NEWCASTLE COLLEGE LIBRARY
02317818

ART AND DESIGN

EMILIO AMBASZ
NATURAL ARCHITECTURE
—
ARTIFICIAL DESIGN

INTRODUCTION BY

TERENCE RILEY

DIRECTOR OF THE DEPARTMENT OF ARCHITECTURE AND DESIGN

OF THE MUSEUM OF MODERN ART, NEW YORK

ESSAYS BY

TADAO ANDO

MARIO BELLINI

MARIO BOTTA

FUMIHIKO MAKI

ALESSANDRO MENDINI

ETTORE SOTTSASS

Electa

Fabula Rasa

The little village was in the grip of fear; fear of Divine rages and of human passions. One of the men started to build a construction, circular in plan, cylindrical in volume, and with a dome-like roof. He used stones, wood, and mud. His travails finished, he came back to tell the group the building he had erected was in the shape of the Universe and inside dwelt the Universe's Gods.

Then, using a rod he had taken from the temple he made a circle around the village and with the help of others he encircled it with a high wall built of earth and stones. In the center of the village, next to the temple, he erected a large hut, which he then covered completely, except for the entrance, with a mound of earth. On top of this mound, he vertically placed six large stone slabs. That, he called: his home. The others called it: the palace.

When he died, his body was laid down inside the hut he had called his house, together with all his belongings and his son covered the entrance with the large stone slabs he removed from the mound's top. Some people say this was how architecture started.

E.A.

Emilio Ambasz: The Landscape Of The Marvelous

Terence Riley*

More than a few essays on the work of Emilio Ambasz note the difficulty in analyzing his work within familiar critical boundaries. Despite what might be referred to as a continuous presence in and around the architectural milieu of New York in the 1970s and 1980s, Ambasz' work has consistently portrayed a lack of concern with the issues which have defined that milieu.

Ambasz has been referred to as a "fabulist;" in his own words, he cites the importance of "wonder," "mysticism" and "myth-making." In the words of Alessandro Mendini, "he answers to no predictable pattern."[1] As a way of piercing the veil which surrounds Ambasz' work it should be noted that even if a pattern is unpredictable it does not deny that a pattern exists nor does it deny the possibility of scrutiny as to method and intent. In this regard, the words of René Descartes are relevant:

> *When our first encounter with some object surprises us and we find it novel, or very different from what we formerly knew or from what we supposed it ought to be, this causes us to wonder and to be astonished at it. Since all this may happen before we know whether or not the object is beneficial to us, I regard wonder as the first of all passions. It has no opposite, for, of the object before us has no characteristics that surprise us, we are not saved at all and we consider it without passion.*[2]

The preceding was quoted in an introduction to an exhibition organized by the Hood Museum, Dartmouth College, titled "The Age of the Marvelous." The title refers to the sixteenth and seventeenth centuries, a period when rapid advances in the sciences, voyages of discovery, and both archaeological and paleological excavations produced an explosion in Europe of exotic and heretofore unknown objects, instruments, specimens of flora and fauna, fossils, and fragments of antiquity.

Perhaps the most interesting manifestation of the marvelous was the Kunst-und Wunderkammer, a collection of sometimes hundreds and even thousands of holy objects assembled and displayed in a specially designed room. Holy relics, stuffed birds and animals, glass prisms, quartz crystals, tobacco leaves, butterfly wings, and ancient Roman cameos would be mounted side-by-side to produce an overwhelming display of the earth's and mankind's most unusual production.

The Wunderkammer provides a model for an attitude which is quite apparent in the work of Emilio Ambasz, an endless juxtaposition of the unexpected, the exotic, and the unfamiliar calculated to evoke a specific response on the past of the viewer. In the Banque Bruxelles Lambert project in Lausanne, the view confronts in the main reception room a miniature version of the bank's facade, through which he had just entered. Surrounding it are trompe-l'oeil reconstructions of the surrounding mountainous landscape. The sense of the unexpected is then heightened: the miniature bank facade is actually the door leading to the vault. Like the Wunderkammer, the objects of wonder are not limited to the man-made: The Houston Center Plaza is composed of a pool from which emanates an unending cloud of steam surrounded by a rationalized topiary garden.

The twentieth century has produced its own version of the Wunderkammer in the fantastic world of the surreal. In addition to exploiting the more classical forms of the marvelous such as perspective, Ambasz' affinities with the work of Magritte, Duchamp, and others is undeniable. As an example, the water-covered, stepped facade of the Grand Rapids Arts Museum is tilted to reflect the sky, producing the sort of irreconcilable perception that delighted the Surrealists. The Mercedes Benz showroom project, which is designed so that automobiles seem to float on a glass-block roof, more successfully exploits the surreal moment even in a highly commercialized landscape where tractor trailers on top of truck stops and gigantic fiberglass donuts fail to attract notice. Ambasz' taste for the surreal is not limited to his own built work. His surrealist interpretation of the work of Luis Barragán,[3] with its references to De Chirico, Delvaux, and Magritte, is so convincing that is almost impossible to see it otherwise.

Despite his ongoing search for an architecture of extreme unfamiliarity, Ambasz' stance as an artist apart is in itself not singular. His work can and should be analyzed in relation to various external cultural phenomena. of his contemporaries, John Hedjuk has followed a similarly distinct path, carefully allying himself with various cultural currents within and without the canonical sources of architectural synthesis. Earlier in this century, Frederic Kiesler and Paul Nelson defined similar territory for their work by assiduously developing contacts with the avant-garde artists of Paris and New York, ironically the Surrealists.

Perhaps what most distinguishes Ambasz' work from this model of "the artist apart" is the range of his influences. While all the architects cited can be read through the broader concerns of their self-selected contemporaries, Ambasz' sources are distinctly ahistorical and cross-cultural. In this sense, it is not coincidental that the architect has gravitated from his native Argentina to the seemingly polarized cultures of North America and Italy. Despite their respective dissimilarities, these two cultural referents provide the defining framework of his architecture.

Both cultures have developed distinctive methodologies of resistance to a prevalent architectonic culture, which was defined principally by the critical positions developed in France, Germany, and the Netherlands. In America, the resistance to northern European, particularly German, modernism is perhaps best represented in the work of Frank Lloyd Wright. As opposed to the Neue Bauen, which was fueled philosophically by the tensions between a static, historically defined culture and an emergent culture of invention and innovation, Wright's work represents the Emersonian conviction that American culture was, in its essence, innovative and dynamic. The traditional American sense of ease with the unfamiliar is, thus, rooted in Emerson's seamless vision of the New World's destiny: in its inception, its present and its future, America represented a continuum of progressivity.

Ambasz' Schlumberger Research Laboratories is particularly emblematic of this phenomenon. The beneficence of technology is so wholly embraced that it ceases to be an issue; it is not only assumed but also subsumed into a broader range of issues. One of the central design features is the conception of the office units as self-contained, 9' x 9' modules, installed and relocated if necessary by a forklift. Yet this highly articulated vision of assembly line production and deployment is absent from the external image of the architecture, which itself is wholly deferential to the landscape

In the instance of the Schlumberger project, the House project for Leo

Castelli, the Manoir d'Angoussart, and others, the landscape can still be seen as essentially American in its unbound expansiveness; a condition described by Henry James in his novel The Europeans as the "uncastled landscape." Yet in these instances, the landscape imagery and its relationship to the technological world is decidedly more complex. If it is American in the Emersonian sense it is also a vision of the New World as imagined by Europe during the waning years of the Enlightenment. Annie-Louis Girodet-Trioson's allegorical painting of 1808, The Entombment of Atala is illuminating in this sense. Based on a popular romantic novel by Chateaubriand, the painting depicts an Indian warrior embracing the body of Atala, a baptized Indian maiden, in the presence of berobed European monk. The painting, in a very literal way, portrays an idyllic vision of American where the "primitive" and the "civilized" are combined in a persona of virgin innocence.

The allegorical quality of Girodet's painting is not far removed from Ambasz' landscape projects wherein highly technological components are sheltered beneath archaic mounds, in plan appearing as ancient glyphs inscribed on a vast plain. The glass-and-steel pyramids of the Lucille Halsell Conservatory similarly evoke a conscious effort to overlay images of technology and primitivism.

If pre-industrial Europe sought a vision of a potentially utopian future in America, it also imagined an equally utopian past in the pastoral innocence of classical Arcadia. As northern Europe moved into the industrial age the ruins of Roman and Greek antiquity assumed an ever-increasing importance as compensation for the transience of post-classical culture. If it is accepted that Ambasz' work alludes to eighteenth- and nineteenth-century Europe's romantic aspiration for an idyllic American future, it must also be accepted that it represents Europe's anxiety, ever-increasing as the Industrial Revolution progressed, about the modern present. In the Schlumberger prophecy, the architect cites the need for a "sense of permanence,"[4] clearly a response to both the Spenglerian sense of cultural doom and the existentialist angst of the pre- and postwar periods.

As noted previously, it is not coincidental that Ambasz' career has been divided between America and Italy. If America has been largely resistant to the ideological modernism of northern Europe, Italy has been no less so, albeit for different reasons. The attitude which had characterized Europe's attitude towards history in the eighteenth and nineteenth century reverberates in Italy today. The Italian architect and historian Vittonio Lampugnani writes:

> *The world we live in is changing more and more, in an ever-greater hurry. And since design culture has to give shape to this world, it cannot fail to measure up to the changes. Yes, but in what way? By supporting and representing it in its evolution, acceleration, restless dematerialization and confusion? Or by opening this frenzy and embodying a stability which, as reality or even just as an image, acts as comforting compensation for a collapse witnessed in dismay? We have said already and repeat: we favour the second option.*[5]

Ambasz' sense of permanence is closely related to Lampugnani's call for stability. Nevertheless, Ambasz' work retains a melancholy not to be found in Lampugnani's activist position. If the architect's design for the Center for Applied Computer Research displays a sense of resistance to the frenzy of contemporary consumption, it is also cognizant of the inescapable transience of the man-made world across a broader time frame: an explicit awareness that the utopian future, in time, will become the utopian past. Foretelling its eventual demise the architect predicts: "Ultimately, only the silent walls and a single barge, turned into an island of flowers, would remain." In this sense, Ambasz' structures are elements of a future vision of a ruined landscape, a yet unrealized Hypernotomachia Polyphili.

A final reference to European Romantic painting is illuminating: Baron von Gros' Portrait of Christine Boyer of 1800 portrays a solitary woman contemplating a rose being carried on the current of a brook; a wistful scene capturing the melancholic essence of the transient moment. On the scale of the landscape, Ambasz portrays a similar mood, the future ruins arranged along an ever-moving stream or bathed in an evanescent cloud of steam.

If the mainline development of European modernism found a cultural resistance in both Italy and America, an analogous development can be seen in Germany itself Without overtly criticizing the technological underpinnings of the modern movement, the so-called Expressionists presented their own alternative vision of anxiety towards an increasingly material world. Ambasz' Nichii Obihiro Department Store with its faceted crystalline structure present immediate affinities with the mystical Glasarchitektur and Alpine architecture of Bruno Taut and Paul Scheebart:

> *The face of the earth would be much altered if brick architecture were ousted everywhere by glass architecture. It would be as if the earth were adorned with sparkling jewels and enamels. Such glory is unimaginable. All over the world it would be as splendid as in the gardens of the Arabian Nights. We should then have paradise on earth, and no need to watch in longing expectation for paradise in heaven.*[6]

Likewise it is difficult to avoid comparing the Fukuoka Prefecture International Hall with Hans Poelzig's submission for the competition to design a "House of Friendship" in Constantinople in 1916, a huge construction with five levels of terraces planted with greenery as the "gardens of Semiramis." The works by Taut, Poelzig, and Ambasz are not so much landscape works as highly emotive constructions of a hyper-natural condition, not unlike terraria on an architectonic scale. Within this attitude is a desire to address the perceived distress of the man-made world, to infuse the material with the spiritual qualities believed to rest inherently in the "natural" world.

If the Nichii Obihiro and Fukuoka projects resonate with certain historical precedents they and other recent projects are also potentially, the most controversial. Any architect in the late twentieth century who aspires to the humanist values which characterize Ambasz' work must contend with, or at least neutralize, the commercial nature of the world we live in. In this regard, that Helmut Jahn was able to convince his client to ban advertising from the new United Airlines terminal in Chicago seems astonishing. Ambasz has also been particularly skillful at addressing his client's needs without sacrificing his own goals. If the imagery of the Mercedes Benz showroom is an intellectual commentary on surrealist composition, it is also a clever advertising ploy (though apparently too subtle an approach for a car dealer).

In Ambasz' recent projects the tension between the artistic goals and the commercial goals is heightened; unsurprisingly, considering that many of them are in Japan where land values exert an enormous pressure on all aspects of the design, construction, and use of the built environment. A number of these works, such as the New Town Center in Chiba Prefecture, Mycal Cultural and Athletic Center, Worldbridge Trade and Investment Center, and Kansas City Union Station projects, portray the architect's ongoing development of a landscape of the marvelous. However, in these cases the landscape is within, or, at least, of the building rather than the opposite. The inversion needs no justification on an intellectual level. Indeed, traditional Japanese gardens can be described in precisely the same terms and the Fontana di Trevi is not far removed.

Even so, the construction of a hyper-natural situation which is, in effect, acting as compensation to the citizen of an increasingly "unnatural" world must be recognized as a potentially global strategy and addressed in those terms. There are certain aspects of the human experience which are of such great universal experience that even private interests must approach them with the utmost caution. The relationship between the individual and nature is one such experience and in the context of Ambasz' recent work, very much to the point.

The "compression" of the natural landscape in the New Town Center in Chiba Prefecture, Union Station and other projects certainly tests the limits of Ambasz' ability to balance intellectual and commercial interests. In fact, the architect appears to not only have recognized the increased tension in his strategy, as must all architects working in our commercialized environment, but devoted considerable attention to exploiting it.

If the architect has pushed his work to the edge, so, too, has the American artist Jeff Koons, albeit from the opposite perspective. In his recent work entitled "Puppy," the artist has assiduously courted the viewers' most extreme sense of sentimentality. The sculpture simultaneously invokes multiple standard kitsch strategies. It is, at once, diminutive and gigantic. It is both an endlessly repetitive and a monumentally singular glorification of the archsymbol of accepted beauty: the flower.

How is it that Ambasz' work can so closely parallel Koon's manipulation of our denaturalized relationship to the environment and achieve such different results? Both artist and architect are aiming at different parts of our intellectual and emotional sense of memory, however, they both speak from a singular position: a convincing sincerity which overcomes any sense of irony. In Jeff Koons words:

> *I've tried to make work that any viewer, no matter where they came from, would have to respond to, would have to say that on some level "Yes, I like it." If they couldn't do that, it would only be because they had been told they were not sup-posed to like it."*[7]

If, for a moment, the reader can disassociate the above from Koons' work and re-associate these words with Ambasz' position as an iconoclast architect with a sense for the marvelous, the membrane that separates the banal and the wonderful in the late twentieth century appears increasingly porous. To press the point, the reader is invited to compare the words of Descartes cited at the onset of this essay with Koons'.

Despite these cultural parallels, the work of Ambasz and Koons should not be confused: if the architect's work has been characterized by its resistance to any accepted notions of modernism it can also be characterized as resistant to any definition of a culture that did not include the potential for the Good, the True, and the Beautiful. Philip Johnson referred to Frank Lloyd Wright as the last great architect of the nineteenth century. Emilio Ambasz must be considered the last great architect of the Enlightenment.

*Terence Riley, Director of the department of architecture and design of the Museum of Modern Art, New York.

[1] E. Ambasz, *Emilio Ambasz, The Poetics of the Pragmatic*, New York: Rizzoli International Publications, 1988, p. 14.

[2] R. Descartes, *Le passioni dell'anima e lettere sulls morale*, trans. by Bari, Laterza, 1954.

[3] E. Ambasz, *The Architecture of Luis Barragán*, New York: The Museum of Modem Art, 1976.

[4] E. Ambasz, *Emilio Ambasz...*, cit., p.126.

[5] V. Magnago Lampugnani, *Progetto e durata*, «Domus,» n. 737, April 1992.

[6] P. Scbeebart, *Architettura di vetro*, trans. by Milano, Adelphi 1982, p. 35.

[7] J. Koons, *The Jeff Koons Handbook*, New York: Rizzoli International Publications, 1992, p. 112.

Tadao Ando

Mario Bellini

Mario Botta

Fumihiko Maki

Alessandro Mendini

Ettore Sottsass

Mario Botta

Lugano, Switzerland

The colored vapors of Houston Center evoked other squares for me. While Emilio Ambasz, in his office, was illustrating the magic and enchantment of this extraordinary project, images and memories of familiar, friendly squares, came back into my mind. But suddenly I noticed their daily and millenary faces. Those colored vapors conjured back pictures for me of spaces once dearest to me (those linked to memories of a hope); distant images, images of a past that belonged to me but of which I could now perceive only a far-away echo. Fading with the vapors of Houston was the idea of a piazza, that of the village where I grew up, so rich in memories yet so poor in architecture. So full until yesterday, of affections, nostalgias and hopes, and so empty today as a residual space. The image of the "square," swallowed up by the stretched geometric horizon of Houston trees; the square which during my student years in Venice had played such an intense part in my existence, likewise vanished. So lavish in its architecture, it used to fill me with anxiety and fear; it made me feel uneasy whenever I had to cross it, as if it were a forbidden territory, a room too sumptuous and foreign for my Lombard world and its remoteness from Byzantine pomp. The Houston project and its magic spell jolted me back into a confrontation with the problems of our time. Its presentation as an "exterior" infringes the very laws whereby the square is expected to be an "interior," a defined "empty space," in relation to a surrounding area The skillful contrast between the geometric rigor of the labyrinth area below and the fluctuating of steam vapors above creates a new dimension of imagination, play and hope of which an immense need is felt today.

Mario Bellini

Milan, Italy

Emilio Ambasz, born in Argentina, lives now in the USA, but works everywhere: Europe, Asia and in the Americas. He is an architect and designer of unique relevance on the international scene.

•

Ambasz practices his many professions in a manner completely different from that of the traditional American architectural office, which is generally segmented into very specialized and circumscribed fields. His activities range from architecture to landscape design, from graphic to industrial design, from interior design to the creation of furniture, while encompassing among other activities, urban planning, criticism, and an expertise in industrial manufacturing processes. This is not meant to imply a dilettante's approach or a diluted concentration. On the contrary, one may say that in practically every one of his fields, Emilio has achieved extraordinary results, as evidenced by his many accomplishments and awards. Some time ago, I learned that he had even been awarded the Prize Jean de la Fontaine for his Working Fables. It does not surprise me; the subtlety and intelligence of his criticism and written comments are well known.

•

I still remember the surprise that arose from his unexpected entry into the Italian design scene when, in June of 1976, we learned that, in addition to his already renowned image as Curator of Design at the Museum of Modern Art, he had also achieved the distinction of being the creator of the "Vertebra" chair, a chair he originated, and then designed and developed in collaboration with G. Piretti. "Vertebra" was the first automatic, articulated office chair in the world.

On this occasion, as with the many others which followed, the importance of his achievement was immediately evident to all. We were not being presented with yet another variation from the large catalog of ergonomic office chairs; "Vertebra" represented a new and important re-evaluation of the notion of ergonomic seating, to such an extent that it dated everything that had been done before. It was, and is, the reference point for everything that has been designed in its field since.

•

Even today, after so many years, we still perceive the importance of the changes brought about by "Vertebra." The chair has become the standard for any other further improvement in design. It has influenced even the all-important German school of ergonomic seating design, which has in the last years seen a shift form the user manipulated chairs, characteristic of the last years of German seating design, towards Ambasz's notion of automatic and flexible seating. (This latter change is due in no small part, to the collaboration of the German furniture industry with non-German designers.)

•

In the same way that Ambasz's basic concept for Vertebra impressed us all, it seems to me that we will never get used to his omnipresent design ability. It is both stimulating, as well as slightly frightening, to know that at this very moment, in some part of the world, he is designing yet another project or inventing another concept that may, once again, bring about a momentous change in design.

Ettore Sottsass

Milan, Italy

I believe that such a quest for a constant state of fluidity, such a perception of existence as an ever-changing process pervades everything that Emilio has ever designed.

In his architectural work, for example, there are almost never objects plainly resting on earth, as is usually the case in more conventional architecture where buildings are just a statement, and that is all. Emilio's architectural creations are a bit outside the earth and a bit inside it. They are like stone slabs emerging from the earth, or fissures cracking the earth open, rather than attempts at controlling the universe by means of logic or agreed upon signs. His is an architecture seeking, almost always, to represent the internal and eternal movement of an all encompassing planetary geology while at the same time respectfully reflecting local pulses, explosions, contractions, tempests, and deeply welled mysteries.

•

Visions come to my mind when looking at his building for the San Antonio Botanical Garden, his terraces and entry lobby for the residential zone near Lugano, his project for a Cooperative of Mexican-American Grape Growers, his project for the Center for Applied Computer Research, his house in Cordoba, as well as many other projects Emilio has created.

Looking at all the projects, I have come to think that Emilio's buildings cannot with certainty be called monuments, even less can we call them literary exercises. By that I mean to say they are not exercises in architectural composition; they are not even intellectual conceits or objects; still less can one catalog them as attempts at technological rhetoric. I would call Emilio's architecture propitiatory designs seeking to invoke the presence of architecture. Each element of his edifices is a bit like a talismanic instrument of a wager, of a hidden ritual to fascinate some immense natural divinity. Maybe they are aspects of a liturgy, performed to obtain forgiveness for the scars we inflict everyday on the planet, or, maybe they are part of a magic ritual performed to establish harmony with those strange celestial rotations which Indians and Greeks, a long time ago, had already intuited.

This is a unique way of imagining architecture, certainly a new way, a special way, or perhaps it is a very ancient way. Perhaps this new way belongs to those very ancient times when, in order to found a city, an animal, perhaps a bull, was let free and the city became established wherever the bull, after a while, stopped to drink.

•

Maybe the difference between the ancient propitiatory processes performed to secure the planet's good will – and then that of the whole universe – and those new propitiatory processes Emilio performs today lies only in the different technologies and in the different gestures he uses to satisfy archetypal longings.

•

Like very few architects, he senses the emergence of a technological mythology, fully aware of all that it may bring with it. I also know that he is very knowledgeable and that with relentless precision and painstaking patience he pays attention, as do very few, to the possibilities of technology as an irreplaceable device for bringing about that rare existential event that is architecture.

All of this I know already, and all of this that I know is also known by others, as we also know that it is not here that the root of his great originality lies. The thing which is original with Emilio, and which is very rare, is that technology is for him an instrument for suggesting architectural presences; that for him it is the architectural event, when it occurs, which serves as the magical instrument to bring about that still larger and even more complex event which is our spiritual existence. I imagine Emilio Ambasz to be a man outside the norm, at the cutting edge, and therefore, under surveillance. I see him as an imaginative and illuminated young man, resembling some old Chinese priest who, first for months, and then for years, uses ancient techniques to polish the surface of a great disk of green jade which will allow him, perhaps, to penetrate beyond the daily, or better still, which will allow him, even for just an instant, to place daily existence in some exalted architectural domain.

Fumihiko Maki

Tokyo, Japan

I first became acquainted with the work of the architect Emilio Ambasz in the 1970s and 1980s when several of his projects won a number of awards, given annually by "Progressive Architecture." The projects for the Grand Rapids Art Museum, the Mercedes-Benz Showroom, and the San Antonio Botanical Center, the last one constructed several years ago, were quite original in their design and concept, and I remember being impressed by them.

•

What was so impressive about them? I believe it was the fact that those projects, like other projects he has subsequently made public, represent a continual search for what is primary in architecture.

Ambasz, has paid little heed to postmodernism and has always been concerned with the more elementary aspects of architecture: that is, domain and boundary. He has offered clear and bold proposals that define domain and boundary by means of walls, roof, ground, water, space, and light. For example, the inclined plane introduced between two existing classicist buildings in the Grand Rapids Art Museum project was not just an addition to a pair of old structures. The intention was to create something new through a violent opposition.

What is sought by the above described procedure is not harmony, but tension through opposition. The plane may be functionally a stairway, a roof, or even a cascade, but in maintaining that tension it serves as a device that transcends time and continues to generate diverse meanings. Again, in the Mercedes-Benz Showroom, a single black wall and a white, wavelike floor create a place of ceremony in the same way that the Japanese tokonoma does. The static wall and the dynamic floor are in a relationship of tension. At the same time, these basic elements create a microcosm.

•

My feeling that there was something about Ambasz's work that recalled an earlier time in architectural history became stronger when I learned that he has always had a strong interest in the role of rituals and ceremonies in architecture.

•

In looking at the plan of the conservatory in San Antonio, I was reminded of cave house of Le Queu. The San Antonio Conservatory was the first of a series of realized projects by Ambasz using earth architecture. Other projects include a number of proposals in Japan now under construction, such as his projects for Shin-Sanda and Fukuoka.

Employing vegetation and earth, Ambasz has developed a new architectural vocabulary that one might call "meta-architecture."

His architecture also reveals an intense awareness of boundaries.

•

We are today in an age that offers many parallels to the Enlightenment of two hundred years ago. However, it should be added here that Ambasz rejects the kind of cultist stance that eighteenth-century architects were apt to assume. When Ambasz declares the need for rituals and ceremonies, he means it within the context of people's everyday actions. In that sense, he remains a humanist and populist.

Today, architects in our society must invent new programs. Contemporary architecture must not simply supply containers for conventional functions required of existing lifestyles. Ambasz appears to recognize that, on the contrary, new programs must be conceived if a new architecture is to be created. The conservatory in San Antonio is an excellent demonstration of the correctness of such an approach. Through the deeply inspirational architecture and environmental art that he has created, Ambasz continues to expand our own horizons.

TADAO ANDO

OSAKA, JAPAN

In recent years, consciousness has indeed grown in matters regarding nature and environmental issues. It was Emilio Ambasz who first called our attention to nature and the environment at quite an early point in his career, and ever since he has striven to achieve a fusion of nature and architecture.

•

The use of abstract forms and modes of expression in architecture, while it is the most remarkable concept born of the twentieth-century modern movement, contains the germ of contradiction. I have become convinced that architecture cannot be derived from the purely abstract as long as a function is demanded of it. The concrete aspects of nature, climate, and tradition, are intrinsic to the existence of architecture and it is not possible to ignore the extremely concrete demands of daily life. By drawing nature into abstraction and giving it expression within his method of architectural creation, Ambasz, in his brilliant insight, appears to have embarked – and he brings us with him – into a realm of architecture previously unknown to human experience. To underscore this marriage, I may want to call it "environmental architecture." Someday we may just call it, again, Architecture. Ambasz's quest has started a trend in the recent work of many other architects.

By using nature on a massive scale, Ambasz presents us with the entire environment as a constellation from which architecture draws its essential being. There is, I believe, no prior example of nature governing architectural creation with such power and haunting seduction.

•

The scope of his vision and the depth of his insights overcome immense differences in scale, traveling freely between poles of macro and micro, thoughtfully contemplating the intervals between the abstract and the concrete. This is especially true of his attempts to transform nature into architecture on a grand urban scale. We can see in his most recent work, like the Fukuoka project, that we are presented with ideas and images which are quite outstanding.

The journey on which he has embarked is as yet uncharted, as is always the case of original design, and it is one where he shall surely encounter immeasurable difficulties. Nevertheless, as we all now stand before the twenty-first century, I have high expectations that the results, which his endeavor and findings are yielding, will prove truly substantial. He has taught us to see a dimension where nature and architecture are inseparable a realm extending from the God-given to the man-made nature. His work promises an ample domain where the found and the made, the natural and the artificial, coexist joyfully. He has fulfilled the promise of his early projects and has indeed shown us the way to a re-beginning of architecture.

Alessandro Mendini

Milan, Italy

I do not know who Emilio Ambasz is. As a man and as an artist, he answers to no predictable pattern. He is, first, a man who shields his tenderness with a transparent layer of aggressiveness plated over his substantial timidity. Second, he is an exquisite poet. Then he is a land artist, then a farmer-cum-engineer cast in a biblical mold, a boundless juggler, a pioneer who suggests the image of mythologies yet to come, and finally the designer of sophisticated terrestrial paradises for modern times. Perhaps he is a prophet, obstinately setting himself in the role of "Anti Master." Or, maybe he is an empirical genius, as seen, for example, when he endowed the office chair with vertebra. Or maybe, his projects, as large as unpolluted continents or as small as flexible ballpoint pens, play a game of running away from us and from themselves, resembling a bed of self-deprecating narcissuses that project on the water their own alternately self-admiring and self-amused image.

•

The seduction by Emilio's projects comes about from relegating man-made culturally conditioned forms and colors to the background in order to favor meadows, lakes, valleys, flowers, sunsets, suspended gardens, and skies with colors like Tiepolo's white, blue, green and gold. Thus, not only as an artist but also as a person, Emilio is a "unique case." Impossible to pin down, impossible to classify, he continually appears in a different guise. An inexhaustible inventor of metaphors – blue prince of his own fables – he is also the mystical master of ceremonies of a ritual, of a liturgy, and of an astrology created by himself. Thus, Emilio grants us only one certainty: that of his absolute singularity and originality.

•

This gentleman, this playwright actor, this cat, this child, this little bird, this nest – as a person, as a professional, and as an artist, is then, I must repeat, a unique case. His very large and complex body of work doesn't seem to seek the friendship of the academy, nor care to contribute to the canonical history of architecture, design and language. It seems, rather, to be born from an obsessive search for primary principles, from a careful and wise observation of the surrounding reality, from an identification of humanistic problems perceived as only a hypersensitive instrument such as Emilio can – a sort of imaginative and scrupulous sensing device that gets to the core of all questions, that strives to satiate the essential quest for techniques and images that men have, that registers the original sense of beginning as well as the anthropological and ecological tremors of a modern world. The passion and pragmatism he offers in opposition to prevailing simple minded functionalism is, perhaps, a result of his direct lineage or ancestral connection the the Cyclopean anti-monumentalism of Buckminster Fuller. The work of Emilio is not post-Modern. His vocabulary is not made up of references and ready-made sentences; thus it is free of the decay of fashions and styles. He proposes as principle and method, as archetypal idea, pure and primary, the notion that an arch is an arch, is an arch; that a man is a man, is a man; and that a house is a concept that should contain both its past and its future, the beginning and the end of all our dwelling memories, whether true or imagined.

•

Perhaps, after the efforts we have undertaken to understand him, a few strong messages begin to emerge from our magician's top hat. Emilio believes deeply that architecture and design are mythical acts. He proposes a different, emotional method, a passionate and sensual mode of existence. His thoughts and images are based on the primitive but eternal process of being born, falling in love and dying – those things that have always moved the world, those irreducible drives that always return. Who then is Emilio? He is a dreamer who dedicates heart and soul to men while longing for angels. Or perhaps. . . .

MYCAL CULTURAL AND ATHLETIC CENTER AT SHIN-SANDA, 1990
SHIN-SANDA, JAPAN, P. 2

A cultural and athletic center in the new town of Shin-Sanda benefits not only its employees but also the growing local community. The difficult challenge of this project was to accommodate the immense massing requirements of the building's 450,000 sq.ft., while sympathetically acknowledging the serene open landscape beyond. Ambasz was able to return to the town of Shin-Sanda virtually all of the greenery that this enormous building footprint would normally have taken away. The Japan Housing Authority was very pleased with the overall benefit that such a design solution would bring to the community, and as an incentive and ultimate reward for this inventive solution, they reduced the cost of the land by almost two-thirds.

THE PHOENIX MUSEUM OF HISTORY, 1989
PHOENIX, ARIZONA, USA, P. 14

The Phoenix Museum of History – a pilot project in an ambitious revitalization program for downtown Phoenix – nestles within a sloping earth-covered ramp and beneath a new public park. Located on a site traditionally used for many public gatherings, it gives back to the city virtually all of the land that would have been taken away by a more traditional design. Rising to a height of fifty feet, the earth-covered ramp in front of the museum shields from view an existing convention center located directly behind the site while providing unobstructed views from the park of the surrounding scenery. Visitors relaxing in the inclined garden can see a collection of existing historic houses which face the museum, as if these houses had been naturally integrated as one of the museum exhibits.

Structural elements beneath this slope are shared by the museum and the 800-car parking garage, helping to keep costs within a very limited budget. Little visible architecture protrudes above the slope except for one sinuous wall delineating a free-form L-shape like a sheer cliff face. Triangular wall sections functioning like buttresses of the indigenous adobe architecture protrude into the center courtyard and leave openings that allow light to flood into the gallery space behind. The two sides of the oval courtyard meet in a circular double-height lobby where visitors experience the sensation of descending deep within the earth as the slope rises around them.

Fukuoka Prefectural International Hall, 1990
Fukuoka, Japan, p. 22

The city of Fukuoka, Japan desperately needed a new government office building, yet the only available site was a large garden park in the center of town. When news emerged that this potential new structure would be located upon the last remaining green area in the city, the citizens of Fukuoka erupted in protest. Ambasz was awarded this commission for successfully achieving a reconciliation between these two opposing desires: doubling the size of the park while providing the city of Fukuoka with a powerful symbolic structure at its center. His design utilizes a series of garden terraces stepping up the facade of the building, thereby giving back to Fukuoka's citizens virtually all of the land that the building would subtract from the city. The result was immediate community approval and the complete avoidance of any construction delays due to community challenges.

San Antonio Botanical Center Lucille Halsell Conservatory, 1982
San Antonio, Texas, USA, p. 32

The Lucille Halsell Conservatory is a complex of greenhouses located in the hot, dry climate of southern Texas. Unlike northern climates where traditionally glazed greenhouses maximize sunlight, the climate of San Antonio requires that plants be shielded from the sun. Mr. Ambasz's design uses the earth as a container and protector of the plants, controlling light and heat levels by limiting glazed areas to the roof. This innovative design concept significantly decreased the need for expensive mechanical systems, thereby reducing the overall cost of the building by over 20%.

Museum of American Folk Art, 1979
New York, USA, p. 44

This project provides enlarged quarters for the Museum of American Folk Art within a midtown Manhattan high-rise building. A freestanding, eighty-foot-high portal – designed as a proscenium – creates an identity for the museum as well as the office tower above. Through this monumental entrance, two symmetrical stairways lead to the glass foyer of the museum – a skylit, three-storied, stepped section inner court. The tower entrance penetrates the staircase axially at street level. Visitors proceed down to the exhibition floors via a series of ramps or stairways designed as a central, elongated spiral. The tower above is divided into three blocks of office floors suspended within a prismatic frame, or envelope. The blocks are set forward as they ascend, expanding their horizontal span as their supporting piers grow taller and thinner. The arrangement of the blocks reduces the tower's perceived mass; increases its resistance to wind pressure; allows greater penetration of light through the building; and expresses the functions of its interior while still presenting a unified face to the street.

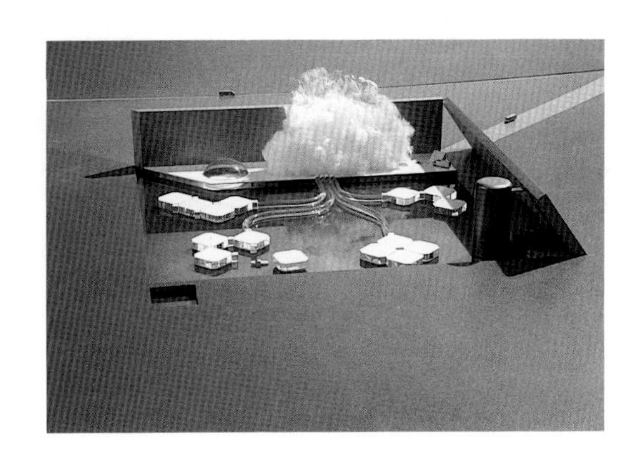

Center For Applied Computer Research, 1975
Mexico City, Mexico, p. 46

On the outskirts of Mexico City, the Center for Applied Computer Research and Programming offers advanced computer-programming services to public agencies and private organizations. As the flagship corporate headquarters, the building establishes the site's primary reference point. Because Mexico City and its surroundings are built on the land-filled site of an ancient system of canals, the building is designed to float within a large water basin which drains the soil and thereby prevents foundation problems. To take full advantage of this basin, the building's office/workspaces are designed as barges that float freely until secured in place. Behind the design of this environment is the premise that nobody should have to work. At worst, one would work at home and not need a large building but rather a small one to simply house a computer and receive messages. The building has been conceived, therefore, as a set of elements that can be progressively reconfigured and recombined as the needs of the office vary over time. Ultimately, elements can be removed as the need for on-site space diminishes. Only the silent walls – leaning upon one another like the traditional walls of a Mayan temple – and a single barge, turned into an island of flowers, would remain.

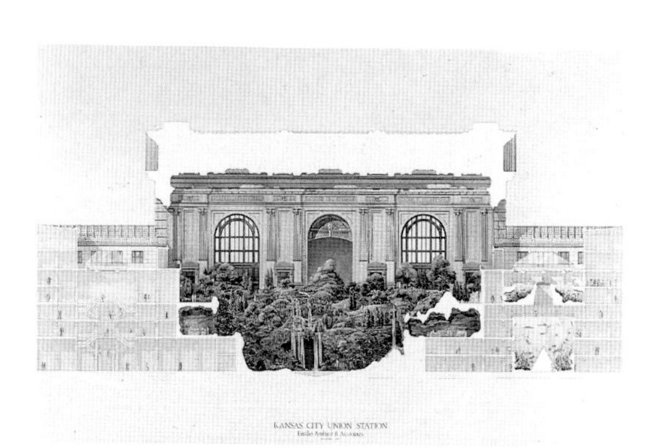

Union Station, 1986
Kansas City, Missouri, USA, p. 50

The planned reuse of Union Station attempts to preserve an historical building by making it a vital part of the community, both in its urban presence and economic contribution. In the course of prior urban expansion, building and landscape were severed. The key element of this design is the reconnection of Union Station to the magnificent green carpet of Liberty Park. The park continues into the former station's vast public areas, converting them into wintergardens. The Grand Hall – once the station's waiting room – becomes a botanical conservatory, illuminated and warmed by winter sunlight pouring through the south-facing windows. Use as a wintergarden drastically reduces heating and cooling costs, making the space an economic asset rather than a liability. Open to the conventionally cooled and heated wings, the conservatory's temperature would be moderated by the continuous influx of conditioned air. The plants therefore survive nicely without the expense of any mechanical heating or cooling. Rather than obscure the hall's elegant architecture, the conservatory is located in a series of terraces set within a "quarry" which step down deep into the floor of the high space. The main hall thus becomes a connecting focus between the wings of the building which house various cultural facilities – museums and theaters – while supporting a grotto and an additional wintergarden on the upper level.

Grand Rapids Art Museum, 1983
Grand Rapids, Michigan, USA, p. 54

This proposal revitalizes the city's downtown area by reusing a distinguished, but now vacant, former government building and by adapting a number of abandoned or underused neighboring structures for additional cultural functions. The main 1908 Beaux-Arts building is upgraded to meet codes and programmatic requirements. A relocation of the entrance from the closed side of the U-shaped building to the open courtyard establishes a direct physical relationship between the museum and its surroundings while creating a single, unified entrance. Covering the courtyard, a translucent inclined plane of stairs has an opening at its center that leads to the Grand Foyer which is a shared space for the different departments. Water descends slowly and evenly from the top of the stairway along a channel carved into the plane, leaving a clear pedestrian path for those walking up the stairs. This water is, in fact, the very water used for the air-conditioning system. In this way, the need for an unsightly cooling tower has been eliminated and this silent cascade becomes the focal point of the city's newest public space.

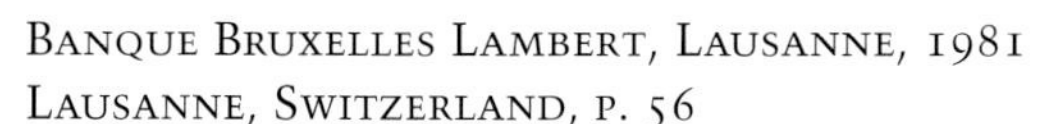

Banque Bruxelles Lambert, Lausanne, 1981
Lausanne, Switzerland, p. 56

Lausanne evokes images of mountains, snow, lakes, steep streets and quaint façades. However, this landscape of the mind rarely matches that of reality. The nearby peaks are often hidden by Lausanne's buildings. There are few open spaces from which to contemplate the buildings at a reasonable distance. Nevertheless, these images are pervasive ones and are therefore recalled in the design of the bank's interior in the hope that those who enter will, for one ineffable instant, discover the view of nature that once was possible. At the bank entrance, a serene, quietly luxurious atmosphere envelopes the visitor. The surrounding scene of far-away mountains is deeply, even ambiguously, reflected in the midnight-blue lacquered ceiling and filtered through a soft, hazy light. On closer inspection, the haze reveals itself to be a curtain of fine silk threads hanging in front of a trompe-l'oeil representation. With this illusion, inside suddenly becomes outside. The inversion is carried further: bronze perspective lines along the floor converge on the façade of a model of the building, while two Carrara marble tables – used for banking transactions – are positioned to emphasize the illusion of vanishing perspective. Minimal in line and pure in volume, these objects assume an architectural role as contrapuntal elements in the landscape.

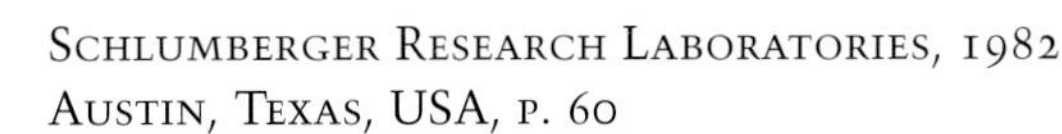

SCHLUMBERGER RESEARCH LABORATORIES, 1982
AUSTIN, TEXAS, USA, P. 60

Located on the edges of Austin, Texas, the site for this computer research facility was valued by the surrounding neighbors as a beautiful green parkland. Because the site therefore warranted a design that harmonizes with the landscape rather than standing out against it, the project was divided into a series of smaller buildings with earth berms built up against them to help integrate them into the landscape and simultaneously reduce energy costs. In this way, neighbors see a beautiful man-made landscape rather than intrusive buildings. The buildings and recreational facilities are arranged casually around a man-made lake, in the manner of an English landscape garden. Because of the earth berms, the architecture blends into the surroundings while providing a pleasant atmosphere for employees wishing to take advantage of the pleasant vistas. The laboratory design consists of a large, undifferentiated space in which the researchers' offices – enclosed mobile units – are placed in any variety of configurations. The units can be moved quickly and easily to accommodate changes in research-group size and to facilitate dialogue within and between the groups. The laboratory design incorporates the best characteristics of the open office landscape – flexibility and ease of communication – with those of the traditional office: reduced noise levels, individual control of the environment and privacy.

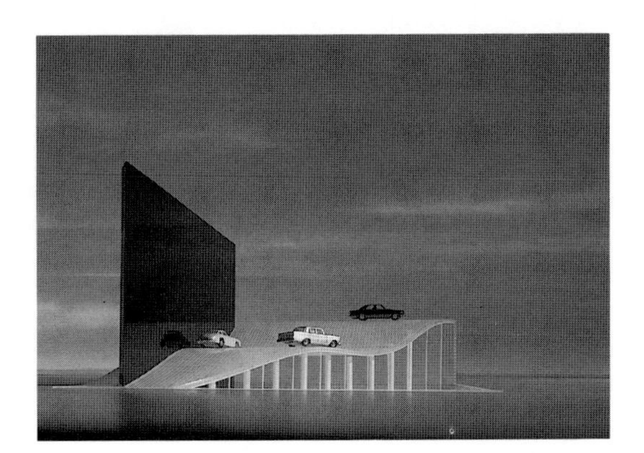

MERCEDES BENZ SHOWROOM, 1985
NEW JERSEY, USA, P. 66

A showroom for new Mercedes Benz automobiles is located on a small site with typical commercial buildings to its left. In order to hide an unattractive diner from the view of people approaching the showroom via the primary road access, a tall wall of polished black stone is erected. This reflective, black wall serves as a display board as well as a backdrop for profiling the new vehicles. The exhibition spaces – both exterior and interior – are conceived as continuous ramped surfaces that lead up and down from the road level and evoke the feeling of movement. The enclosed showroom consists of translucent glass-block floors that are inclined so that visitors are able to view cars from above and below as well as laterally. The cars – presented as sculptures – suggest acceleration and deceleration; and the translucent glass plane, dematerialized by the light passing through it, emphasizes the poetic aspect of movement.

NICHII OBIHIRO DEPARTMENT STORE, 1987
HOKKAIDO, JAPAN, P. 70

Climatically similar to Siberia, Hokkaido's image is that of an ever-present winter in which snow seems somehow to fall horizontally. Ambasz's design for a new two-and-a half acre department store creates a perpetual spring-like garden within this otherwise bleak environmen where people can gather under a glass skylight to look at plants, listen to water cascades and stroll along walkways in the courtyard around which the building is organized. The local building ordinance only typically allows 23,000 sq.ft. of retail in this area but, as a reward for the civic-mindedness of this building, the client was allowed to construct over 58,000 sq.ft. of shopping, thereby significantly increasing the profitability of the entire venture.

Worldbridge Trade and Investment Center, 1989
Baltimore, Maryland, USA, p. 74

In planning a new international trade center, the client intentionally did not want a building that referred to any particular building style, rather one that synthesized Eastern and Western traditions. In addition, the neighbors would not allow any construction that would affect their view of the countryside. The structures that comprise Worldbridge, therefore, create two inviting features on the Maryland topography: one, the office complex, is the seeming result of an orchestrated up-lifting at the earth's surface; the other, the exhibition hall, is an implosion forming a carefully-wrought cavity. Both buildings are, in fact, composed with the graduated stacking of organically-shaped floor plates. At the office complex, gardens are cultivated where one plate extends beyond the next. The monumental atrium – a truncated cone – is dramatically lit from above by an oculus. At the base of the interior space, a two-story bowl carves into a highly-detailed and episodic landscape of rock, moving water and abundant plant life. Adjacent to the extroverted superstructure of the office complex, the sunken exhibition hall – dedicated to revolving trade shows – offers a sense of spaciousness through the controlled use of vegetation and the introduction of natural light from its hollowed core.

Residence-au-Lac, 1983
Lugano, Switzerland, p. 78

The terrace of Residence-au-Lac, a 1950s resort hotel converted into luxury apartments, overlooks Lake Lugano. Visible from the lakefront avenue and gardens, it commands a panoramic view of the surrounding mountains. This view of mountains, sky and clouds is recreated theatrically on the terrace and in the building's lobby. It is composed of rough-finished granite slabs alternating with strips of polished white Carrara marble. These marble strips emerge at irregular intervals, forming jagged shapes that echo the mountains on the horizon. Inside the lobby, the walls are painted a gradient blue suggestive of the distant sky. The marble landscape takes on a surrealistic air under translucent silk "clouds" hanging from the blue top-lit ceiling and spaced at intervals corresponding to those of the marble forms. Architecture, sculpture and landscape merge into a dreamlike whole.

Focchi Shopping Center, 1990
Rimini, Italy, p. 80

Due to a changing economy, the owners of the Focchi factory building decided to retain the historic structure and adapt it to a new use more suitable to the current marketplace. Ambasz was given the commission to enlarge the facility and transform it into a new shopping center with a compelling and inviting image. Ambasz partially covered the roof of the structure with a grid of mesh covered in planted ivy, helping to naturally cool the interior while enriching the appearance of the raw concrete. Portions of the roof vaults are left uncovered to establish a clear distinction between the decorative and structural portions of the building while visually uniting the old and new sections. Along the sides of the building, the ivy covered mesh cantilevers past the structure, echoing the barrel vaults, to create translucent verandas reaching the building. Like a symbolic cornice, this overhanging mesh provides shade as well as an atmosphere of welcoming arrival to patrons. Six ribs of the central vault were removed to create an internal garden courtyard. Surrounded by an open arcade that links the various shops and cafes, this courtyard provides sunlight and fresh air to the interior.

House for Leo Castelli, 1980
East Hampton, USA, p. 84

Practical and wholesome, this house is at peace both with nature and the man-made environment. Two large earth berms serve as gates, shielding the house from the street and framing a grand entrance. Another gently terraced earth berm insulates the north-facing rooms and walls of the house, making it an integral, unobtrusive part of the landscape. Carefully planned cross-ventilation eliminates the need for air-conditioning, while the combination of passive and active solar design reduces heating requirements. Large south-facing windows capture the sun's rays while the thick walls of the house absorb and store heat. The open layout of the rooms allows for a flexible organization of the interior spaces. All rooms open onto an arcaded courtyard which also functions as a graceful entrance court.

Casa de Retiro Espiritual, 1978
Cordoba, Spain, p. 88

Outside of Cordoba, in the middle of a rolling wheat field, this house is the weekend retreat of a childless couple. Inspired by the traditional Andalusian house, it has a central patio onto which all rooms open. The house is insulated by the earth which keeps it naturally cool in the hot, arid southern climate. Two tall, rough stuccoed white walls meet at a right angle to herald the entrance of the house which is otherwise contained by sinuous walls. Water cascades within both high walls along the handrails, creating a great amount of noise at the bottom of the stairs but becoming increasingly quieter as visitors ascend to the top. The large, continuous living space of the interior area is defined by smooth cavities excavated into the floor and echoed in the ceiling above. Small, inlaid glass tiles are washed by the soft, diffused light that descends from the skylights and filters in from the breezy patio.

Manoir d'Angoussart, 1979
Charleroi, Belgium, p. 94

Taking advantage of its site – a twenty-acre estate which is completely flat except for a ravine – this house is built into the earth so that its largest windows open south onto the ravine. Masses are defined by geometric volumes and earth berms that shield and contain the house to ensure a pleasant temperature year-round. Texture and ornament are provided by grass, plantings, and the moiré patterns created by the window frames. The design relates to the earth and is rooted in the architectural tradition of the Low Countries. Freestanding façades of double lattice are covered with ivy in order to maintain a discreetly low profile when viewed from afar. As one approaches the house, the full height of the façade emerges, giving it a majestic presence. The house and garden are conceived as a single, indivisible entity; diverse architectural elements are treated as integral parts of the landscape to suggest an expansive domain of ever-changing perspectives.

Private Estate, 1991
Montana, USA, p. 98

On a magnificent 1,400-acre site along a secluded river valley, this complex of three buildings houses a family, estate caretaker and private art collection. The siting creates a dynamic sequence of surprise and discovery as the natural landscape envelopes each building mass and leaves only the bold façades exposed. The landscape appears to flow down the hillside completely covering each building, in the historic Montana tradition where log cabins typically had roofs covered with earth. And like buildings which have now come to symbolize the traditional historic American Western town, each façade plane extends well above and beyond the volume of the building which it masks from view. The only major visual element is the façade. These ivy-covered trellis façades lean against colonnades of bare logs – a rustic classical reference which employs the primeval vocabulary of the forest. At the top of each log is a gilded capital to match the building's bronze cornice which captures the lingering reflections of the setting sun. The strong, inward curve of the main residence embraces a raised court d'honneur. The art gallery, set in a quiet meadow across a pond, is accessible only by foot and curves gently outward. Beyond the houses and the art gallery, a small meditation pavilion peeks enigmatically out of the wilderness.

Casa Canales, 1991
Monterrey, Mexico, p. 116

Projecting out from a steep, rugged mountainside, this house commands a panoramic view over the valley that contains Mexico's industrial center. From the shade of the elongated porte cochère, the visitor is left to discover and contemplate the beauty that lies below. Deceptively cantilevered far out over the mountainside, an edgeless triangular plane of cool, aquamarine water appears suspended weightless in space. This reflecting pool doubles as the water-cooled roof of the house. A temple-like observation pavilion, perched at the farthest corner of the pool, tempts visitors to a refuge that seems supported by the sky. A monumental set of stone steps leads enigmatically into the depths of the cool areas below, like an ethereal portal to an underwater world. Directly below, the house seems to emerge from the very stone of the mountain. Overlapping layers of open lattice walls, colonnades and glass allow breezes to filter through softly and create rich patterns of light and shadow. Each room within opens completely onto broad covered terraces and balconies to the outdoors which capture the natural breezes and create cool, shaded areas for the family's daily activities.

Financial Guaranty Insurance Company, 1984
New York, USA, p. 122

The design for the Financial Guaranty Insurance Company office is composed of a series of landscape vignettes. Curtains of gauzy silk strings that transmit light are used as partitions, screening the views while hinting at a presence beyond. Veiled behind, a blue wall covering shimmers and gives the illusion of a distant horizon. The need for flexibility in the workplace is accommodated by an open-office environment into which self-contained, modular, movable work units are inserted and configured. Experiencing this space is like walking in an English garden: glimpses of each scene hint at what lies around the corner. One leaves this series of visual episodes with the impression of quiet and functional elegance – of spaces only partially seen and understood, yet part of a carefully choreographed drama.

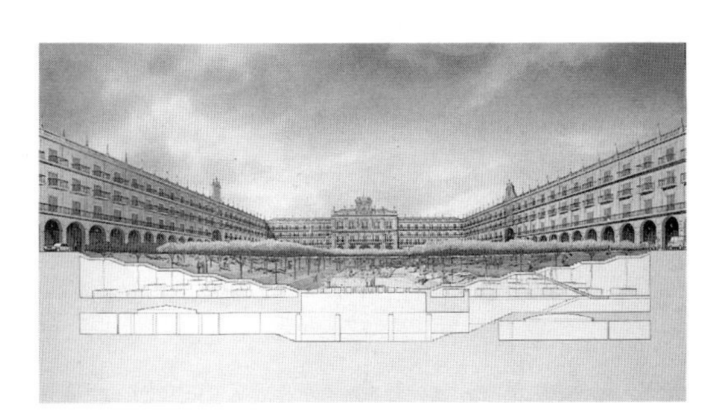

Plaza Mayor, 1982
Salamanca, Spain, p. 128

Salamanca's Plaza Mayor is, in true Iberian fashion, the center of the city's commercial and cultural activity as well as a place of repose. Around the plaza is a four-sided baroque façade designed by Alberto Churriguera – one of the glories of Spanish architecture. The plaza itself, however, is a flat, barren place, unsuitable for sitting or gathering. This design uses concentric squares stepping down toward the center of the plaza to create a sheltered, tree-shaded space which in no way impedes views of the magnificent surrounding façade. The plaza's former ground level is maintained by the tops of the trees, which create a metaphorical ground cover of leafy clouds. As one descends into the plaza, the tree trunks emerge into view. These columnar trunks and the green canopy overhead allude to the arcaded loggia beneath the surrounding façades. Below this forested plaza are cinemas, theaters, gymnasiums, community offices and – in keeping with one of the plaza's traditional uses – a dance hall naturally lit through metal gratings in the plaza paving above. An air pocket between the steps and these public functions acts as a plenum, trapping shaded air to cool the plaza in summer. In winter, when the leaves have fallen, the sun strikes the exposed plaza, warming its stone steps. Growing physically as well as symbolically from the very stone of the city, the new plaza offers a quiet, shaded retreat, while maintaining the integrity of the historic surrounding architecture.

Houston Center Plaza, 1982
Houston, Texas, USA, p. 130

The most outstanding feature of Houston is the continual urban grid. This site – one square city block – finds in the purity of the grid a perfect foil to urban interventions. Thus, the grid of the city became the grid of the plaza, with a rough edge on the outside representing the incomplete nature of the growing city, and the square pool in the center representing the plaza itself. The plaza is also meant to portray various aspects of Houston on a metaphorical level. The culture of the city is embodied in the theaters, galleries, restaurants and social amenities below the reflecting pool. Most important of all, though, is the spiritual quality of the space. The ground slopes down from the edge of the plaza to the large square pool in the center, with its circular opening above the atrium. The taller trellises toward the center serve as gazebos, with portals and seating. The plaza engages all the senses: colorful, fragrant flowers grow between the vines and mist emanates from the top of each enclosure, thereby cooling the surroundings and producing a soothing hissing sound. The gazebos offer spots for relaxation from office work and from the heat of the city as well as places for socializing or quiet contemplation amid green shade. The effect is reminiscent of Islamic Mogul gardens: the sounds of falling water, the lush vegetation and the cool shade form an oasis in a growing city.

Pro Memoria Garden, 1978
Ludenhausen, Germany, p.134

The people of a small town south of Hannover – completely rebuilt after World War II – were determined that their children should not forget the horror, grief and destruction of war. This project – intended to impart the lesson of peaceful cohabitation – draws on the Lower Saxon tradition of providing pensioners with a plot of land to cultivate vegetables and flowers. Each garden in the composite consists of irregularly-sized, unique, one-fifth-acre plots defined by hedge walls and separated by narrow paths. Assigned at birth, each plot contains a marble slab inscribed with the newborn's name. Children are taught the rudiments of gardening to prepare for a lifetime of responsible cultivation. When the plot owner dies, the land is reassigned and a new marble slab is placed next to the first. The new gardener is taught to respect the inheritance received, maintaining as well as improving it. The townspeople's implicit hope is that, eventually, the gardeners will clip away the hedges between the individual plots to create one large communal garden.

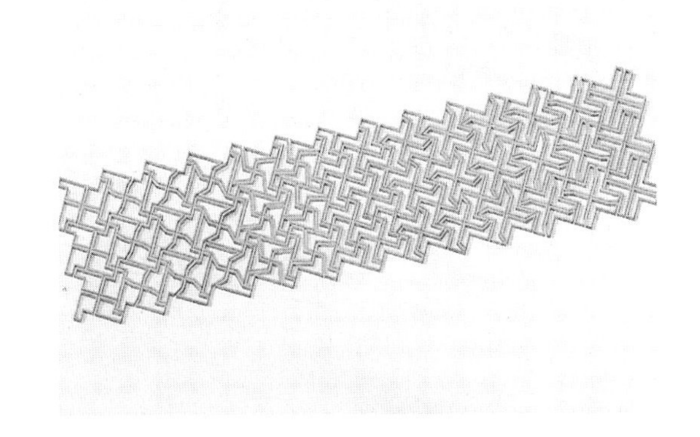

Emilio's Folly: Man is an Island, 1983, p. 136

(A Fable): No, I never thought about it in words. It came to me as a full-fledged, irreducible image, like a vision.

I fancied myself the owner of a wide grazing field, somewhere in the fertile plains of Texas or the province of Buenos Aires. In the middle of this field was a partly sunken, open-air construction. I felt as if this place had always existed. Its entrance was marked by a baldachin, held up by three columns, which in turn supported a lemon tree. From the entrance, a triangular earthen plane steeped gently toward the diagonal of a large, sunken square courtyard which was half earth, half water. From the center of the courtyard rose a rocky mass that resembled a mountain. On the water floated a barge made of logs, sheltered by a thatched roof and supported by wooden trusses that rested on four-square-sectioned wooden pillars. With the aid of a long pole the barge could be sculled into an opening in the mountain. Once inside this cave, one could land the barge on a cove-like shore illuminated by a zenithal opening. More often, I used the barge to reach an L-shaped cloister where I could read, draw, or just think, sheltered from the wind and sun. The cloister was defined on the outside by a water basin and on the inside by a number of undulating planes that screened alcove-like spaces. Once I discovered their entrances, I began to use them for storage. Although I am not compulsively driven to order and thrive, instead, on tenuously controlled disorder, I decided to use these alcoves in an orderly sequence, storing things in the first alcove until it was full and then proceeding clockwise to the next one. The first items I stored were my childhood toys, school notebooks, stamp collection and a few items of clothing to which I had become attached. Later, I started moving out of my house and into the second alcove gifts I had received while doing my military service as well as my uniform. I became fond of traversing the water basin once in a while to dress up in it, to make

sure that I had not put on too much weight. Not all the things I stored in these alcoves were there because they had given me pleasure, but I could not rid myself of them. In time, I developed a technique for using these things to support other objects. I often wondered whether I was going to run out of space but somehow always found extra room, either by reorganizing things or because some objects had shrunk or collapsed because of their age or from the weight of the items that had accumulated on top of them.On the diagonal axis passing the entrance canopy, but directly above it, an undulating place was missing. Instead of a storage alcove, there was an entrance to a man-height tunnel that led to an open pit filled with a fresh mist. I never understood where this cold-water mist originated, but it never failed to produce a rainbow.

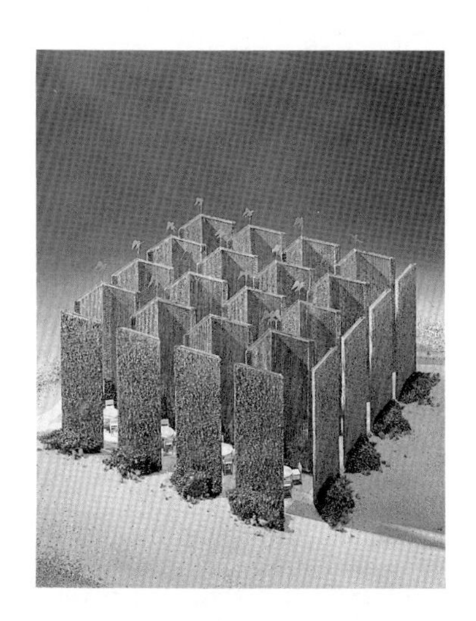

Hortus Conclusus—Centre Georges Pompidou, 1989
Paris, France, p. 138

Hortus Conclusus creates a quiet, secluded sanctuary – a contained orchard – for dining in the center of a dense metropolis. As part of an exhibition on gardens, the museum's roof terrace is transformed into a temporal realm of mystery and delight. Enveloped in mist that animates its forms with shifting light and shadow, a garden rises from the hard surface of the terrace at the end of a serpentine path. Vertical mirrors veiled by trellis panels covered with ivy and arranged in a grid alternate with open space to define the garden's perimeter wall, giving the effect of a forest of infinite depth and delineating semi-enclosed private dining spaces. Above each trellis tower, a gold-plated weather vane gleams, simulating bird songs when the wind blows and calling to mind the expanse of nature beyond the city. Crisply pruned hedges abut the panels to evoke the boundaries of a fortified city – forbidding from without, protecting from within.

Columbus Bridge, 1989
Columbus, Indiana, USA, p. 140

This design for a bridge in Columbus, Indiana, commissioned to commemorate the five-hundredth anniversary of Columbus's arrival in America, creates a sequence of experiences leading one across a river and into the town that makes it an unprecedented urban gateway. The approach road is designed as a gently sweeping curve, gradually rising upward as it crosses the bridge until the two most important vertical buildings in the city – the County Courthouse Tower and Eliel Saarinen's First Christian Church – are perfectly framed by the outstretched columns of the new structure. This dramatic first view of the city begins with the traveler passing through a dense arbor of trees that conceals all views to the city. The splayed columns spread further and further apart, rising taller and taller, in anticipation of a final crescendo as the city skyline ultimately bursts into view. Simultaneously, the splay of the columns allows guide wires to parallel the neutral axis, thereby increasing the efficiency of the bridge and significantly reducing its cost. The irregular paving of the road intensifies the anticipation of approach through punctuated and percussive sounds. At its peak, the curving road becomes smooth and quiet, descending axially in a straight line between columns to suddenly offer a spectacular view of the city skyline.

Master Plan for the 1992 Universal Exposition, 1986
Seville, Spain, p. 150

The history of world expositions demonstrates that most of them left behind ruins: roads, buildings, bridges, monorails, etc. This project – covering several hundred acres of an island in the Guadalquivir River – won the First Prize in an invitational competition for its unique solution to this problem. The master plan proposes three large lagoons on which most of the activity takes place. All of the exhibition pavilions are floating. And after the exhibition finishes, they can be taken away leaving only a magnificent garden park that would belong to the city long after the exposition is gone. This approach totally avoids having to construct in-ground foundations and the last-minute rush to build extra roads and facilities. Against this backdrop, the only buildings erected are those which ultimately become a much-needed administrative center for the University of Seville. The water symbolizes the indispensable communications link between Spain and the New World. The grounds are cooled by numerous shade trees and by cold-water mist dispensed from arbors high above the ground. A significant departure from traditional exposition design, this plan resonates with the theme of the 1992 fair – The Era of Discovery – and its emphasis on innovation throughout history.

Eschenheimer Tower, 1985
Frankfurt, Germany, p. 154

A new walkway connecting the Eschenheimer Tower to a nearby Frankfurt pedestrian area relieves the severe and chronic traffic problem created by the nineteenth-century ringstrasse or "ring road": the heavily traveled vehicular passage is too hazardous for pedestrians to cross, while the tunnels beneath it are too squalid and the overhead tram too inconvenient. A barrel vault, which encloses the road, is covered with earth to create a gently rolling hill that, in turn, serves as a pedestrian bridge. The hill becomes a landscaped mini-park uniting inner and outer sectors of the city and reconnecting patches of greenery previously separated by the ringstrasse. In series, these bridges produce a greenbelt around the city to be enjoyed by pedestrians, joggers, bicyclists and drivers alike. The walkway not only defines the edge of the urban fabric but also reinstates the Eschenheimer Tower – the city gateway – as a garden foyer.

Banque Bruxelles Lambert, 1979
Milan, Italy, p. 158

The Milan branch of the Banque Bruxelles Lambert is housed in a nineteenth-century building. The renovation leaves the original ornament intact but rigorously tones it down to create a subdued background – a visual basso continuo. Rooms are treated as ready-made stages. Columns and pilasters – pure, minimalist volumes – are lacquered black and highly polished to emphasize their form and reflect the existing ornament. Functional, yet symbolic, they are the architectural elements in the bank's urban spaces. One simple, bold device – a curvilinear, free-flowing handrail in the entrance's grand staircase, wide, straight, black ribbons placed flat on the floor and tall slender, black floor lamps in the offices – works as a black cordoning line, defining the edges where new and old meet. Each room is perceived as being enclosed or enveloped by a prism of edges.

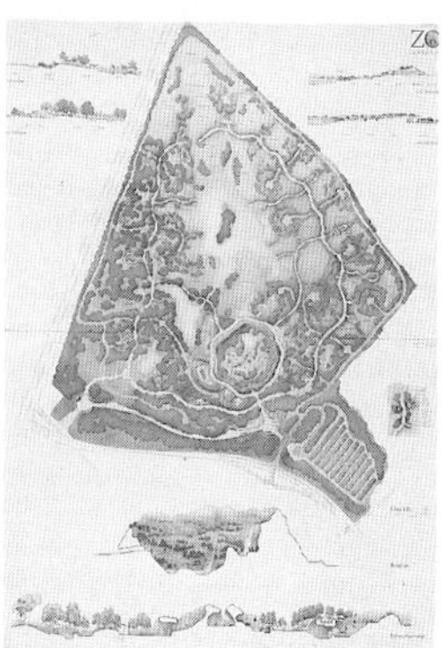

Frankfurt Zoo, 1986
Frankfurt, Germany, p. 164

This 160-acre zoo is conceived as a succession of natural habitats in which landscape and architecture are integrated as an organic whole. Each animal habitat maintains typological affinities amongst herds and follows a topological sequence: forest blends gradually into bush and eventually returns – full cycle – to the forest. Herds are separated from one another and from visitors by means of natural barriers such as waterways, ditches and berms. The animals' enclosures blend into the landscape of their habitats. Two walking paths circle the zoo. Both routes include a stop at the central Paradise, a metaphorical Garden of Eden contained within an extinguished volcanic crater that is populated by flying, walking and swinging animals that live in a sylvan valley of rocks, caves, ponds and waterfalls. The paths of the long and short journeys form constantly undulating lines through the zoo in order to minimize crowding and maximize the sense of being alone with the animals in their natural setting.

Cooperative of Mexican-American Grapegrowers, 1976
California, USA, p. 166

The climate of this site necessitates the use of a southern European grape-growing technique: an elevated grid of wires allows the vines to branch out, creating a dense roof of leaves that shades the grapes from the sun and frees the ground below for the cultivation of other crops. Within the gridded vineyard, a small, excavated open-air chapel has a cross inserted in the earth at water level. The square plots are laid out in a formal pattern reminiscent of early Hispano-American towns. New inhabitants will continue living in their trailer houses, assured of privacy by walls of hedges which separate each family's land and are the only urban reference in the valley. It is hoped that eventually these walls will be clipped away and that a more communal way of living will develop – a metaphor for the eternal wish for all walls to wither away and for man to live in peace under the vineyard's shade and off of its precious grapes.

New Town Center, 1989
Chiba, Japan, p. 172

Taking advantage of a high-speed rail system which brings distant agricultural areas of Japan within easy commuting distance of Tokyo, a leading developer acquired property to develop an entirely new suburban town. Ambasz was faced with the challenge of creating a commercial center that feels complete and welcoming even before the remainder of the town is built. The unique multi-tiered vertical garden gives a sense of arrival and enclosure, establishing a visual definition of the city center, yet screening the surrounding agricultural land which will take another ten years to develop. Simultaneously, it establishes a uniform concept for new façades of a town which needs a clear identity from the beginning. This open, ivy-covered structural grid contains a potted plant in each module, creating a natural, transparent transition between urban center and landscape. Four different varieties of flowering trees are used to heighten the drama of the changing seasons with new fragrances and colors. A huge Torii gate comprised of office towers announces the city to arriving trains which pass through it into a new elevated station. Its vegetation-hung scaffold veils the structure's glass curtain wall, visually relates to the gardens below and recalls the roofed verandas of traditional Japanese houses, becoming a dynamic symbol for a new city in search of an identity.

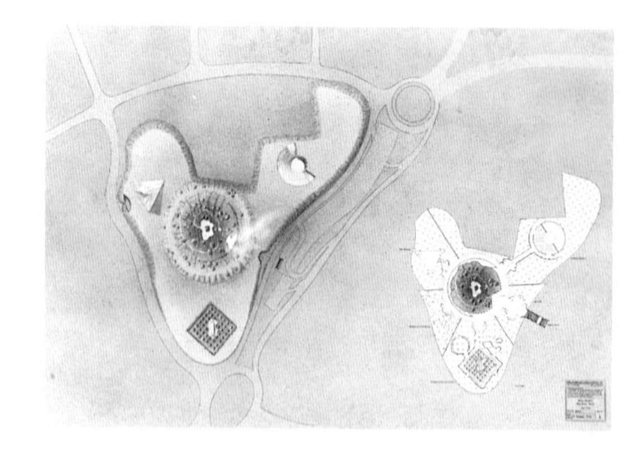

Realworld Theme Park, 1989
Barcelona, Spain, p. 182

No longer believing that a modern theme park should be restricted to themes of adolescent nostalgia and childhood cartoons, Peter Gabriel, together with Brian Eno and Laurie Anderson, conceived of a new, personal theme park for the modern age emphasizing contemporary wisdom, technology and visual delight. A great mountain plateau of stone and verdant landscape rises against the surrounding urban fabric. At the top, a glass pyramid and revolving wheel stand sentinel as symbols of innovation and the promise of technology. Visitors enter a tunnel – a sensory transition space – in the mountain, leaving the familiar world behind. Light beckons from the center of a large outdoor crater where two vaporous tornadoes form a gateway to the undiscovered world. A stream of water – Living River – begins at the top of the crater and circles the sloping sides until it reaches a green sea. Along the way, ever-changing exhibits and chambers of exploration are open to view. In this unknown realm of a garden paradise, each image denies a preconception and evokes the existence of a new, real world.

Marine City Waterfront Development, 1988
Otaru, Japan, p. 184

At Otaru, this commercial development rebuilds and improves the facilities of the port town. New structures are covered in earth and landscaped providing an extensive urban park at the harbor's edge. A canopied promenade – in the Japanese tradition of arcaded shopping streets – organizes the design by linking the existing railroad station to the new commercial, residential, cultural, educational and recreational facilities. The glazed, sinuous arcade, called the "Maritime Promenade of the Four Seasons," simultaneously offers protection from the cold winter winds and opens out to constantly changing colorful interior vistas of the wintergarden: flowers and plants exist year-round to maintain the organic connection with the natural landscape. Near the water's edge, an open-air boardwalk symbolically reflects the protected indoor arcade as it moves out toward the sea. By placing a landscaped park above the building facilities – putting the green over the gray – architecture and nature are harmoniously united.

Rimini Seaside Development, 1990
Rimini, Italy, p. 188

The redesign of nine miles of beachfront along the northeastern coast of Italy proposes to recover the myth of Rimini as an elegant seaside resort. A shaded garden promenade, or Green Dune, greatly increases the width of the existing planted walkway and runs the length of the beachfront, offering unobstructed views of the sea. The raised promenade creates a public park in which freshwater pools, bars and cafes operate around the clock, four seasons a year. Beneath the seaward edge of this garden promenade, beach-related functions and service areas are located, visible and accessible only from the beach. It has gentle ramps which extend to give direct access to the beach, connecting the town to the water. Piers project into the sea, continuing important avenues of the town across the beach and into the water. At the far end of each pier is a floating pavilion that accommodates additional concessions and serves as a sunning platform during the mild seasons. Once out at sea, the view of the city and the hills behind is enhanced by the green frame of the garden promenade.

European School of Hotel Management, 1995
Omegna, Italy, P. 194

This international school, containing classrooms, administrative offices, auditoriums, gymnasium, two restaurants, a bar and a training kitchen, is located in Omegna, Italy, at the top of a mountain peak on the northern tip of Lake d'Orta. It is oriented to take maximum advantage of views of the water and the beautiful mountain ridges which frame the southern tip of the lake. The only portion of the building complex visible from the city is a tall, circular observation tower which points in the direction of the lake and appears as an inviting landmark from the city below. This tower marks the public entry to the school. Visitors approach from the northern ridge of the site and only upon entering the tower do they suddenly perceive the spectacular view of the lake beyond. From this entry/reception area, visitors then descend to either the classrooms to the north or the dining facilities to the south. Residential units step gently downward in successive terraces following the natural contours of the site, so that all units have lake views and private gardens, without being visible from one another.

Nuova Concordia Resort Housing Development, 1994
Castellaneta, Italy, p. 204

Set outside the town of Castellaneta al Mare, in the province of Puglia, Italy, separated from the Ionian Sea only by a protected forest, Nuova Concordia is a 1,500,000 sq.ft. resort community of villas, shopping complex, international hotel, sporting facilities, meeting center, golf course, health spa and night clubs. To preserve the primal beauty of this site, Ambasz embraced the architecture and the roadways with landscaped berms and flowering trellises to make them appear like natural, undulating hillsides within the landscape. In this way, a medium density development is virtually turned into a park by the sea and one hundred percent of the land covered by the buildings is returned to the community as greenery. The complex is organized around free-form artificial lakes, the largest of which forms the nucleus of an international hotel. A landscaped island of shopping arcades is accessible by pedestrian bridges or by boats from the open public entry plaza. Surrounding the international hotel is an assortment of housing types and private villas, becoming lower and lower as they approach the forest, thus maximizing views and diminishing the overall impact on this spectacular site.

Baron Edmond de Rothschild Memorial Museum, 1993
Ramat Hanadiv, Israel, p. 212

This new Memorial Museum for Baron Edmond de Rothschild, the Founding Father of Israel, hopes to speak eloquently to the heart but with a soft and whispering voice. Drawing upon the Eastern tradition of shaded gardens, the new museum creates an encompassing sense of peace and serenity while leading visitors to the gateway of the existing European-style gardens. Arriving at the first circular entry court, visitors walk quietly past a single leafy and shading tree, representing Israel's Founding Father's historic commitment to agriculture. Visitors emerge into an inviting ambulatory surrounding a rectangular, shaded courtyard and reflecting pool. An array of columns bearing olive trees appears to grow from the very waters of this pool, subtly arching to a high point at the center. The most important challenge within the Memorial Museum was to address not only a need to gently guide the first-time visitor with varied choices, but also to stimulate the repeat visitor who may wish to bypass the exhibition areas and proceed directly to the gardens. In this design, the needs of both types of visitor have been reconciled.

JARDINS DE LA FRANCE, 1992
CHAUMONT, FRANCE, P. 222

Chaumont, a large hill facing the Loire River, in France, is famous for its Gardens Biennale, an exhibition that every two years allows invited participants to build their image of an ideal garden. This is done on a very small piece of land. When summer ends, these gardens are dismantled and the land becomes available for the next gardens, two years hence.
The winds blowing on the hill have pernicious effects on the plants. To avoid the wind-related problems, Emilio Ambasz imagined his garden as a grass-covered amphitheater excavated into the hill. As expected, a new microclimate developed within this earth excavation and the plants were able to grow robust, covering all the ground available. This happened because the plants were protected from dehydration within the amphitheater's confines. The visitors, sitting on the steps created within this excavation, could enjoy an unexpectedly warmer climate. They could also choose to remain in the balmy open sky amphitheater, or to enter the small grotto excavated on the steep side of the amphitheater. As the plants were able to extend their roots well within the earth, the vegetation originally planted became an exhuberant "jungle." This garden became such a popular place that the organizers decided not to dismantle it at the end of its exhibition period but to transform it into a permanent installation.

LA VENTA OFFICE COMPLEX, 1993
LA VENTA, MEXICO, P. 224

Located on the far outskirts of Mexico City, the region of La Venta contains a man-planted forest of pine trees, once used for paper production and now abandoned. Mr. Ambasz was commissioned to develop an architectural concept for La Venta which would maintain the very nature of this forest, while providing development income to fund its preservation and that of its surrounding woodlands. A series of nurseries was designed as stepped or sloping terraces, covering the roof of an office building below. The income from these offices then can be used to help finance the continual future preservation of the forest. The office buildings blend completely into the existing forest landscape as the heights of each new structure seek to match the surrounding tree canopy. Positioned on the downhill slopes of the site, the buildings are invisible from the highway as well as fully insulated from the sound of the highway's traffic.

Clean Coal-Powered Electrical Plant, 1994
Hekinen, Chubu Prefecture, Japan, p. 230

Power stations can be considered one of the greatest anomalies of the twentieth and twenty-first centuries for they provide power that is vital to every aspect of our everyday lives but often appear as blights upon the landscape, viewed universally as adversaries of the environment. In an attempt to overcome the negative public image of the power station in Hekinen, Emilio Ambasz has invented a new look for this building which makes it appear both inviting and "environmentally friendly."
A simple veil of cables and wire mesh extends above the roof line in a series of peaks behind which landscaped planters are attached to the entire façade. The ultimate effect is that of a verdant hillside shimmering in the breeze. Three high smoke stacks stand between the main power plant and the harbor and Mr. Ambasz infills the diagonal mainstays with wire mesh to create the appearance of a great mountain upon the horizon, appearing softly beyond the green hillside that veils the main plant. As smoke wafts gently from the stacks, the effect is like a gentle cloud embracing the top of the mountain. The original buildings are not changed structurally in any way and for minimal expense a major transformation has occurred which visually benefits the entire community.

B-ticino Exhibition, 1994
Venice, Italy, p. 236

In conjunction with the '95 Venice Biennale, B-Ticino commissioned Emilio Ambasz to design an interior exhibition for the company entitled "Living and Light." Mr. Ambasz's design visually transforms this space into a metaphor for living, a passage through time. From the entry portal to the exit portal, an arching bridge spans the entire length of the space, connecting one portal to the other. The entry portal is over-sized and off-axis, as if leading to an alternate world and dimension. The floor below appears as softly illuminated white mist, wafting like a gentle cloud in response to the movement of people above. The ceiling appears like a surreal "sky," brightly illuminated yet sharply delineated by the boundaries of the bridge. In the distance the diffused image of a window is suspended in space as if undefined by any wall. Through this window the hint of a future world is visible, the new Green Town where gardens merge in harmony with architecture, the "Green over the Gray." Upon the bridge stand four gateways, each appearing to be sequentially transformed and evolving from the previous one. At eye level upon each gateway is a switch plate and when each switch is touched, soft words of wisdom are spoken. Then silence returns until the next gate is entered. Switches are Keys to Other States...

Air Force Memorial, 1994
Arlington, Virginia, USA, p. 240

Emilio Ambasz was one of eight architects selected to compete for the design of the new United States Air Force Memorial to be located near Arlington National Cemetery on axis with the Lincoln Memorial, the Washington Monument and the U.S. Capitol. Emilio Ambasz's design is composed of a building and a public plaza whose forms symbolically unite to represent the image and achievements of the US Air Force. The main building is in the form of a sleek elongated triangle, thrusting upwards from the earth. Its massing simultaneously suggests the innocence of a child's paper airplane and the high technology of a stealth bomber. The space inside is a museum of the history of the United States Air Force. The public plaza in front is a fog-covered circular depression with stepped edges that can be used as an amphitheater for special events and the center of the circular plaza uses computer-generated illumination to create the image of the earth turning on its axis. When viewed from above, the plaza takes on the appearance of Earth itself as seen from a shuttle in outer space about to return home from a vital mission. The image is simultaneously one of highest technology united with the natural embodiment of Mother Earth, harmonized by the imagination and the technology of the United States Air Force.

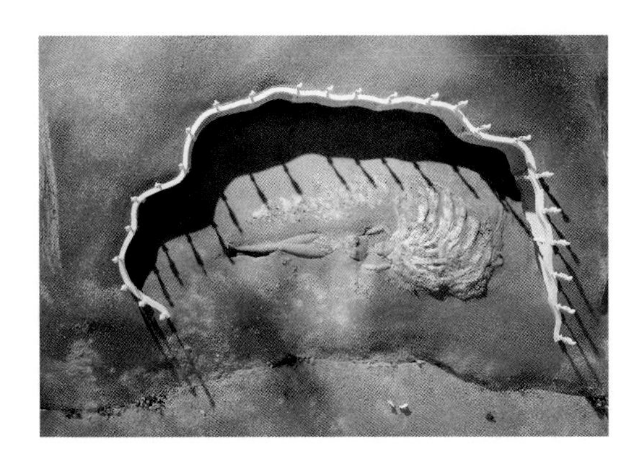

Barbie Knoll, 1995
Pasadena, California, USA, p. 244

"Barbie Knoll," commissioned by Mattel for an international traveling exhibition, was conceived by Emilio Ambasz as an ultimate commemoration of the Barbie doll. "Barbie Knoll" is the ecological embodiment of the "Barbie Doll," the transformation through the years to the ultimate icon. It rests partially visible in the meadow like a half-covered ancient Greek statue, immersed in greenery having become an integral part of the primordial garden itself. Families visiting "Barbie Knoll" arrive at a sacred entry courtyard reminiscent of a Greek temple. Passing through this open-air entry court, visitors arrive inside a grand exhibition space where each column is a large-scale Barbie sculpture showcasing the evolution of Barbie herself. The grand interior space is a permanent display exhibition showcasing the history of Barbie for children and adults to see and enjoy. As children pass beyond the grand exhibition space to the outdoors, they finally see "Barbie Knoll" silhouetted on the horizon, covered in grass and wildflowers. Children can climb upon her, tumbling and playing, making her become not only an ageless symbol of youth but an enchanted playground. Lying protected in a green, outdoor courtyard, Barbie lies resting in the sun, an icon to the happiest times of days gone by and days to come.

Lisbon EXPO '98 Commercial Development, 1995
Lisbon, Portugal, p. 250

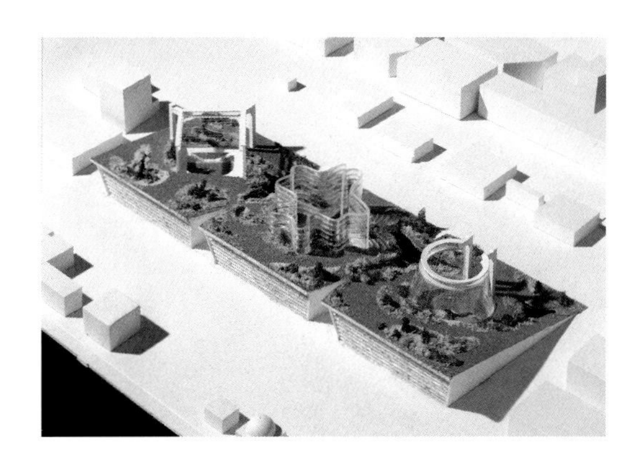

In preparation for Expo '98, the city of Lisbon selected a site between the marina and the Calatrava train station for the development of 300,000 sq.ft. of new commercial office space. The new site slopes downward toward the marina from the main commercial street above and Mr. Ambasz took advantage of the sloping site to suggest a highly unique solution. He proposed that a "garden platform" be extended outward from the upper edge of the site, completely covering the sloping area below. This "garden platform" becomes both the roof of the new building and a new garden park for the surrounding community. The resulting enclosed volume is more than adequate for the commercial needs of the development. Each of the three main commercial tenants desired a unique corporate identity, recognizable even from a distance within the city, and Mr. Ambasz designed three distinctive glass structures inhabiting three open courtyards. With "walls" of clear glass punctuated by a grid of "windows" composed of open voids, the three structures appear transparent: recognizable urban structures, yet completely ethereal in nature. The city is given a gift of a new community garden park, accented by flowering plants and trees, pools of cascading water and uninterrupted views to the sea beyond.

Utrecht Residential and Commercial Development, 1995
Utrecht, The Netherlands, p. 254

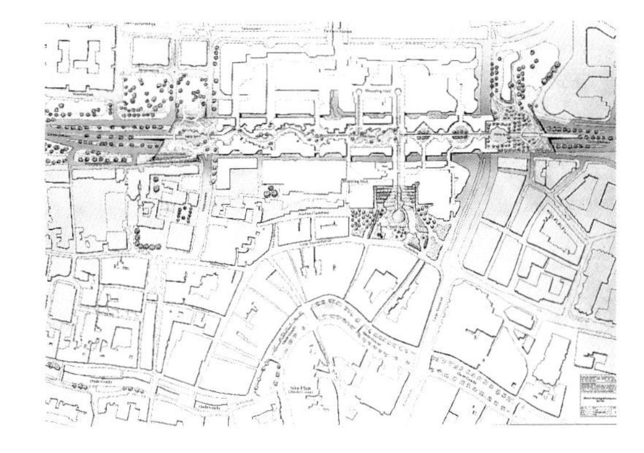

A continuous garden park and ring canal, the Catharijnesingel, still encircle virtually the entire old city of Utrecht. Within the ring lie a charming medieval city of elegant boutiques and cafes; outside the ring lies modern shopping centers and housing developments. This sharp contrast is most evident at the one area where the Catharijnesingel has been filled in by the city for a new highway. This once vital location now appears like a gaping wound in the heart of the historic old town. The city of Utrecht realizes the need to reclaim the original image of a continuous canal and garden park surrounding the old city, yet simultaneously acknowledges the enormous value of this location for new commercial space. To solve this dilemma, Emilio Ambasz proposed that an elevated pedestrian garden platform be constructed above the highway, thereby hiding the traffic below and creating a garden park surrounding the old city once again. At the center of the platform, Ambasz placed a channel of water to symbolize the old Catharijnesingel, flanked by shops, cafes and housing made of brick and stone in the character of the old city. These buildings are articulated to form a series of intimate courtyards and in this way the original character of the old city returns once more, using modern materials and a modern vocabulary.

KANSAI-KAN OF THE NATIONAL DIET LIBRARY, 1996
KANSAI SCIENCE CITY, JAPAN, P. 258

The new Kansai-kan of the National Diet Library, the national central library in Japan, will be the core facility of the new town of Kansai Science City. The Kansai-kan will be composed of stacks, service system, library cooperation system, operational system, administrative system, common departments and training center. As the principal archive of the National Library for all of Japan, it is important that the symbolic form of the new library embraces the philosophy, spirit and history of the Japanese people. For this reason, Emilio Ambasz uses the form of the ancient earth mound, one of the earliest architectural representations of sacred wisdom in Japan, as a powerful architectural prototype for the new national library archives. The outer form of the library appears from the front as a natural mound of earth, serenely landscaped in the tradition of Japanese meditation gardens. Embraced by a reflecting pool, it suggests the fundamental beauty and serenity of the natural environment, protected from the chaos of the modern world, a place of wisdom and reflection. An undulating stone wall slices through the mound of earth, transforming it on the opposite side into a five-tiered series of man-made, flowering terraces. The natural hillside on one side and the man-made hillside on the other co-exist, inseparable and in harmony.

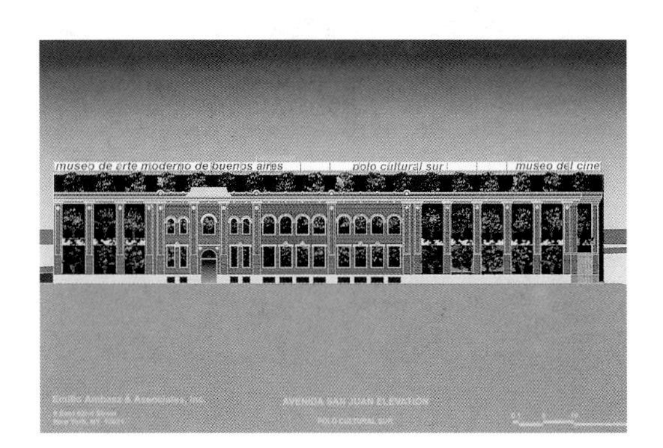

MUSEUM OF MODERN ART, AND CINEMA MUSEUM (MAMBA), 1997
BUENOS AIRES, ARGENTINA, P. 262

Situated in an historical district, the two independently operated museums form a single cultural center known as the Polo Cultural Sur. Each require separate entrances, collections, workshops, auditoria, archives and administrative facilities. These two buildings have little in common spatially or stylistically: one an aging modernist concrete frame office building (Cinema Museum), the other a handsome nineteenth century factory building known as the Tabacalera (Museum of Modern Art). The solution is found through the use of Grand Façades. The front and side façades are conceived as a complex façade of "hanging gardens" which dually shelters the interior spaces from the direct north light and donates to the street a lush, green backdrop. The rear façade, which faces onto an elevated highway, is conceived as a giant video projection, a kind of "moving billboard," to provide an appropriately scaled presence for the zooming traffic overhead.

With the façades modulating the exteriors, the interiors are designed to be highly functional and archivally sound spaces for each museum. The Tabacalera building's well-detailed masonry and concrete structure preserved as loft spaces offer the greatest flexibility and protection for the extensive and extremely varying spatial requirements of contemporary art. The new gallery building provides huge, unobstructed space for large installations. New spaces created below ground level provide storage, archives and a multimedia auditorium. The Cinema Museum's concrete frame holds three screening rooms, the largest of which occupies the lower portion of an existing courtyard, as well as gallery spaces for the permanent and rotating exhibits, gift shop, library and reading room, and administrative offices.

Marina, 1998
Bellaria, Italy, p. 268

In designing the new Marina for Bellaria, one of Italy's most frequented shores, Emilio Ambasz further developed those principles which have successfully guided other designs destined to improve the environmental quality of beach resorts, such as the nearby beach of Rimini or the tourist center in Castellaneta al Mare, in Puglia. This approach – which substantially consists of returning, in the form of gardens, every square inch of earth which has been subtracted to nature by the building's footprint – translates itself, in this specific case of Bellaria, into the design of a generous vegetation-covered "roof," which hides under such a plane all the necessary facilities for a marina, such as parking, boat moorings, repair shops, restaurants and clubs.
This "roof" is like an embroidered, undulating carpet. It extends from the edge of the residential area towards the port. This roof/garden's skyline resembles the wavy surface of the sea towards which it offers a belvedere. Located near a remarkable maritime colony for children built in the 30s, this variegated garden fits panoramically the vision of a marina while at the same time offering it a verdant terrace.

Office Complex, 1998
Hilversum, The Netherlands, p. 272

Hilversum has a well-earned reputation as a "green" city. This is due to the generous areas destined to forests which surround it. The project takes place on the edge of this natural forest reserve. Here the city possesses the last available land for building. On this extension of approximately 12 acres Emilio Ambasz has had the commission to design a complex of office buildings which will accomplish the miracle of adding 600,000 sq.ft. of new office space without taking away even one square inch of greenery.
As it is now, Hilversum is the city of Willem Marinus Dudok, who worked as Director of Public Works and Architecture for the City of Hilversum in the first decades of the twentieth century. The city enjoys a high reputation for the splendid architectural quality of its schools and especially its municipal building. Appreciating the healthy number of visitors who come to admire Dudok's work, the city has become sensitive to the importance of the quality of architecture in daily life, as well as a source of citizen prestige. It is therefore logical that, even in this case, the city should have sought an architect capable of providing such a large amount of square footage while at the same time maintaining the environmental qualities of this much-prized area. Ambasz's proposal foresees the construction of an office complex of approximately 600,000 sq.ft. placed within three buildings which from the outside resemble hills cultivated in a terraced manner. A thin glass plane cuts these artificial hills in half, following the grounds of a once-existing hippodrome, transformed on this occasion into a walkway that connects the three buildings.

Argentine Pavillion at Biennale Venezia, 1994
Venice, Italy, p. 278

Organized as a year-round complex of international cultural pavilions, the Venice Biennale is the premier showcase of exhibitions representing the arts and the technology of major countries from around the world. Because Venice is densely urbanized, the few remaining green areas are highly valued and it is now almost impossible for countries not yet represented at the Biennale to find a site for new pavilions. Argentina approached Emilio Ambasz with this seemingly impossible dilemma and Ambasz proposed an ingenious solution. Ambasz's unique design for the Argentine Pavilion utilizes his concept of the "Green over the Gray" to create a very large new pavilion, while returning to the city of Venice virtually 100% of the green landscape of the original site. Ambasz selected a prime location at the Biennale that presently is a green open space where people enjoy gathering outdoors. Ambasz began by designing a landscaped circular courtyard which is the entry to the pavilion. The groundplane cants slowly upward toward the far corner of the site, creating an enormous space within. Covered in greenery and flowering plants, this canted roof remains a lush garden park for visitors to the Biennale, yet completely hides the fact that a large pavilion space lies below.

Concert Hall on the Waterfront, 1993
Copenhagen, Denmark, p. 280

The waterways of Copenhagen make it one of the unique cultural cities of Northern Europe. Located on the site where Frederiksholms Kanal joins Inderhaven, this design for a new concert hall facility celebrates Copenhagen's waterways by actually allowing musical and cultural events to travel throughout Denmark upon its waters. The grand Concert Hall stands as a bold anchor upon the quays while the Auditorium and Copenhagen Philharmonic are each housed on floating barges. These barges can travel to any other port in Denmark, providing an extraordinary ability to introduce live music and cultural performances to communities which would otherwise never have such an opportunity. During the primary "music season" for the city, these two barges remain secured to the quay, embracing each side of the main Concert Hall and becoming formal symmetrical "wings," much like the great traditional concert halls of Europe. For other occasions, however, the barges have the opportunity of turning sideways and engaging the Central Square in a variety of diverse configurations. In this way, the concert hall becomes a unique dynamic structure, ever changing throughout the year, even allowing one or both barges to take the great musical events of Copenhagen to distant regions lacking music halls of their own.

Thermal Gardens, 1996
Sirmione, Italy, p. 284

Located on a lush peninsula at the southern tip of Lago di Garda in northern Italy, the medieval town of Sirmione is a lake resort that has been famous for its hot sulfur springs since ancient Roman times. Sirmione has developed strict regulations limiting the extent of new construction, yet ever-increasing numbers of tourists are expected each year. Mr. Ambasz was therefore commissioned to develop an architectural concept for a series of indoor and outdoor thermal pools, restaurants, bars, changing rooms and additional service amenities to accommodate up to 1,500 additional visitors a day, while not affecting the character of the existing garden landscape. Mr. Ambasz proposed a unique design solution that he refers to as the "Green over the Gray." The assorted pools and changing facilities are bermed with earth and then landscaped so that the new structures virtually disappear and the garden-like character of the original site is completely maintained. Only a few vestiges of freestanding colonnades are allowed to remain partially visible, like modern translations of ancient ruins standing enigmatically within the landscape. In his concept of the "Green over the Gray," Mr. Ambasz places a very high priority on nature, and the total number of trees on the site actually increases by 15%.

Environment Park, 1997
Turin, Italy, p. 292

Environment Park is situated in the center of an extensive post-industrial landscape on the periphery of the city of Turin, Italy. It is an area of approximately 20 million sq.ft., which the city master plan designated as an area for a green park. At the same time there are a number of ongoing attempts to transform the abandoned factories and sheds of this area into an urban concentration of tertiary industries, service and research as well as lightweight industries. Environment Park is the first technological park in Europe entirely dedicated to the problems of the environment. It is an ambitious architectural project seeking to further Ambasz's idea of "green architecture," which aims at establishing a strong relationship, technical as well as symbolic, between "green" and "gray," between garden and building.

This project creates a unitary landscape which also seeks to rediscover and give value to the adjacent River Dora. Thusly Environment Park is a project with a deep-seated ecological and environmental commitment, which is quite aware of the need to control resources and reduce energy consumption by means of natural and sophisticated techniques for the management of the buildings.

The complex of buildings of Environment Park consists of 600,000 sq.ft. organized into several blocks of compact buildings built on three levels. They present themselves with roofs covered by gardens which are bisected by walkways integrated into this new artificial green landscape. In addition, the buildings' vertical walls are covered with a dense net of nylon on which ivy is to climb. So, in addition to the horizontal gardens, this project also provides "vertical gardens." The public can use the roof of the buildings, starting directly from the sidewalk ascending a very gentle ramp as if it were a public park. This process of turning Environment Park into a "natural park" is also extended to the façade of the buildings facing the River Dora. With trees placed on the different balconies protruding at each level, they constitute a "green skin" of trees.

ENI HEADQUARTERS, 1998
ROME, ITALY, P. 296

"ENI: the Palace of Vertical Gardens" is the result of a radical transformation of the existing headquarters of the ENI Company outside Rome. ENI is the petrochemical cartel of the Italian government. This building, built in 1963, has proved insufficient in its functional arrangements and its curtain wall façade has become defective over the decades. These insufficiencies were at the basis of the design of this new architectural complex. Its principal characteristics lie in the redesign of the façades facing east and west which confer to the building the image of a veritable "Vertical Garden" twenty stories high. The designer, in addition to solving ecological problems as well as the problems of reducing heat loss, also sought to associate the image of this colossal petroleum company with the idea of a dynamic industry that is sensitive to the problems of ecological equilibrium. The new project also goes a long way to show that it is possible to create tall buildings which establish a much more human relationship with the surrounding urban context and with the hopes of society.

The image of ENI as a company in harmony with the environment will be further reinforced when the plants on the façade change seasonally from leaves to flowers and from flowers to branches. This metaphor and icon, offering flowering energy, informs the new image of ENI, rendering this company an important forerunner of an industry concerned with maintaining and improving the quality of life through an architecture that respects the human and natural environment.

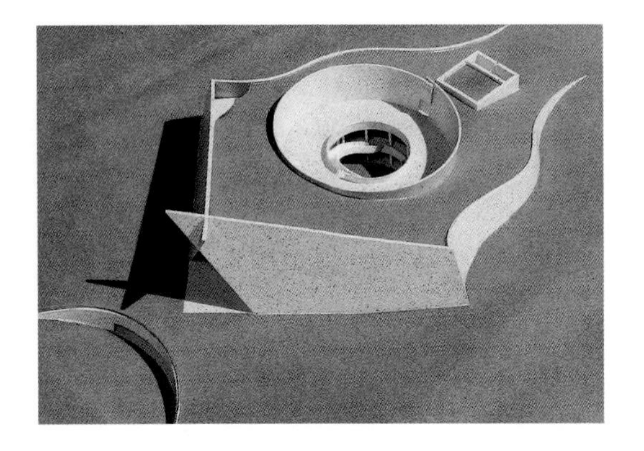

GLORY ART MUSEUM, 1998
HSIN-CHU, TAIWAN, R.O.C., P. 306

Mr. Glory Yeh is an important collector of art of the new Chinese/Taiwanese avant-garde. The importance of this new work is now beginning to become recognized even in the world of Western culture. Important exhibitions of this emerging mode of expression have been held in the USA and Europe. In order to make his private collection more accessible to his compatriots, Mr. Glory Yeh has commissioned Emilio Ambasz to design a new museum in the city of Hsin-Chu, the capital of Taiwan's "Silicon Valley."

For Ambasz, the conception of the museum has provided new possibilities to express his longings for an architecture which is reconciled with nature, adding to the palate of his "natural architecture" vocabulary a further element of interest: the theme of the roof/façade. Thanks to this unique idea, the visitors will find themselves immersed in a spatial continuum that revolves tightly inside and outside the museum, integrating the sculpture garden around it with that on its roof/façade. The emphasis on the ceremonial entry to the grand hall for public events gives further proof of the important cultural role that the new museum is called to play in the context of Taiwanese society. With its 100,000 sq.ft., and the great flexibility of the exhibition spaces, this building reveals strong innovative characteristics.

Shopping Center, 1999
Amersfoort, The Netherlands, p. 314

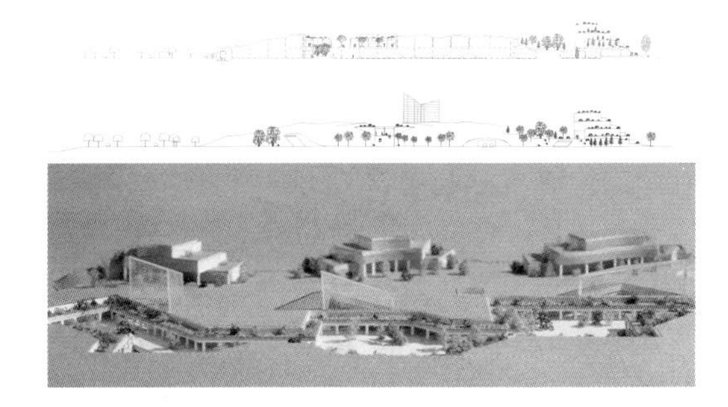

This 17,500 sq.mt. shopping center comprises two supermarkets, one discount store, 30 retail shops and a large amount of covered parking places. In addition, although originally not foreseen, designing this shopping center as a green area permitted the developer to request and obtain a building variance that allows him to build 78 apartment units.
This shopping center has been designed so that the ground level gently berms up to the roof continuing as a green grass-covered blanket on the roof. In this way, the building gives back, in the form of accessible gardens, almost 100% of the ground the shopping center covers. It is for this reason that the building variance was extended to allow for residential housing, thereby greatly increasing this site's rentability.
Vathorst, a new 27,000 people town next to Amersfoort, has been publicized as a green town. The only amount of green this town will actually enjoy will be on this shopping center when built as designed.

Master Plan, 1997
Barletta, Italy, p. 318

The program for Barletta given to Ambasz did not contemplate the restructuring of the old historical center.
The master plan was just to propose a touristic settlement along ten kilometers of coast to be coupled by commercial and residential development. It was one of Ambasz's primordial goals that the project be a green project as far as the newer areas were concerned. Going beyond his brief, Ambasz redesigned the town center with an emphasis of recovering pedestrian access to the sea which had been taken away by truck roadways leading to the port. To do this he took advantage of an abandoned railroad track placed on a deep trench. The trucks are directed to travel along this trench and the trench is to be roofed over with greenery. To allow for natural ventilation, one side of the tunnel has large window like openings. Whenever a level occurred or a street had to be crossed, a series of "wide green bridges" have been devised so as to not interrupt the flow of the green park. In order to further enhance the quality of some derelict areas of the old historical town, a high level promenade was created in the form of a pergola – shaded belvedere – like street. A wide coastal avenue that inhibited the access of the citizens of Barletta to the beaches was covered with trees and transformed into parking places, thusly turning a seldom frequented beach resort into an adornment to the city and an enjoyable place. Open areas that had been abandoned and illegally turned into were trasformed into a very lively park, with sport facilities and an open air auditorium. In all cases, parking was provided above ground but covered with earth berms so that no building is visible and no land is forfeited.
The newer touristic development and residential settlement answer to Ambasz' design principle that the buildings should be so integrated into the landscape that they are invisible to their neighbors and that their gardens be connected to the roofs by means of easily walkable earth ramps. This device will allow to maintain almost all of the Mediterranean vegetation presently growing in the area. Provisions are being made to utilize the value added to the land to finance annual maintenance and improvements.

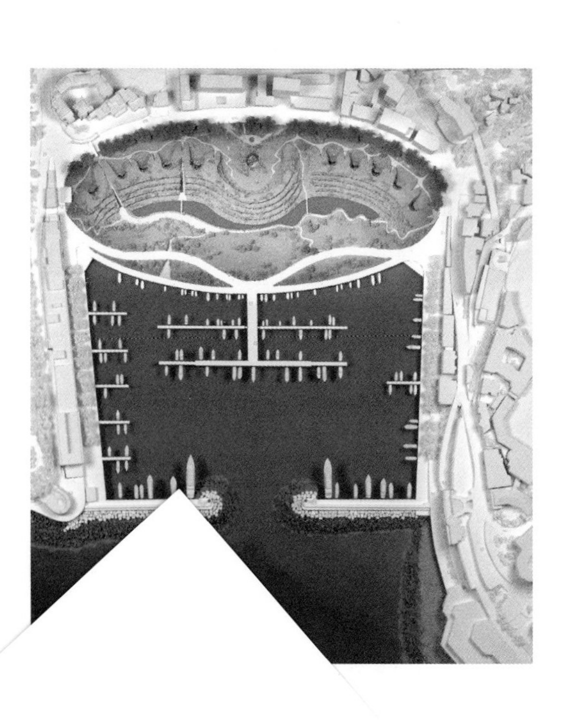

Public Park and Residences, 1998
Old Port, Monte Carlo, p. 326

This project takes into account the fact that Monte Carlo has gradually lost all of its greenery except for some precious and extremely well tended spots. This project intends to give to Monaco a 5 hectare – park plus another 5 hectares of today not existing land for residential and hotel construction. This land would be reclaimed from the sea by a very economical device.
The present port of Monaco is approximately 18 meters deep and a dam made of earth cement of this depth would be constructed. This would create the 10 hectare area. The dam itself would contain parking facilities as well as a road on top to improve the present design of the Formula 1 car races. Along this road on top of the dam there would be temporary for the Formula 1 spectators.
All construction would be on an open air, but below the level of Avenue Albert I. Seven story high apartments and hotels with deep gardened terraces would be built facing a park adorned with ponds and playgrounds.
This idea has been carefully worked out by the project's Dutch engineering consultants who are the world's most experienced specialists in dam construction. The land thus recovered is land created anew since it is gained from the sea. The cost of recovering the land is still lower than the cost of the no longer existing downtown land. This project can be built without engaging in the very expensive and extremely time-consuming process of gaining land by creating an artificial peninsula. The cost of the resulting residences is not any higher

than that of any other residences recently built in the area and the social advantage of having an amenity such as a park would certainly be appreciated by a densely populated community.

Winnisook Lodge, 2000
Catskill Mountains, New York, p. 330

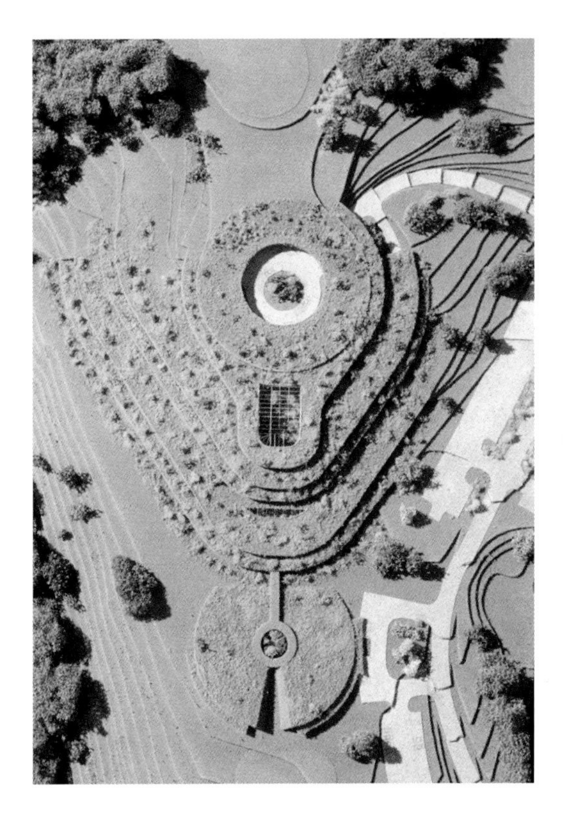

The Winnisook Lodge is located within the Catskill Park in upstate New York on a 2,000-acre private golfing community on the slopes of Belleayre Mountain. This 150-room luxury hotel will be the centerpiece to a resort comprised of an 18-hole golf course, private home and golf clubhouse.

The site for the hotel is a wooded ridge stretching from the summit of Belleayre Mountain just below the 18th hole. It has exposure to the north, east and south sides of the ridge offering panoramic views of the surrounding landscape.

In order to obtain approval with the many state and local governments as well as the many environmental groups opposed to developing the area, Emilio Ambasz was asked to design a 226,000 sq.ft. building that is virtually invisible within the natural landscape while taking advantage of each possible vista to make it a premier hotel destination.

To achieve this, Emilio Ambasz has designed the project into two buildings: the main building for the hotel with the restaurants, banquet facility and support facilities and a second building housing the spa and fitness center and the golf clubhouse. The main building is organized on five levels, each conforming to the natural terrain to effectively integrate the mass into the landscape with minimal excavation while approximating the original profile of the ridge. Each floor conforms to the natural contour of the ridge and steps back at each level creating roof terraces. The entire building is then blanketed with native plants which essentially make it invisible. The spa and golf clubhouse are situated in a one-story structure reached from the lowest floor of the hotel.

Ambasz has placed the lobby at the highest floor where guests enter into a circular motor court to reach the reception and restaurants. Arriving guests are offered breathtaking views of the surrounding mountains as they check in or while dining in two restaurants or sipping drinks in the lounge. Patios on the roof terrace offer guests the best views.

The lobby continues down one level to a pre-function lounge near the banquet functions. Here, a ballroom for 200 people and three 350 sq.ft. conference rooms are located. From these spaces doors open to the outside expanding the events onto the roof terrace.

At the core of the hotel is a five-story skylit atrium filled with plants and waterfalls. The lower floors are reached by descending through the atrium in glass elevators. The atrium expands towards the bottom into a large space where guests may fully experience this interior landscape year-round.

At each of the four guest floors, elevators open onto generous sitting areas facing the golf club and spa at the base of the hotel. These intimate spaces provide guests a chance to orient themselves within the hotel and to walk outside onto the roof terraces. From each sitting area wide corridors connect the guestrooms that wrap around the atrium.

To optimize views, the guestrooms have lofty ceilings and are divided onto two levels: the sleeping area and bathroom are located a few steps above a sitting area. From their beds guests have unobstructed views of the mountains. Each room has a private terrace facing the view.

At the base of the hotel lie the health spa and golf club in a structure that is reachable both from within the hotel by a covered walkway and by a separate public entrance from a driveway. These functions are housed under a sloped, landscaped roof that is circular in plan.

Monument Tower Offices, 1998
Phoenix, Arizona, p. 336

In the center of Phoenix's downtown core, opposite Patriot's Square where the city was founded, two spectacular 34-story office towers rise from the desert floor with all the wonder and mystery of the rock formations of Monument Valley. In keeping with an environmentally responsive agenda, the buildings' design responds to the climatic demands of the desert in a visually evocative and cost-effective way while also providing distinct and flexible amenities to its commercial tenants.

The entire exterior of each tower is clad using a system of horizontal fins that deflect the sun and shade the interior glass-enclosed floors, thereby minimizing the need for costly mechanical cooling systems much needed in the hot climate of Arizona. The façade components are standardized but flexible to respond to the constantly changing angles and slopes of the walls whereby each panel is adjustable to maintain visual consistency over the entire façade. The horizontally oriented fins are spaced tightly to block any direct sunlight entering the office behind to dually reduce the heat gain and to achieve solidity and monumentality.

The highly unique interior organization addresses the needs for flexible, multi-tenant spaces. Affording multiple tenants with a sense of ownership, each tower has four distinct double-height tenant lobbies directly opening onto the street and discreetly connecting to the ground floor elevator lobby. A common lobby is entered through the plaza between towers for smaller tenants. A mezzanine floor offers multi-function rooms for the tenant's use. Every office floor features a double-height elevator lobby achieved by alternating odd and even floor elevator service. Adjoining these lobbies are atria for major tenants.

With the intent of increasing pedestrian traffic and providing a shady respite from the oppressive heat, a "canyon" plaza between the towers also serves as entrance to the six floors of underground parking and service area.

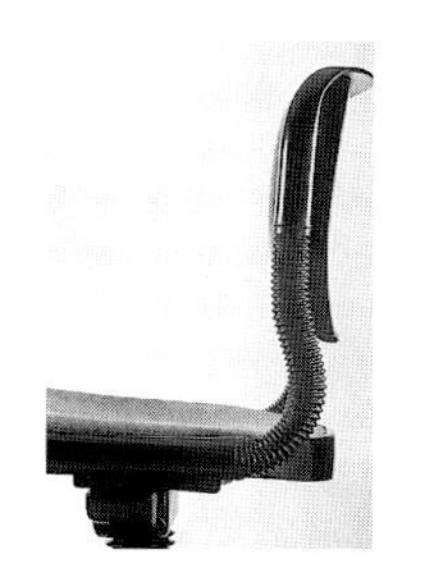

Vertebra Chair, 1974-75, p. 316*

The Vertebra seating system responds automatically to the human body's desire for comfort. Since people rarely stay in the same seated position for more than ten minutes, the chair is designed to keep up with the sitter's every movement – without the use of buttons, levers or mechanisms. Whether leaning forward or back, the chair follows the body's requirements when shifting positions. Recognizing that the ideal chair requires no adjustment whatsoever, Vertebra changes its configuration gently and automatically.

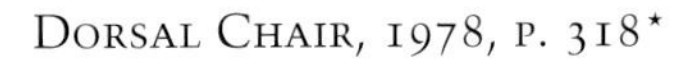

Dorsal Chair, 1978, p. 318*

Like Vertebra, Dorsal was created with the concept that a chair must function as both an ergonomic design and as a cultural artifact. Economically-minded this institutional chair is designed to provide ease of adaptability to the range of the contemporary office worker's seating positions. Its automatically-functioning backrest and overall structure employs both ergonomic and orthopedic research, resulting in the relaxation of the sitter's body. The minimal design becomes a part of the worker in motion rather than an overwhelming visible structure.

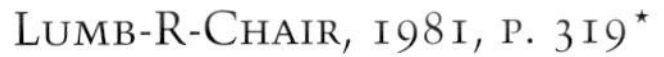

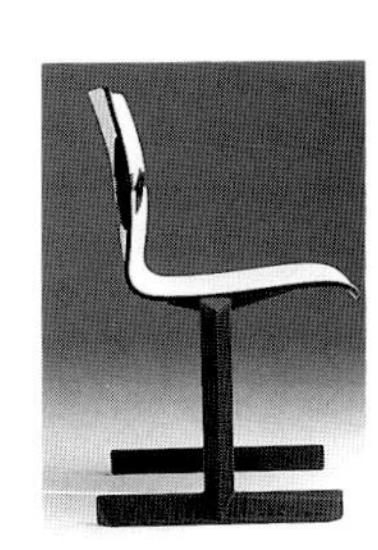

Lumb-R-Chair, 1981, p. 319*

Many office workers lean over their desk for extended periods of time, commonly developing lower-back problems or spinal pains in the lumbar region. This chair incorporates solutions to these work-related injuries into its seat and cut-out backrest design. All movements are automatic, making every seating posture possible without difficult mechanical adjustments. Optimal weight distribution is achieved when the sitter leans back, the center of gravity shifts with him, relieving muscle and spinal-cord strain. The rounded edges of the chair provide maximum blood circulation relief. Designed as a sturdy, attractive line of articulated seating that ensures the worker's comfort, the LUMB-R chair also increases the possibility for more efficient and productive work.

Qualis Office Seating, 1989, p. 320

This is a completely automatic, ergonomically-conceived seating line ranging from computer-operator seating to executive chairs. The chair celebrates its aesthetic and symbolic qualities while underplaying its highly-advanced technology, creating a "soft-tech" look. Adaptable to any decor, the colorful upholstery can be mixed and matched on a single chair, unzipped and washed, or replaced. Blood circulation in the thighs is improved due to the self-adjusting tilt-forward, while the high backrest hinge point enhances back support.

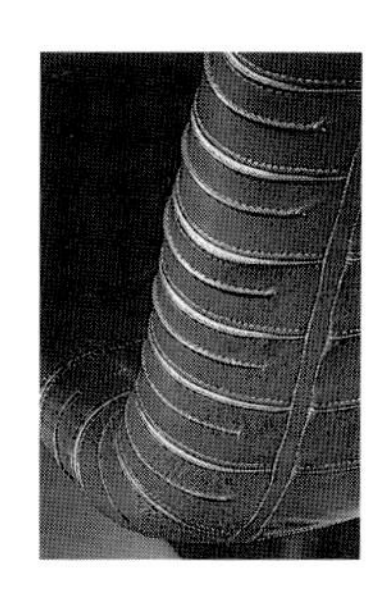

Vertair Chair, 1991, p. 322

This articulated chair conforms automatically to the body's continuously changing posture. A constant flex point in relation to the sitter's body is maintained as the back moves downward while the seat moves forward. The patented upholstery system is made of narrow, overlapping bands of leather that are stitched to elastic bands, formally expressing the chair's flexibility in a soft-tech look. The covering expands and contracts with the sitter's movements while allowing the surface to breathe for greater seating comfort.

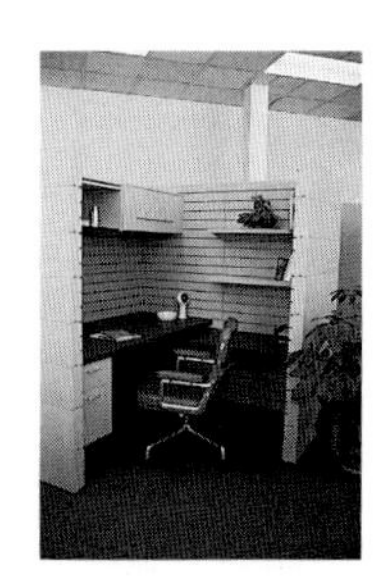

L-System Modular Furniture, 1974, p. 324

The fundamental building block for this modular furniture system is a single L-shaped unit. L-System pieces can be arranged for either office or residential use. Units can be linked together to form a sofa, stacked to form shelving, or organized into work stations or multi-seating areas. The system utilizes only a small group of pins to secure pieces together, unlike other modular systems requiring a myriad of supports and fasteners.

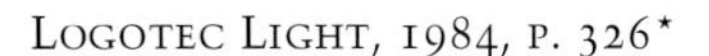

Logotec Light, 1984, p. 326*

Logotec's design combines economic general-service lamps with powerful reflectors in a housing made of two basic fittings – formed by cutting a cylinder at a 45-degree angle – that are backed by a minimum of accessories. Designed for contract and industrial lighting uses, Logotec is capable of meeting all lighting requirements: allowing for both simple controls and complex, fine adjustments such as color correction and image framing. Its camera-like structure allows the light beam to be directed with incremental precision.

Oseris Spotlight, 1985, p. 327*

Named for the Egyptian god, this range of low-voltage, high-range wattage spotlights utilizes a blend of two different small, quartz-halogen lamps as its low-consumption, low-heat-load light source. Able to rotate 360 degrees on its vertical axis, the design is also based on the fact that the intersection of a semi-spherical volume cut two times with a plane generates perfect circles. As the circles are turned against each other, a movement from zero to ninety degrees is described. Oseris works as a system with a comprehensive and expandable collection of accessories including: ceiling mounts, flood lenses, honeycomb screens, infra-red reflector filters, multi-groove baffles, sculpture lenses, fixing rings for both colored-glass filters and anti-dazzle cylinders. Optimal alignment is aided by the scales imprinted on the fixture, allowing precise illumination when multiple units are placed in a row, as in fine art installations.

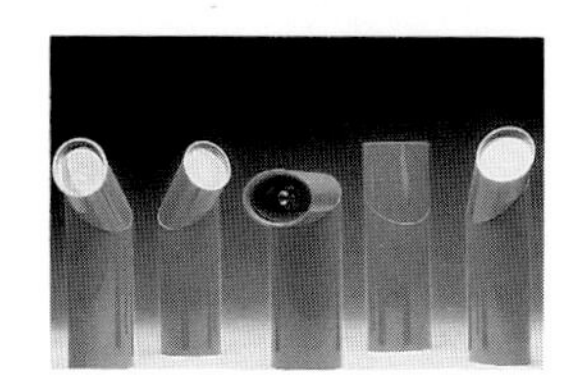

Polyphemus Flashlight, 1983, p. 328

The design for this flashlight is based on sectioning an elliptical cylinder at a 45-degree angle to create a circle on which a rotating head turns. By bending the flashlight into an L-shape, it can easily be carried, and is ergonomically more comfortable to the hand and wrist. Built from separate parts snapped together, the flashlight can stand or lay on any surface while its light beam is aimed in any direction. Additionally, it holds a magnet inside so that it can be attached to a metal surface.

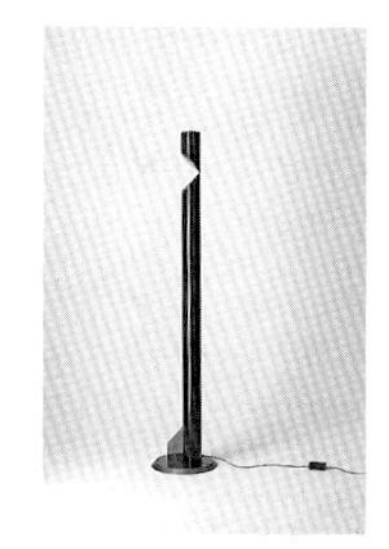

Agamennone Light, 1985, p. 330

This minimalist, black floor lamp, named after the ancient Greek king, is based on cutting an elliptical cylinder twice at a 45-degree angle to create two perfect circles. The circles rotate from a shared point a full 360 degrees while always maintaining a right angle to each other. The lamp's head – driven either by two miniature remote-controlled motors or manually – travels along a circular path through three axes to provide a high degree of lighting freedom. The metal halide bulb produces high luminosity with less energy usage and a long lifespan.

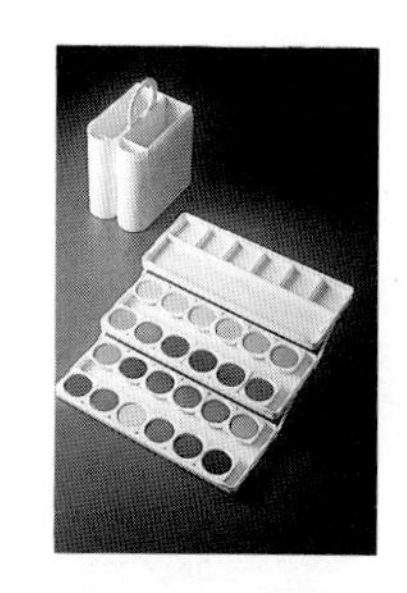

Aquacolor Watercolor Set, 1985, p. 332

The Aquacolor Watercolor Set is composed of three hinged tiers – a palette with two color trays – which are easily removed for cleaning or replacement. The two halves of its self-locking storage/carrying case are designed to function additionally as a segmented water bucket, reducing the number of objects required in order to paint. Innovative in its concern for cleaning, storage and usage problems encountered by both children and adults, the Aquacolor box is easy to carry, empty, and maintain, while providing a distinctive product identity.

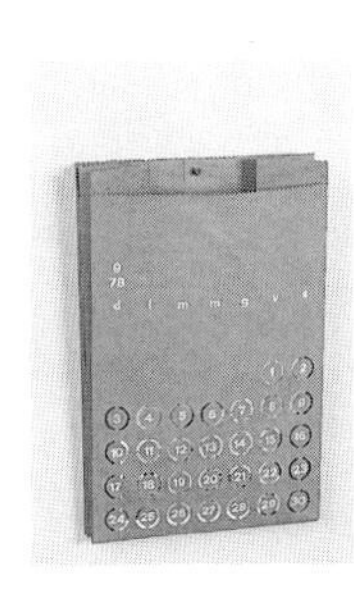

CALENDAR MEMORY BAGS, 1974, P. 334

This calendar system is designed to serve as a reminder of future plans and convenient storage for items from the past on a month-by-month basis. Die-cut circular frames surround each day of the month, providing an aesthetically-pleasing graphic element that allows the user to view the contents of either the brown bag or color-coded inter-office versions without opening. These calendar bags function as a physical reminder of the past as it points to the future.

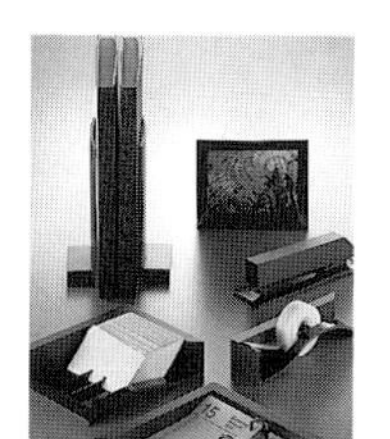

DESK SET COLLECTION, 1988, P. 336

The pieces of the desk accessory collection are designed to organize the home or office while offering a strong emotional presence. The individual articles are united in form and spirit by a gently curved plane evoking paper and its work flow. The shape confers a visual identity that transforms the desk top into a sculpture field. This curve directs the paper to the rear of the letter tray, preventing slippage and facilitating its removal. Front and side loading, the letter tray stacks by means of a molded locking mechanism. The pieces are finished with an applied accent color and with juxtaposed matte and glossy surfaces providing a rich, non-plastic feel with visual and tactile appeal.

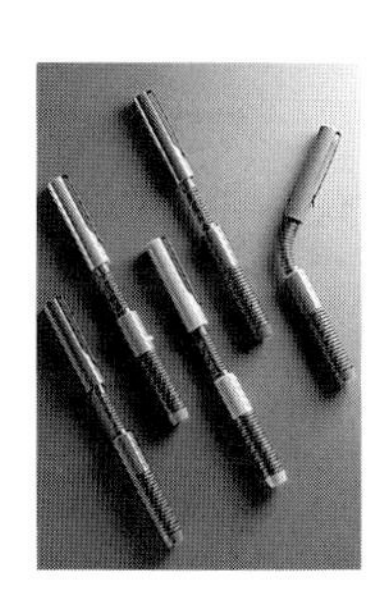

FLEXIBOL PENS, 1985, P. 338

Designed for schoolchildren, these brightly-colored ballpoint pens conform to the elastic movements of the human body. Between the rigid nib and pocket clip housings, is a ribbed portion that flexes with body motion, avoiding breakage when placed in a pocket. The ink runs in a flexible body-conforming plastic tube which runs through the length of the pen alleviating the leaks associated with traditional rigid-body pens. When the pen body is twisted, the lower housing snaps upward to meet the upper housing exposing the pen nib and enclosing the flexible portion in a rigid shell for optimum grasping.

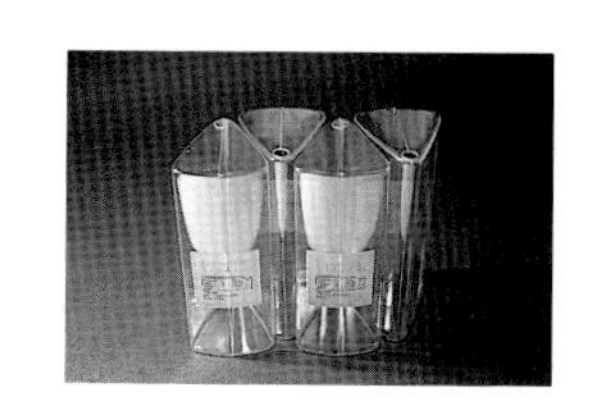

VITTEL WATER BOTTLE, 1985, P. 340

The Vittel water bottle is based on the merging of the bottle's practical requirements and the substance's sensual delight. The bottle has a feeling of quality – a clean and elegant acetic package. A wavy-V texture makes the bottle easy to grip and tactilely pleasant. The pattern suggests both the company name and the ripples of water. The liquid pours smoothly from the mouth of the bottle. They can be stacked on top of each other or packaged in groups of seven to form a circle for a week's worth of drinking. Like a fine perfume bottle for water, the container signals that something of real quality is inside.

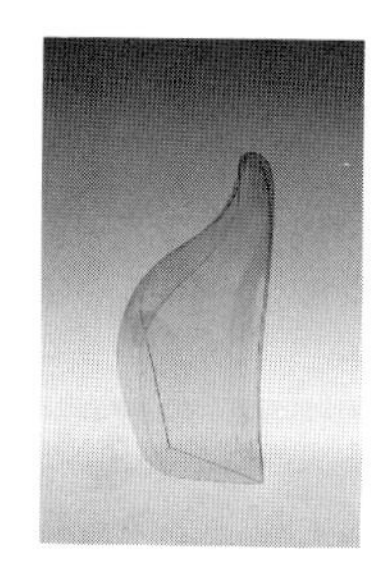

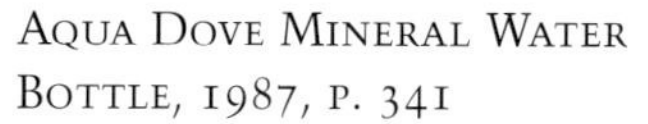

AQUA DOVE MINERAL WATER BOTTLE, 1987, P. 341

The form of this bottle suggests a dove, the symbol of purity. Made of clear plastic, the bottle doubles as an elegant dining-table decanter. Varied presentation effects are achieved by laying it lengthwise or standing it on end. Its narrow beak, serves as a pouring spout with a snap-off cap. Distinguished by their flowing, asymmetrical form from traditional mineral water bottles, Aqua Dove bottles can be arranged in a number of attractive patterns on store shelves.

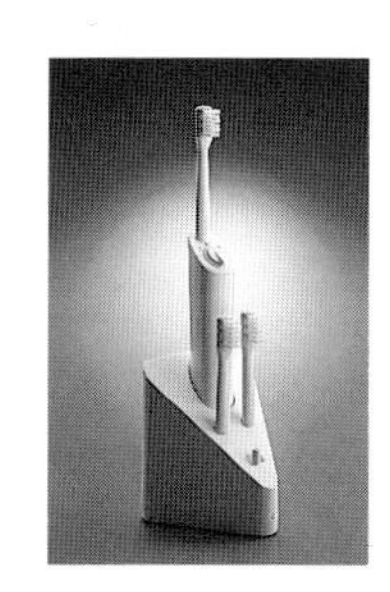

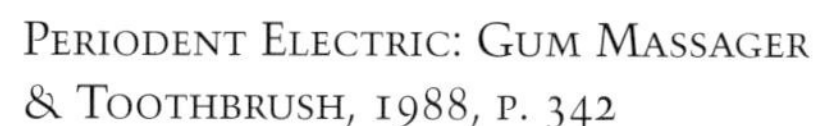

PERIODENT ELECTRIC: GUM MASSAGER & TOOTHBRUSH, 1988, P. 342

Dual-action bristle heads make this dental system unique in design. A crank shaft lowers and raises alternate rows of bristles engaged in a counter swinging motion, simultaneously lifting the gums to cleanse the teeth underneath. This motion massages the gums, removes plaque, helping to fight gum disease while avoiding damage to tooth enamel. The smooth, bulbous head of each bristle improves on the traditional roughly cut blade. The design also reduces the unpleasant vibration of other electric toothbrushes. The small size and rounded triangular shape of the handle make it easy to grip for both children and the aged. Pastel colors in the brush head enhance its appeal for children.

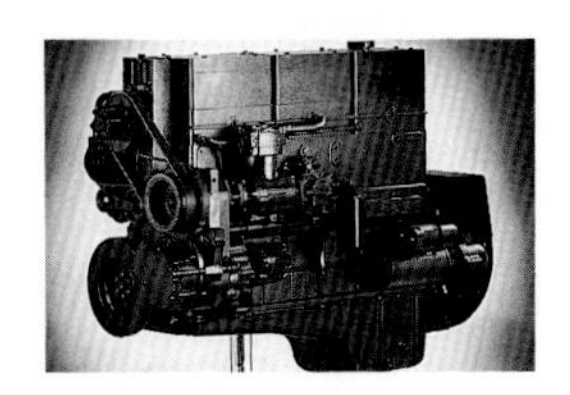

Cummins N-14 Diesel Engine, 1982, p. 344

The Cummins N-14 diesel engine features a revolutionary oil-cooling technique which extends the engine's operating period and allows a reduction in engine weight, resulting in enhanced savings for the manufacturer. The N-14 engine, with models ranging from 300 to 450 horsepower, achieves major improvements in fuel economy, noise reduction, durability, and emissions while retaining enough commonality with current N-type engines to permit use of much of Cummins existing tooling and transfer lines. The arrangements of the Turbo charger, after-cooler, and cross-over on one side of the engine have resulted in a height reduction of nine inches allowing it to fit a greater range of vehicles. The cylinder block and valve covers express power and denote structural strength. The dark blue and black color scheme emphasizes ruggedness.

Cummins X-Series Engine, 1987, p. 345

The five mid-range diesel engines comprising the series range from four to nine-liters, and were developed for automotive, industrial, and marine use. The modular nature of the engine systems and component integration coupled with a commonality of parts results in improved quality-control, cost-efficiency, a reduced parts inventory, and shorter maintenance training time. The clean, uncluttered aesthetic suggests the engine's power and strength. Charcoal gray emphasizes its ruggedness and unique design qualities. A specially-mixed dark blue and black color scheme, complemented with red accents, sets the top-end model apart from the rest of the family. Production is streamlined by producing the engine's fully-integrated electronic monitoring and guidance systems on an advanced variable production line.

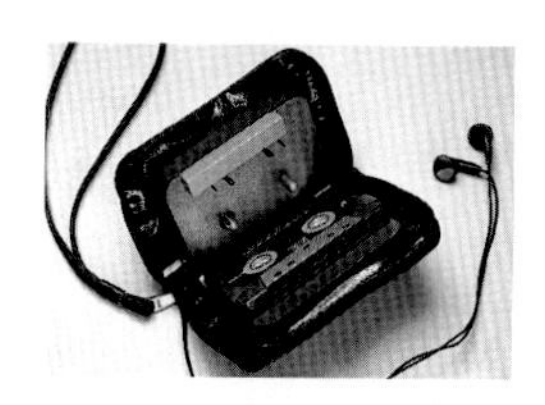

Soft Portable Radio/Cassette Player, 1990, p. 346

One of three items that constitute the Soft Series, all the mechanical and electrical components of this completely flexible tape cassette player with radio are fixed into place in a traditional rigid shell. As in other conventional players, the radio functions with or without a tape in place. The functional elements are then protected while visually and tactilely enhanced with a soft-tech appearance created by a padded leather skin.

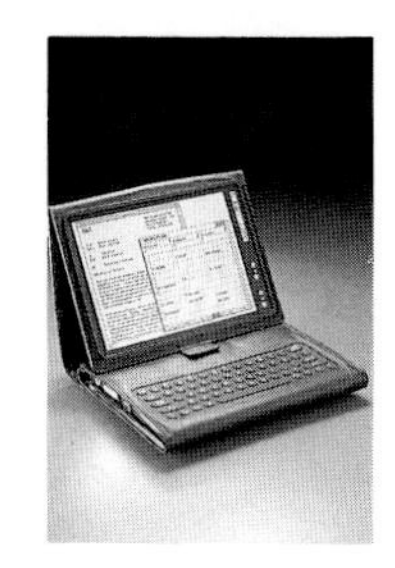

Soft Notebook Computer, 1990, p. 348

The second Soft Series design, the Soft Notebook Computer, is influenced by the traditional paper notebook. Like the fine-quality version, the computer is covered with soft-padded leather to give it a formal elegance and touchable appeal while providing protection for electronic components. Opening the first flap reveals the soft, tactile rubber keypad and thin LCD display screen. Further opening reveals storage for diskettes, pockets for paper and business cards; and an electronic pen for computer sketch pad functions. The screen adjusts like a viewing easel with a velcro tab. Flipping the screen onto its back allows it to be used as a laptop clipboard.

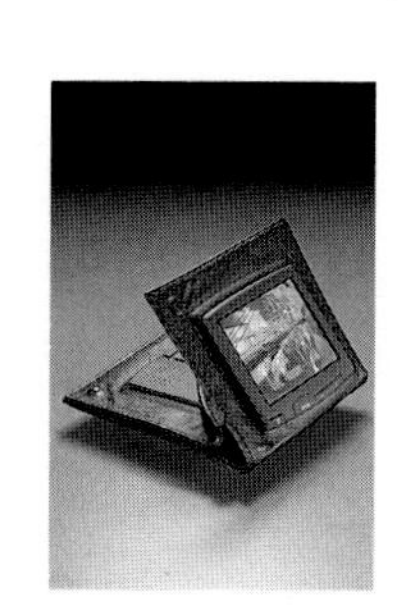

Handkerchief TV, 1990, p. 350

Another member of the Soft Series, the folded-leather Handkerchief TV benefits from the development of flat-screen technology, revealing its technological function only when unfolded. The opened TV has four planes, each with a distinct function: screen, antenna, battery/speaker and external ports. The component cases are made of injection-molded polycarbonate.

Vertebra II, 1985, p. 380

Vertebra II builds upon the epic success of Vertebra I, the world's first automatically adjustable office chair, introduced in June 1976. Vertebra made history generating a whole way of understanding the process of seating, as well as giving rise to a whole new industry.

The office seating system Vertebra II answers organically to the office workers' needs for comfort in a working environment where he/she spend almost a third of their life. Since people remain rarely seated in the same position for more than 8 to 10 minutes, Vertebra II has been designed to follow the user automatically, without any need for adjusting buttons, mechanisms, or levers. By bending both forward and backward the chair follows the user's body demands as he often changes from one position to another. Building on the premise that the ideal chair is that which goes unnoticed, Vertebra changes its shape in a gentle manner. When the user relaxes the seat moves forward and the backrest follows the body as it displaces itself toward the front.

Dorsal II, 1985, p. 382

Dorsal II is the highly improved re-edition of Dorsal I, an economical office chair of high ergonomical capabilities which continues to be sold notwithstanding its 20 years in the market.

Dorsal I was born as a low cost office chair three years after the presentation in June of 1976 of Vertebra I, the first ergonomical chair ever. It is still the only office chair capable of changing configuration automatically. Like Vertebra I, Dorsal I changed configuration according to the positions adopted by the user, without any need for adjustment buttons.

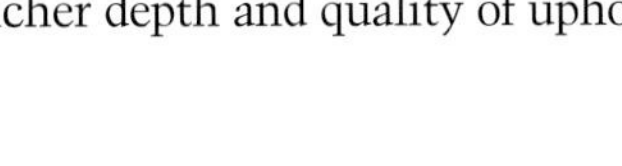

Dorsal II builds upon Dorsal I's good points, greatly improving the original tilt forward function and providing a richer depth and quality of upholstery.

Door Handles, 1996, p. 384

Contemporary design has produced many elegant handles. They are functional, and aesthetically minimal. In their pursuit of the pragmatic, they have overlooked the fact that the door handle is one of the few house building products the user really touches. The present design, to be cast in bronze, follows the contour of the palm of the hand, and has the side in contact with the hand very highly polished, as if it were a terse skin. The underside, where the thumb makes contact, is of a rougher texture. The contrast between the smooth and the rough reinforces the sensorial quality of the handles' very sculptoral shape. Opening the door is therefore elevated to a sensual experience.

Nuage Sofa and Headboard, 1993, p. 386

This design is based on a single element: an 8″ x 8″ (20 x 20 cm) "leather cushion," filled with spongy material of different densities, assembled according to what part of the body it is to support. Each of these cushions, or "leather tiles," are connected to the other cushions by a fastener that secures them only at each corner. The advantages of this construction system are many: each tile can be easily replaced independent of each other; air flows along the edges of the tiles where they overlap; and the hide is fully utilized with very little left over, thusly reducing the cost.

This method of construction requires no expensive craftsmanship. Since the padding in each tile can be of a different density, it allows building a far more ergonomic type of backrest. The traditional construction technique to build a Chesterfield, requiring at least two very skilled craftsmen working for three days, is hereby re-formulated in a contemporary way. The backrest, which is longer than the seat, visually defines the sitting area. No space is wasted since this backrest design recognizes the fact that a sofa always needs side tables. Since the backrest is also well finished, the sofa can be positioned in the middle of a room, serving as a space definer.

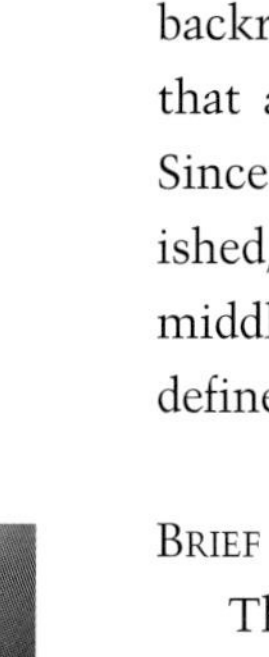

Brief Office Seating, 1995, p. 388

The Brief range of office chairs is an innovative functional solution with an anatomically designed seat and back rest which can be moved around to create endless support positions. Brief can be adapted to suit the size and height of each user, thereby supporting the ever-changing posture the user adopts at work. This feature is a recommended one in leading ergonomic studies, and is a requirement of recent European Union Directives. The arm can be adjusted in height and in orientation to satisfy the complex ergonomic requirements of computer users.
The seat and back are visually connected in such a way that they look like an integrated shell while in reality the seat and back supports move independently from each other. Some of the important features of the chair include height adjustment for the seat, as well as the height adjustment of both the backrest and the armrest. The angle between the seat and the back can be changed either independently from each other or following a program of synchronized adjustments.

Max Operative Seating, 1996, p. 390

The Max office seating range consists of operational, managerial, and executive models. A unique feature of the office chair is that the back is hinged to the seat. More important is the fact that the tilt mechanisms and the support structure, usually two separate, expensive pieces, are integrated into one injection molded piece, greatly reducing costs.
The backrest cushion support goes up and down to adjust for lumbar positions, while at the same time being able to tilt back up to 20 degrees. The armrest can be adjustable in height.

Magic Wand Automatic Roller Pen, 1998, p. 392

The Magic Wand automatic roller pen has no visible movable parts. All mechanical parts are hidden from view and operate automatically. When the user puts the pen in his pocket the roller tip disappears automatically into the main body. When the user takes it out to write, the tip emerges automatically and remains fixed automatically in the writing position until again put away in a pocket.
The clip, which fastens the pen to a pocket, fulfills the very important function of reducing pressure on the fingertips when guiding the pen in its writing function. The clip becomes an ergonomical support to guiding the pen, greatly reducing muscular tension on the fingers.
The pen is equipped with a safety device that keeps the roller tip hidden when not being used as a writing instrument.

Tennis Office Seating, 1997, p. 394

Office Seating system in wood consists of operational, managerial, and executive models. Except for the synchron tilt mechanism of steel and aluminum the entire chair is made of recyclable and/or renewable materials such as; stretchable fabric, and wooden frames.
The back goes up and down to adjust for lumbar requirements being able to tilt back up to 20 degrees. The seat slides backward and forward, both tilting forward 4 degrees and/or tilting back 10 degrees synchronously with the backrests' tilting of 20 degrees.
It is equipped with adjustable and non-adjustable armrests.

Saturno Street/Highway Lighting, 1998, p. 396

Saturno is a vertical truss made up of very thin members making it an extremely light product – a delicate, almost transparent presence on the street or highway. The economical use of materials makes it possible to use stainless steel, which requires a minimum of maintenance since painting and rust are eliminated. Keeping maintenance costs to a minimum is a goal for any organization, but it is particularly important for cities, highway authorities, and other public entities, which often have capital to buy good products but are not always able to allocate funds for ongoing maintenance.

Since Saturno was designed not only to illuminate the street but also itself, the designer chose ultra-violet resistant polycarbonates for the identifying color shields. Shields of different colors are usually used to identify different quarters of the city or upcoming exits or service stations. The bumpers on the sides of the poles are made of an ultra-resistant material made from 100 percent recycled plastic.

Saturno is highly visible, which minimizes the potential of road accidents. In daytime, its matte-polished surface is enhanced by its colored bumpers and light shields. At night or during adverse weather conditions, the more reflective nature of its surface and lighted bumpers help guide drivers. The poles are also designed to bend easily in case they are hit by an automobile. Moreover, the poles can become a billboard of changing signs alerting drivers to changing traffic conditions.

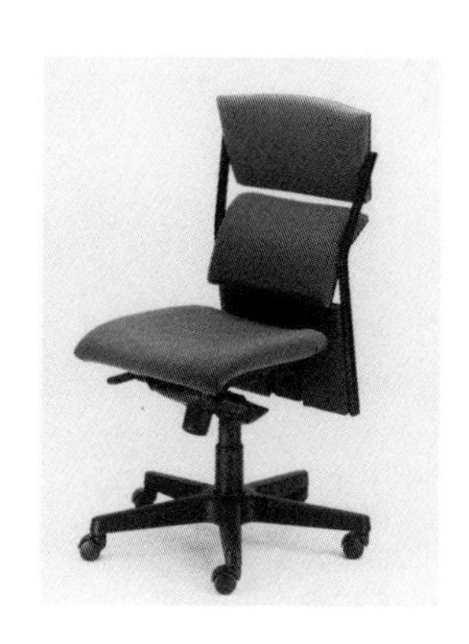

Tris Office Seating, 1998, p. 400

Office seating system: consists of operational, managerial and executive models. It operates using off-the-shelf complex and/or simple synchron systems. It consists of independent spring loaded lumbar and dorsal back supports. The structure holding the dorsal and torsal back supports goes up and down to adjust for lumbar positions, tilting back 20 degrees plus. The seat slides backward and forward, both tilting forward 4 degrees and/or tilting back 10 degrees synchronously with the backrests' tilting 20 degrees. It is equipped with adjustable and non-adjustable armrests.

VoX Contract Chair, 1996, p. 404

VoX is a range of chairs to be used by the general public for periods of time ranging from 15 minutes to 2-3 hours, in schools, waiting areas, auditoria, offices, and factories. This typology of chair must stack vertically for storage and gang safely in rows.

Moreover, to sharpen its competitive edge, the seat of this type of chair must be able to tilt up vertically. This allows increasing the number of rows by reducing the distance between rows, while facilitating the passage of people entering or leaving the rows. Furthermore, this type of chair also needs to be equipped with a writing tablet, for either left or right-handed users. This writing tablet must tilt up to allow the user to get up or sit down, while at the same time, it must also be able to collapse automatically out of the way in case of a panic.

Contract seating is the most exacting type of chair to design. There are very rigorous structural norms to meet, ergonomic requirements to satisfy, and a most competitive price ceiling to respect. Moreover, it must be light weight for ease of stacking vertically or ganging.

It must also be light-weight to reduce cost. At the same time, this seating must be extremely strong to pass the rigorous DIN, BIFMA, and ANSI norms regulating its design. All these requirements must be synthesized in the design of a chair that seeks to be as essential as one of Nature's organisms. That is to say, it must be also elegant and graceful.

The design solution is based on a simple but very strong visual and functional concept: the JOINT. At that joint come together the chair's legs, the support for the backrest, the support for the armrest and writing tablet, the ganging connectors, the stacking bumpers, and the beam supporting the seat. That joint is also the key point onto which elements can be added later, after the chair has been bought, such as arms and writing tablets. The user must be able to acquire and install these accessories after having purchased the chair. The joint also caps the hidden weldings holding the crossing legs inside the central beam.

The structure is made of steel tubes, elliptical in section, conified and ribbed by stamping to increase its strength and reduce the need for material to a minimum. The flexing backrest is made in Delrin™ plastic. The tilt-up seat is produced in a copolymer of polypropylene.

IBM PORTABLE DESKTOP, 2000, P. 408

The Portable Desktop is based on the concept of having a basic infrastructure of connectors housed in a one size fits all container. In this single sized container the components can be changed, added or subtracted at will by the user. The container would remain the same in outer size regardless of whether the screen is 12″ or 16″ diagonal. The user can initially purchase the system without, for example, a DVD and when he decides to acquire it, it would then occupy an empty slot. By this approach, it is hoped to greatly reduce the costs not only of manufacturing, but also of updating the equipment on the user's part.

A very distinctive feature of this design is that the height of the screen can be adjusted to fit the requirements of the user, thereby eliminating one of the problems of present day portable laptops. By this device the laptop becomes truly a portable desktop.

The container is made of a smoked tinted transparent acrylic. When the container is closed the box is black and non-transparent. When the cover is open, its transparency becomes evident. Moreover, a system of lights illuminates the outer edge of the screen making it much easier on the eyes.

The keyboard is a wireless folded device, which fits neatly inside the container when closed, and can be used at some distance from the computer when opened.

A handheld PC device fits inside the container, but its edge is visible when carried in a bag so that the user can observe if there is a message for him. At that moment, he only needs to press the small visible eject button on the container's edge that releases the handheld computer. This handheld PC device will double in the near future also as a GSM portable telephone.

URBAN FURNITURE, 1999, P. 414

This urban furniture is based on a simple concept: to minimize the number of support elements which have to be anchored to a sidewalk. This support element, in essence a column, is based on the design for the Saturno Lighting System. Onto this column all elements are attached, weather protecting canopies, space defining transparent membranes, backlit display units, signs, seats, wastebaskets, etc. Everything attached to the column is raised from the floor to aid cleaning and to avoid breakage.

Each of the units can be decorated to identify the City quarters where placed; metal insignia or glass colored elements, requiring no maintenance, serve to establish the identity of the neighborhood where located.

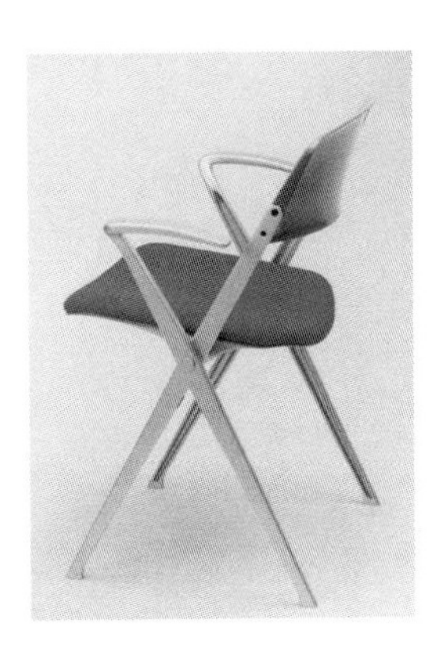

STACKER CONTRACT CHAIR, 1998, P. 420

The Design Solution is based on a simple but very strong visual and functional concept: the CROSS. At that cross come together the chair's legs, the support for the backrest, the support for the armrest and writing tablet, the ganging connectors, the stacking bumpers, and the beam supporting the seat. That cross is also the key point onto which elements can be added later, after the chair has been bought, such as arms and writing tablets. The user must be able to acquire and install these accessories after having purchased the chair.
An important feature of this chair is that the backrest can flex backwards, supporting the user in both the relaxed position when listening to a lecture or an erect one when writing notes. The cross is the symbol of the chair's multifunction mission; reduced to a minimum, it is a design tour de force.
The structure is made of diecast aluminum, c shaped in section. The flexing backrest is made in ABS plastic. The tilt-up seat is produced also in ABS.
Stacker is a range of chairs to be used by the general public for periods of time ranging from 15 minutes to 2-3 hours, in schools, waiting areas, auditoria, offices, and factories. This typology of chair must stack vertically for storage and gang safely in rows. Moreover, to sharpen its competitive edge, the seat of this type of chair must be able to tilt up vertically. This allows increasing the number of rows by reducing the distance between rows, while facilitating the passage of people entering or leaving the rows.
Furthermore, this type of chair also needs to be equipped with a writing tablet, for either left or right-handed users. This writing tablet must tilt up to allow the user to get up or sit down, while at the same time, it must also be able to collapse automatically out of the way in case of a panic.

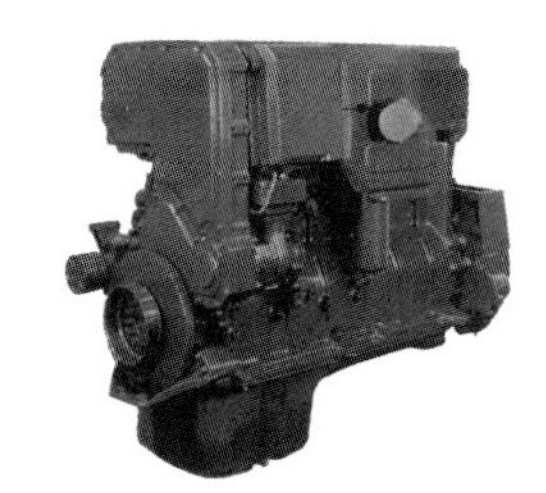

CUMMINS DIESEL ENGINE, 1996-97, P. 424

This totally new engine is the strongest, lightest and fastest in the 600 horse power range. It is also the most fuel efficient, durable, and dependable. A great amount of design effort has gone into making all wire and hose lines absent from view. To achieve this, all the electronics and fluid lines were designed and built as an integral part into the engine.
Another radical aspect of this design, in addition to its very uncluttered lines and ease of assembly and service, is the fact that the number of its parts have been reduced to the point that it has a third less components than its next competitor, thereby minimizing the chances for mishaps. Since the engine is approximately 300 pounds lighter than any other engine in its class, its performance is exceptional in terms of fuel economy. This is also the result of carefully designing integrated parts so as to avoid excess material and structural redundancies
Such remarkable design has allowed the company to give an extraordinary base warranty of 250,000 miles (400,000 kilometers), all items covered, and 500,000 miles (800,000 kilometers) for all major components, parts and labor without any deductible.

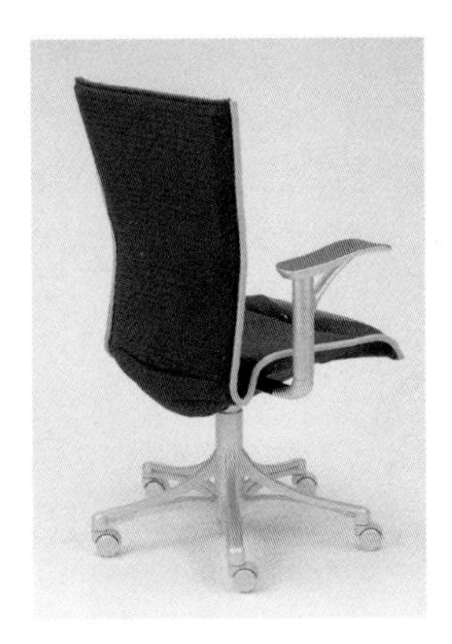

Svelte Office Chair, 1999, p. 426

This office seating range is an elegant and svelte chair, very light and thin, amply equipped with an invisible number of ingenious ergonomical mechanisms. Inside the thin backrest, there are incorporated very sophisticated devices to provide independent torsal and dorsal automatic support. As visible in the diagrams, the backrest support cannot only ascend and descend, but it also tilts at different angles independently, or synchronized with the angle changes of the seat. The seat can slide forward and backward to accommodate different users and has the capacity of tilting forward so as to avoid exercising pressure on the users thighs when leaning on a desk. This chair represents a contribution to a "soft-tech" approach where the emotional qualities of the product prevail over the expression of its technological capabilities.

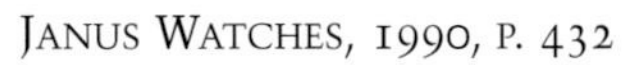

Janus Watches, 1990, p. 432

This collection combines the formal characteristics as well as emotional and symbolic appeal of a fine wristwatch with the efficient functions of a built-in computer for memos, phone numbers, addresses, and calculating. The design enhances the computer's functionality compared to existing models by greatly enlarging the screen and keyboard. The keypads are large enough to be punched with the fingers, as opposed to a pen tip.

In two models with traditional mechanical analog hands, the hidden computer is revealed by flipping open the watchface. Revealing the computing functions in this manner not only allows for the separation of the watch's two distinct functions, but allows the form of each to express its functional and symbolic requirements.

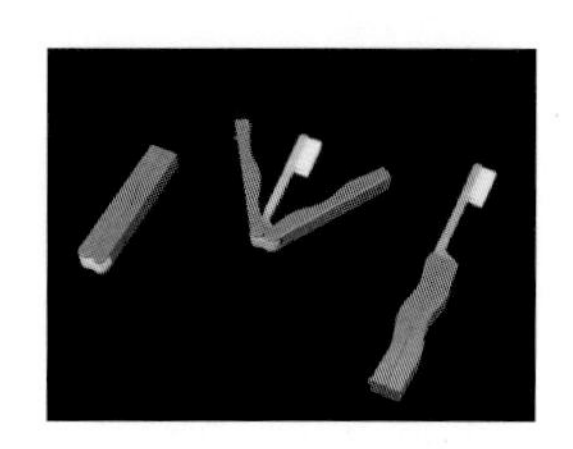

Manual Toothbrushes, 1985, p. 434

Designed for the Japanese market where toothbrushes are a popular item, this line reflects current trends in personal oral hygiene and utilizes advances in brush and bristle design while offering variety and portability to a youthful market. Each molded polypropylene toothbrush responds to different functional concerns, while the playful forms and bright colors unify the product line. The covers of two travel models swing open to form handles, while the brush of a woman's purse-sized model slides out of its handle. A pair of toothbrushes for a couple nestle together when closed and pull apart for use. They share the same geometry, but have differentiating details. Another toothbrush changes the angle of its head to conform to the tooth surface.

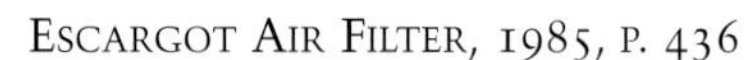

Escargot Air Filter, 1985, p. 436

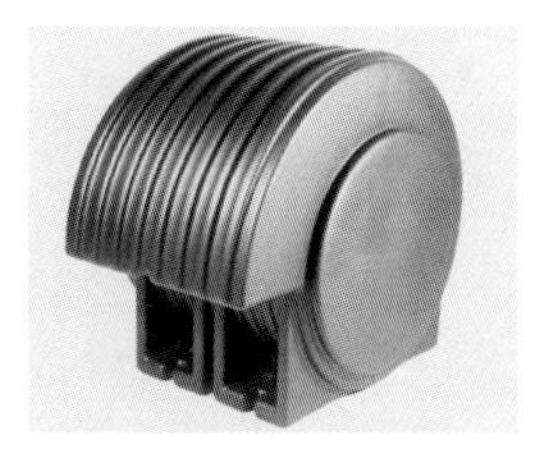

Designed for use with large industrial engines, such as diesel engines or electrical generators, Escargot integrates its plastic housing with the actual air filter of the unit. By moulding the entire Escargot unit integrally, and by making it out of plastic which requires no painting, great savings in manufacturing were realized. Self-contained and having no movable parts, Escargot is essentially a cartridge filter. When its life is exhausted, it is simply replaced, unlike other units on the market. In this way, Escargot ensures that an expensive engine will not be damaged by dirt or dust getting into its innards when the filter is being changed (as is also the case with conventional non-disposable filters).

Escargot is not only named after the snail; it takes its functional clues from it as well. The air inlet – the opening of the snail and the point of entry into the filter – offers a very efficient way of distributing air into and through the filtering surfaces.

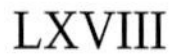

While its appearance is the direct result of functional requirements, the snail-like configuration also operates at a higher level; it is emblematic of an in-out process during which a transformation takes place.

X-Pand Suitcase, 1985, p. 438

We sought to design a line of attaché cases and bags that would expand to a greater capacity within the parameters of the traditional size group range. A patented elastic panel was developed allowing the attaché case or bag to expand or contract according to its contents. This panel is composed of alternating overlapping strips of synthetic leather and stretchable nylon. Not only does this elastic panel fulfill the functional requirements we sought, but it also provides a distinct identity to the line as well as visually suggesting the contents of the bag.

In order to manufacture these panels cost effectively, a special multiple stitching head machine was developed that would first roll the materials' edges over and then sew all stitches whether structural or decorative simultaneously. This produces a continuous panel that can be cut to size as needed.

An unexpected benefit to this stitching process is that it greatly reduced the amount of labor involved in the fabrication of each bag leading to a significant reduction over the initial projected unit cost.

The Attaché cases are formed and machined steel and aluminum frame with diecut plastic side panels covered with a synthetic leather and elastic nylon skin. Synthetic leather was used because it does not have a memory of its shape when distorted and also is easier to feed control for the difficult stitching of the elastic panel. Additional components (leather covered handle, side panels) are injection molded in plastic. The bags are constructed the same as the attaché cases, but without the frame.

GEIGY GRAPHICS POSTER, 1966, P. 352
Both viewer interaction and participation are invited by the poster's dominant die-cut letter. The "G" quickly associates itself with Geigy, a progressive patron of graphic design, while the limitless options in composition remind the viewer of the infinite possibilities for constant discovery in both the pharmaceutical and design fields.

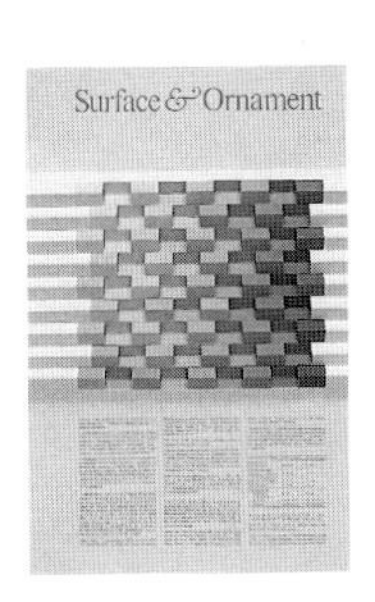

SURFACE & ORNAMENT POSTER, 1983, P. 356
The poster's accordion-like folds and gradient color scheme effectively communicate the two primary benefits of the Formica product, Colorcore: the range of its subtle palette and its ability to create concave and/or convex seams without the tell-tale black line normally associated with plastic laminates.

AXIS POSTER, 1985, P. 357
This exhibition poster, for a retrospective of Ambasz's work, adopts the motif of one of Ambasz's architectural projects. The monochromatic image is enhanced through the use of die-cuts, embossing, and the printing of subtle tonal gradations, while maintaining a simple and evocative image.

ITALY: THE NEW DOMESTIC LANDSCAPE BOOK COVER & POSTER, 1971, P. 358
The poster for The Museum of Modern Art exhibition is composed of thirty-six transparent plexiglas boxes. Contained in each are an identical set of loose die-cut figures representative of the design objects in the exhibition. Sealed in their individual containers, each set of cutouts presents a unique composition to the viewer, implying the infinite range of variations possible within the new domestic landscape. The folded-vellum book cover works as a pocket for the same die-cut elements allowing the objects the freedom to move about in random combinations. Every time the reader opens or moves the book, the graphic relationship is altered, suggesting both the influence and impact that man has over the environments he creates for himself.

LA JOLLA EXHIBITION POSTER, 1988, P. 360
This traveling exhibition poster for the La Jolla Museum of Contemporary Art is based on Ambasz's San Antonio Botanical Conservatory design. Two sheets of plastic encase a silk-screened inner sheet, showing textual information and the plan of the San Antonio Botanical Gardens. The clear outer sheet flows from the flat material into the three-dimensional form of the building, much the same way as Ambasz's architecture springs from, yet remains integrated with, the plain of the earth.

MYCAL GROUP CORPORATE IDENTITY, 1987, P. 362
The logotype, trademark symbol, and identity program developed for this Japanese company draws on narrative symbols from the eastern tradition rather than the more static western symbology utilized in corporate graphics. A cyclical-progression of seasons and life are suggested while the symbol reflects the company's very real concern with the quality of life for its employees and clients.

Chiocciola Logotype, 1988, p. 363

Developed for a client who for years utilized the shell of the snail as its corporate symbol. This new icon represents the growth and expansion of the client's company. The shell, normally associated with the dwelling, is an appropriate association for a real estate company. But, now the snail takes flight, enjoying limitless horizons.

Axis Catalogue, 1985, p. 364

This catalogue accompanied a 1985 retrospective exhibition of Ambasz's work at the Axis Gallery in Tokyo. The dust jacket has been embossed and die-cut so that it can be folded into a three-dimensional representation of two of his architectural projects. Using folded paper to evoke an object is very traditional in Japan, serving as a cultural bridge, and allowing visitors to interact with the exhibition.

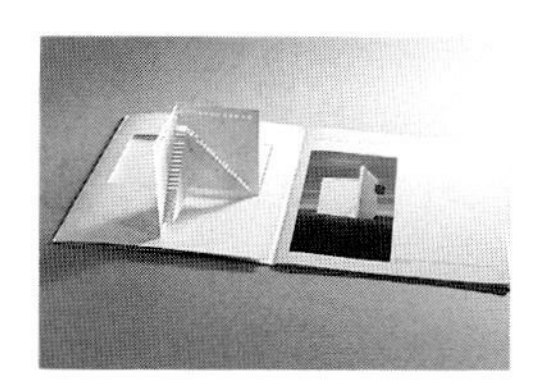

* In collaboration with G. Piretti

Architecture Projects

Mycal Cultural Center at Shin-Sanda

Hyogo Prefecture, Japan

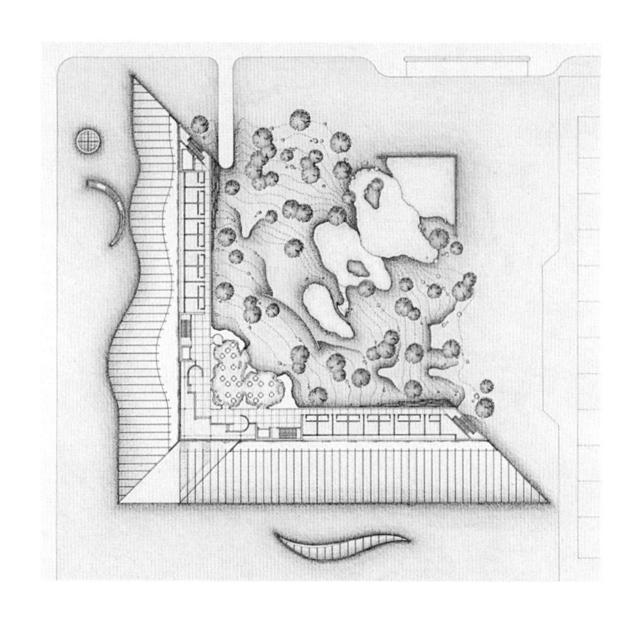

17
15
15
13
15
13
16
17
15
14
14
14
12
10
6
11
5
1
3
9
2
8
4

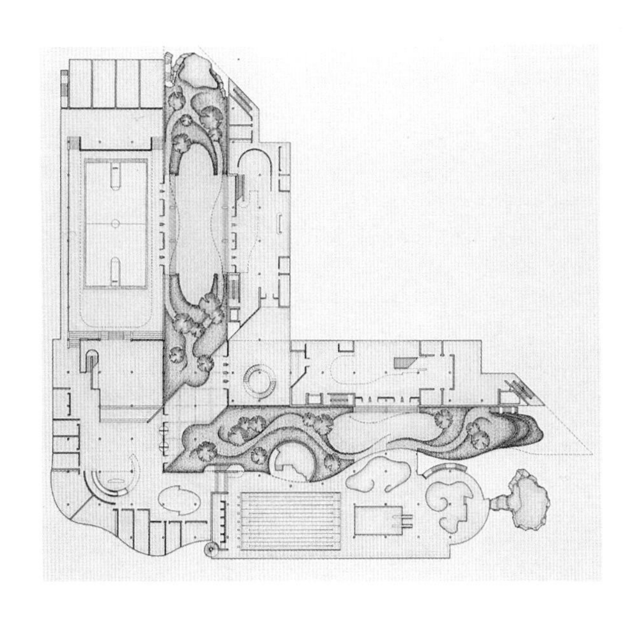

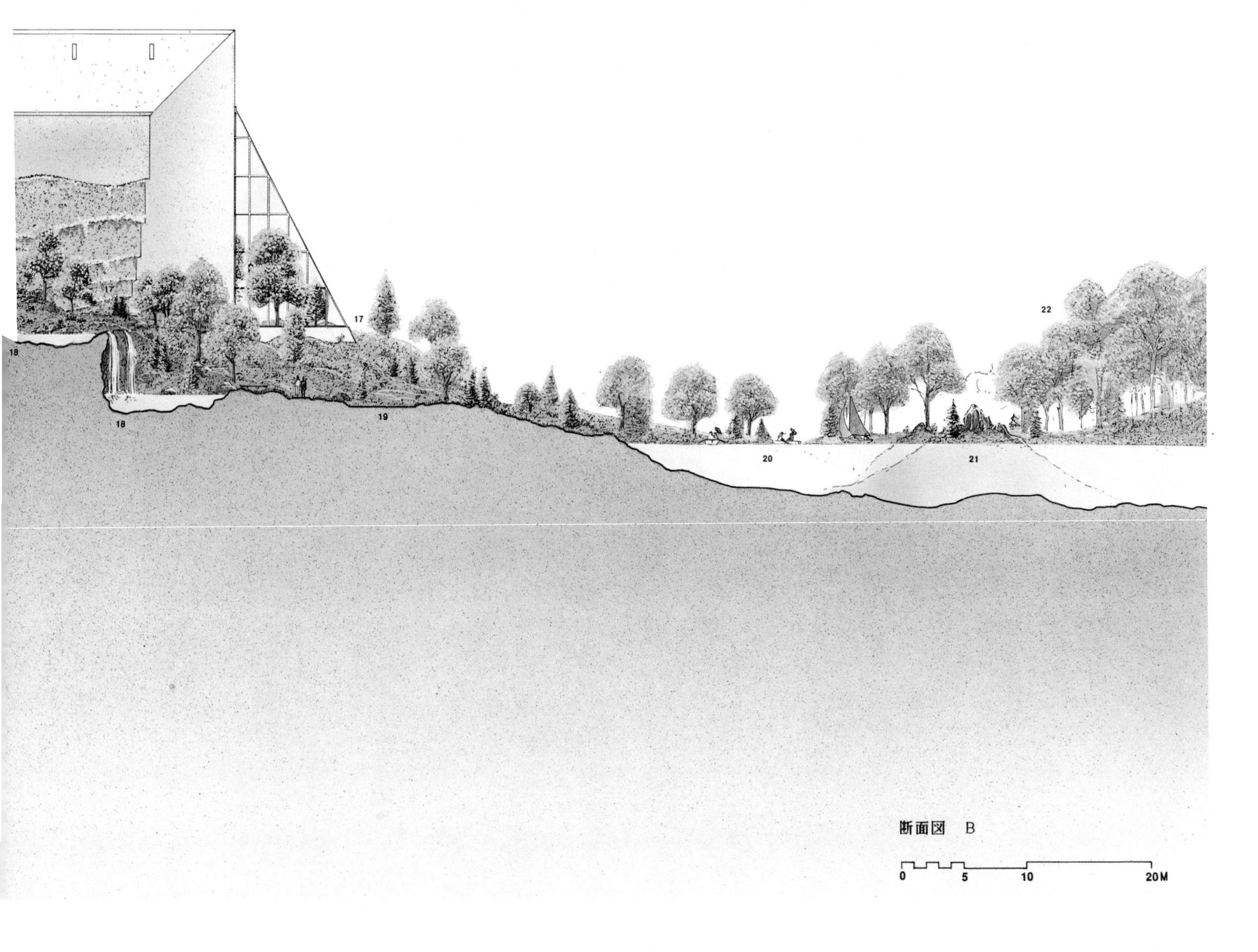

断面図　B

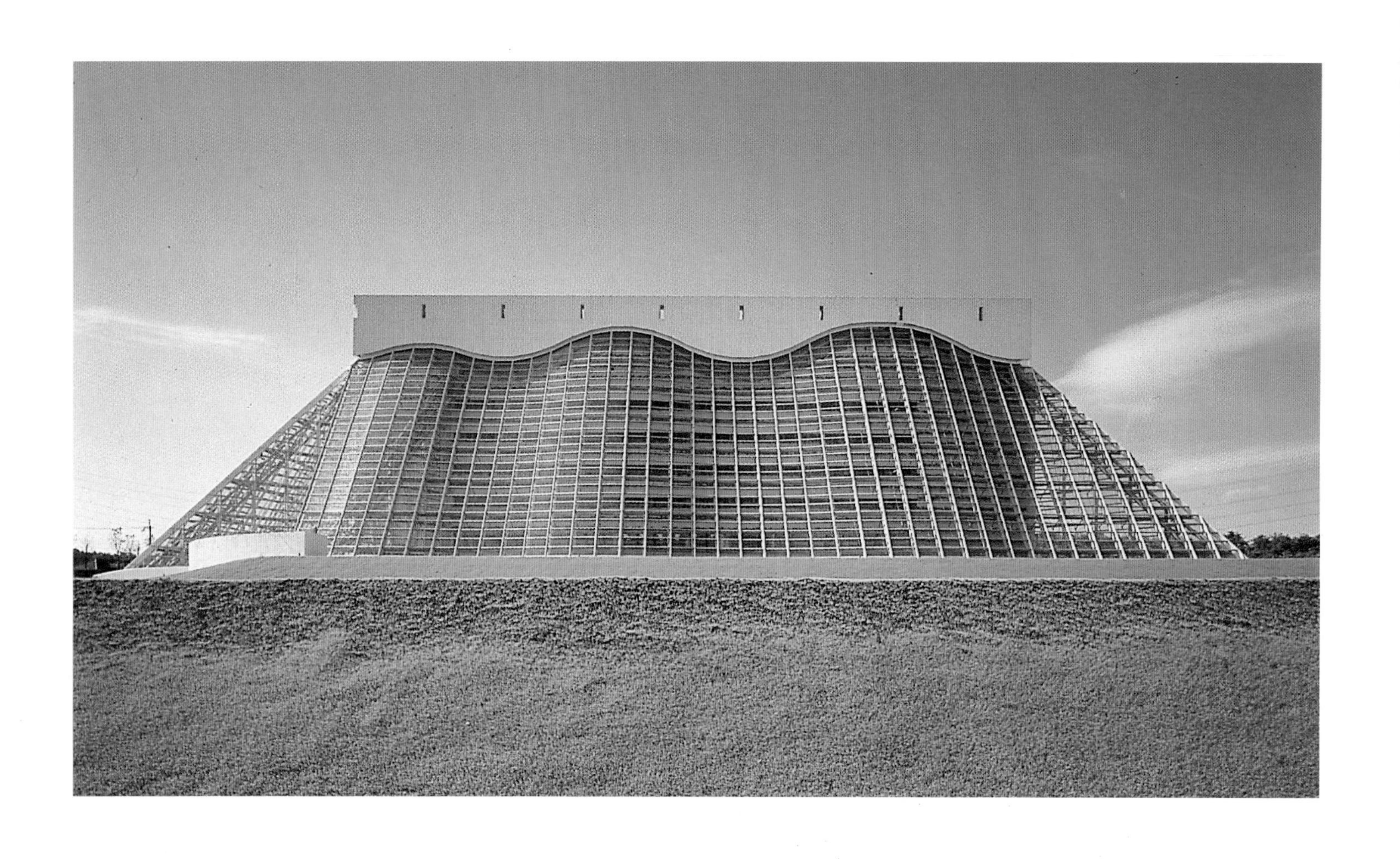

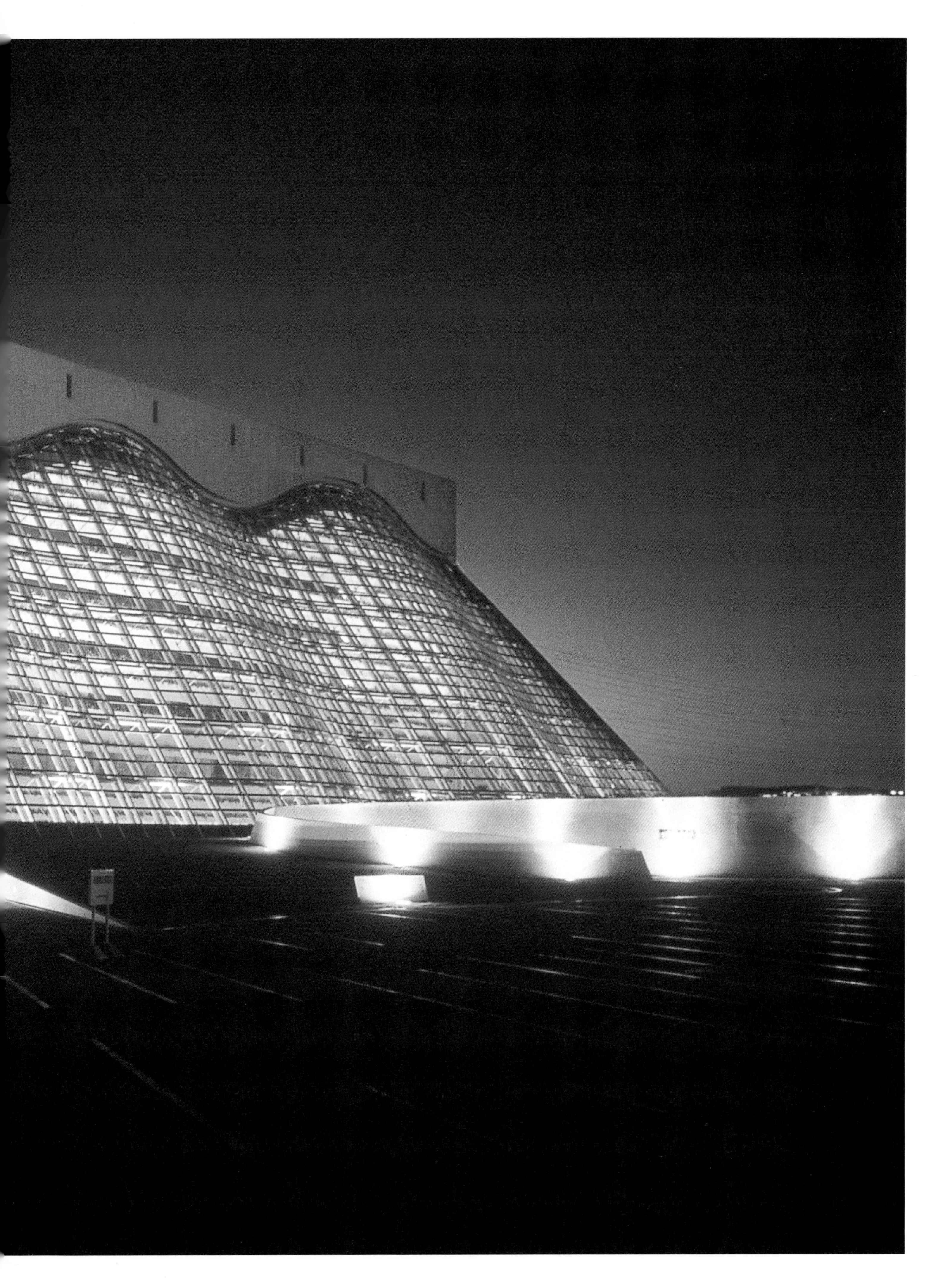

Phoenix Museum of History

Phoenix, Arizona

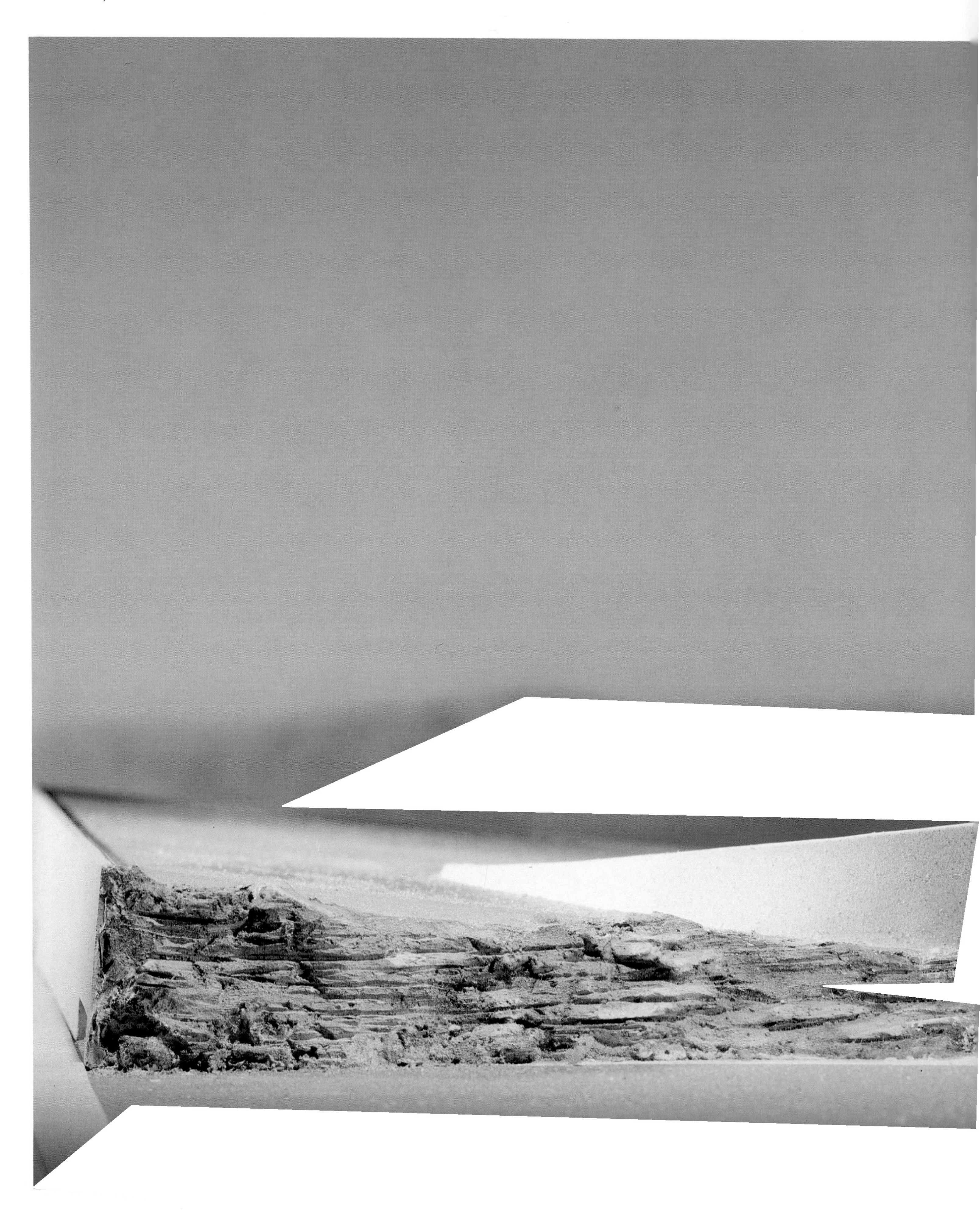

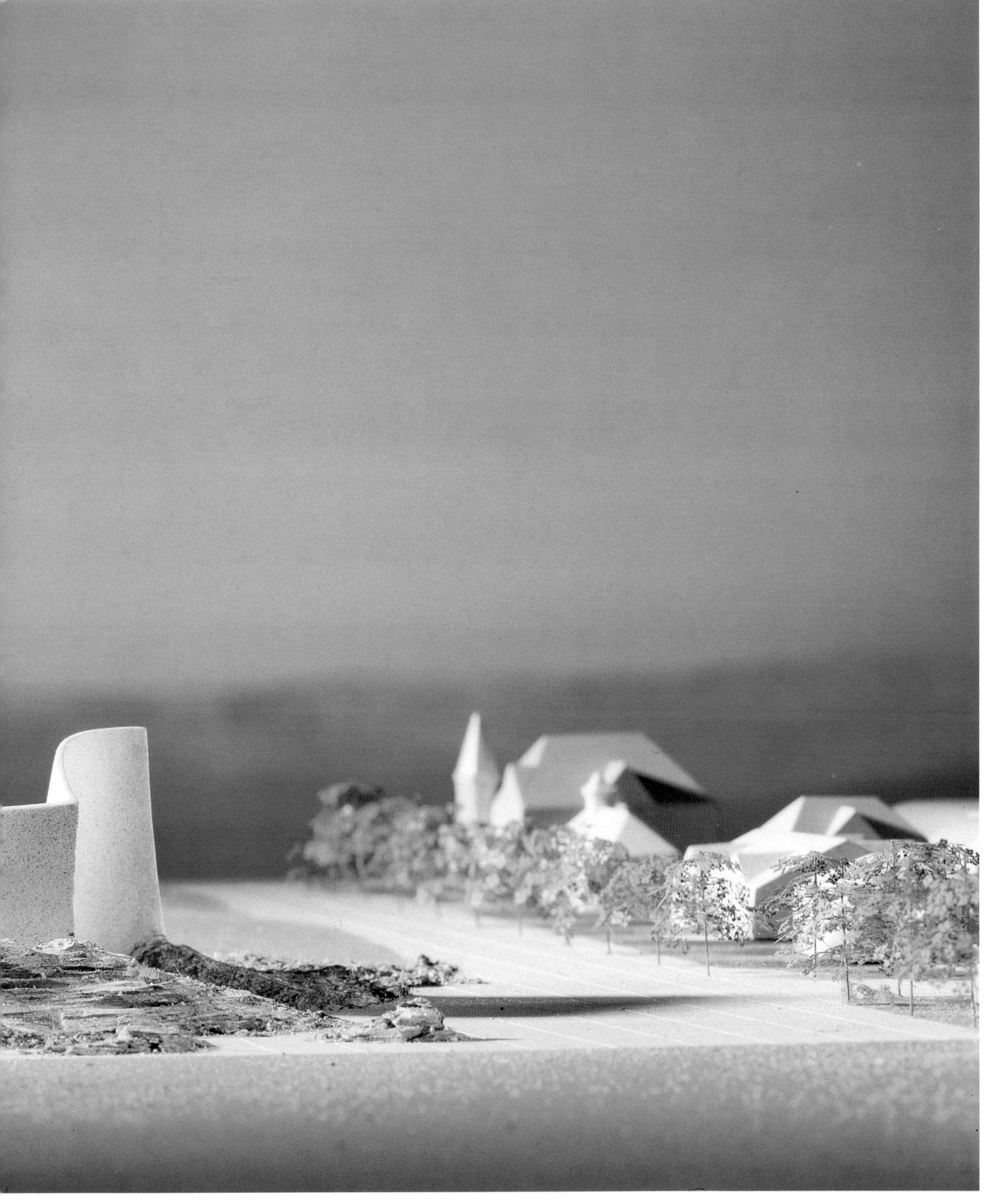

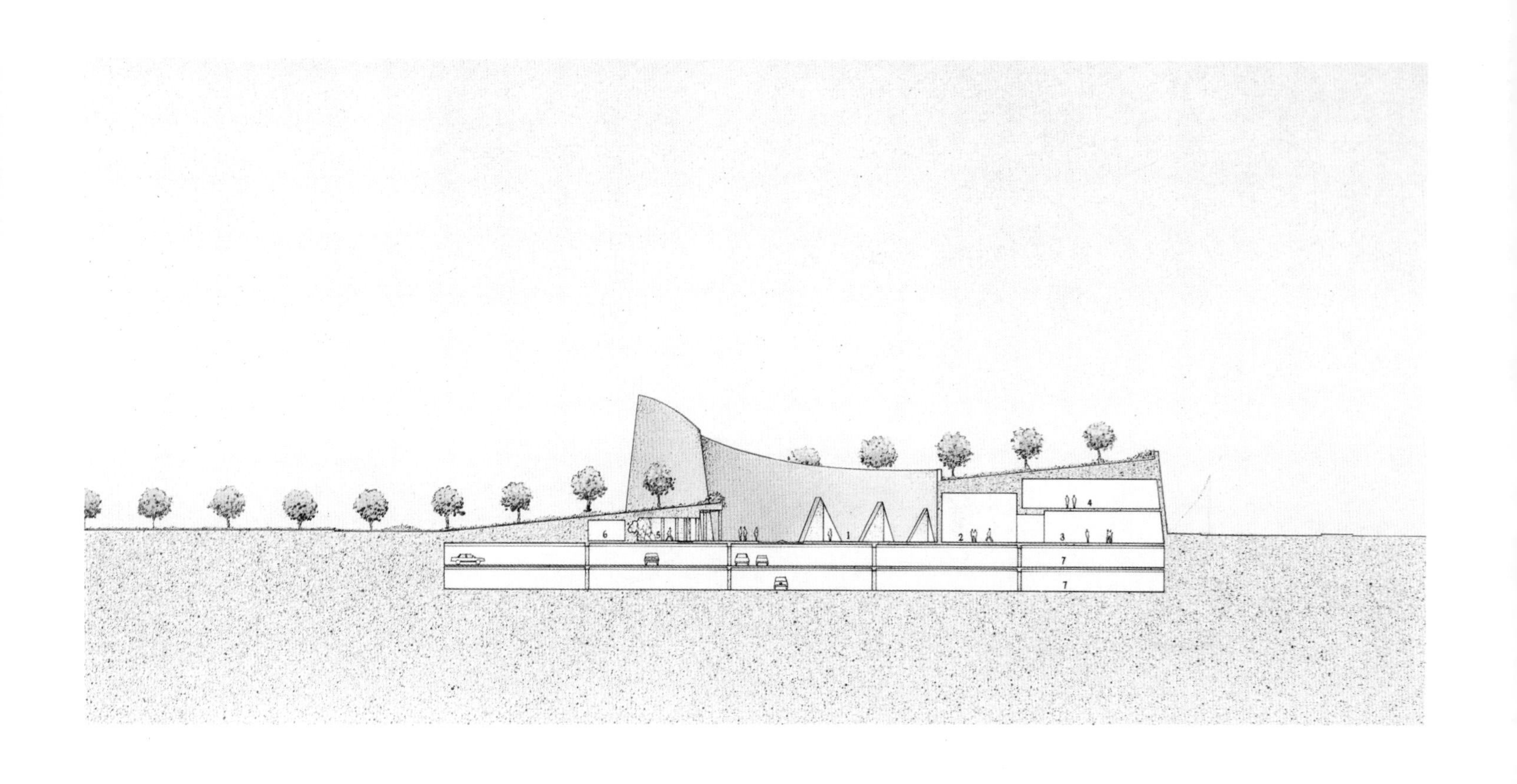
6
5
1
2
4
3
7
7

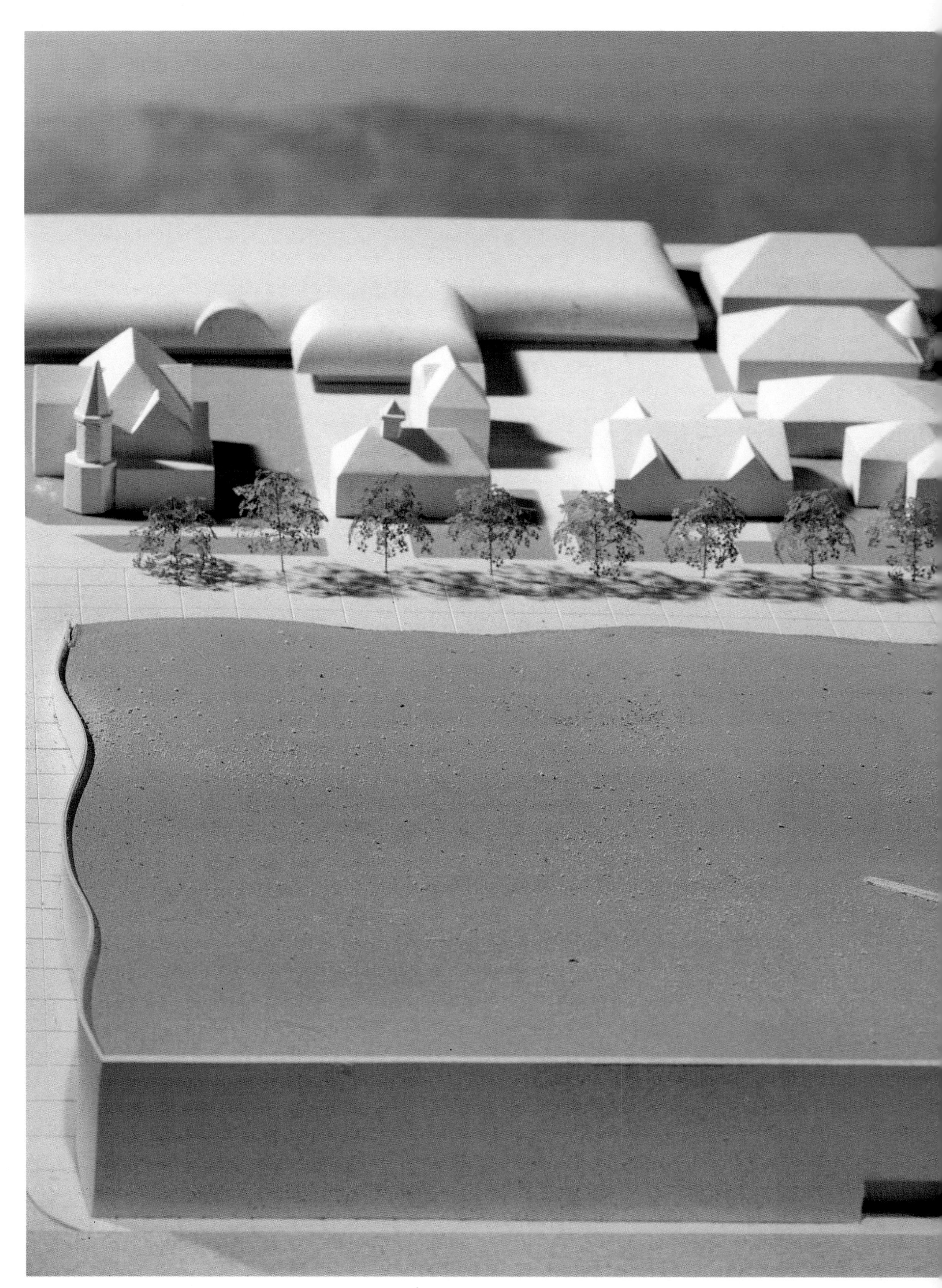

Fukuoka Prefectural International Hall

Fukuoka, Japan

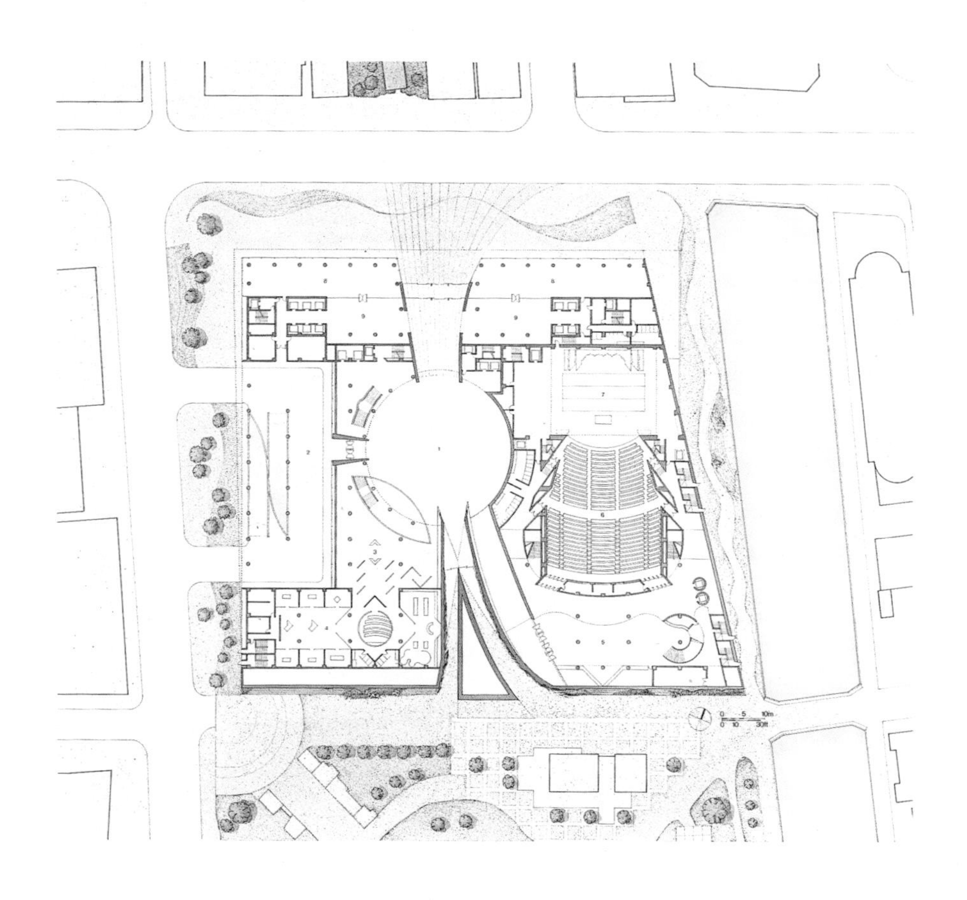

0 5 10m
0 10 20 30ft

Lucille Halsell Conservatory

San Antonio, Texas

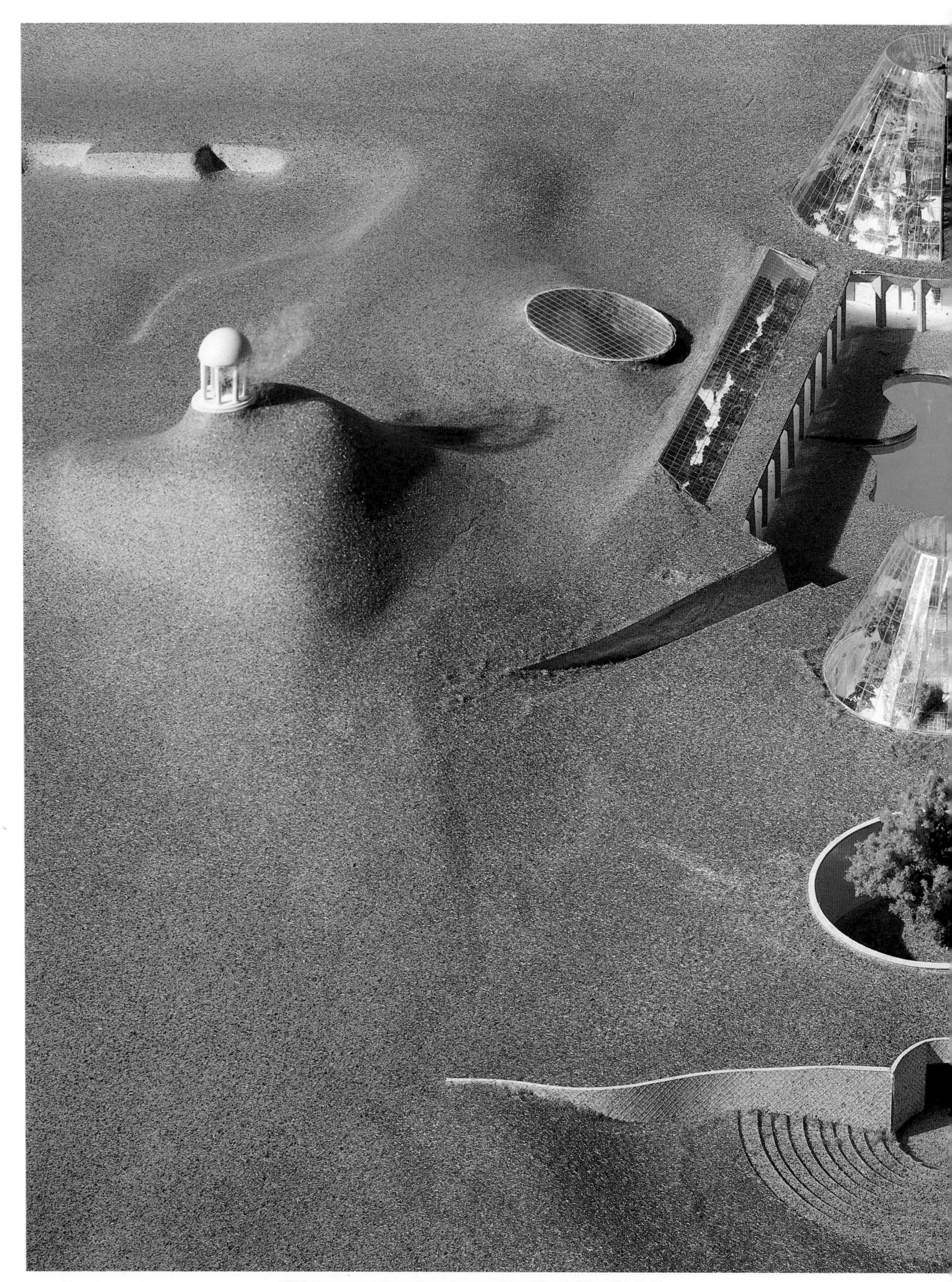

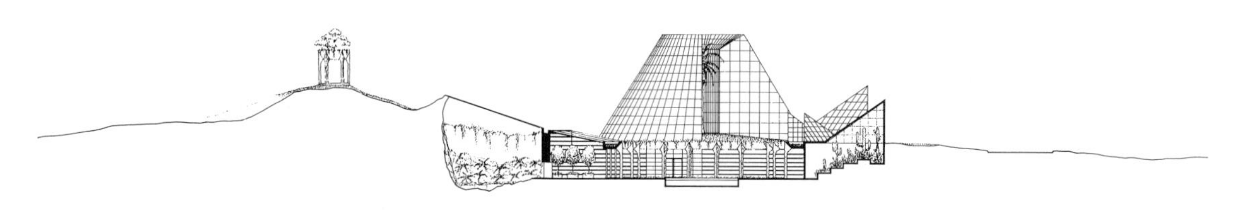

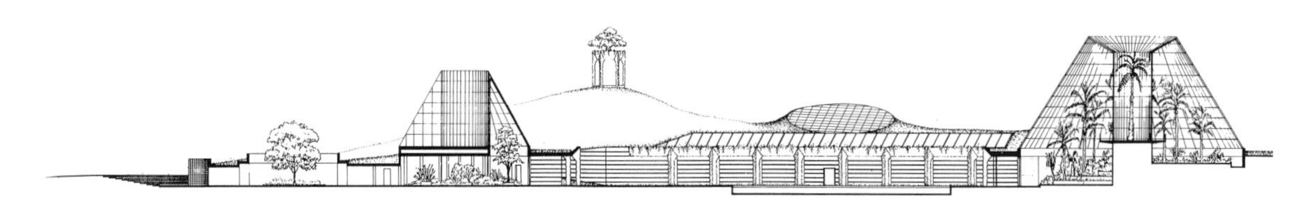

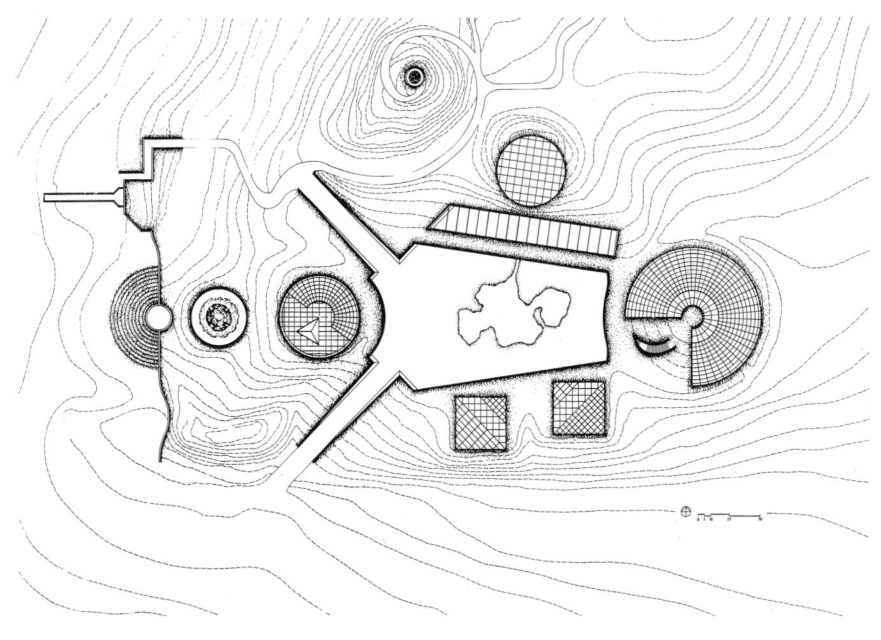

Museum of American Folk Art

New York, New York

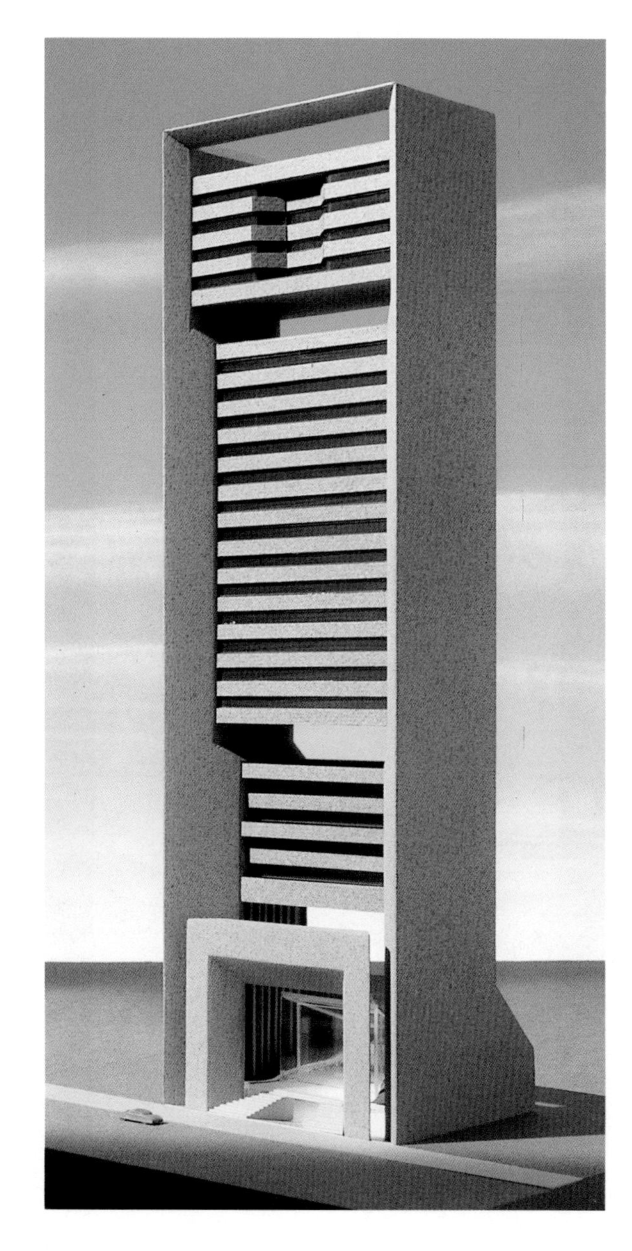

Center for Applied Computer Research

Mexico City, Mexico

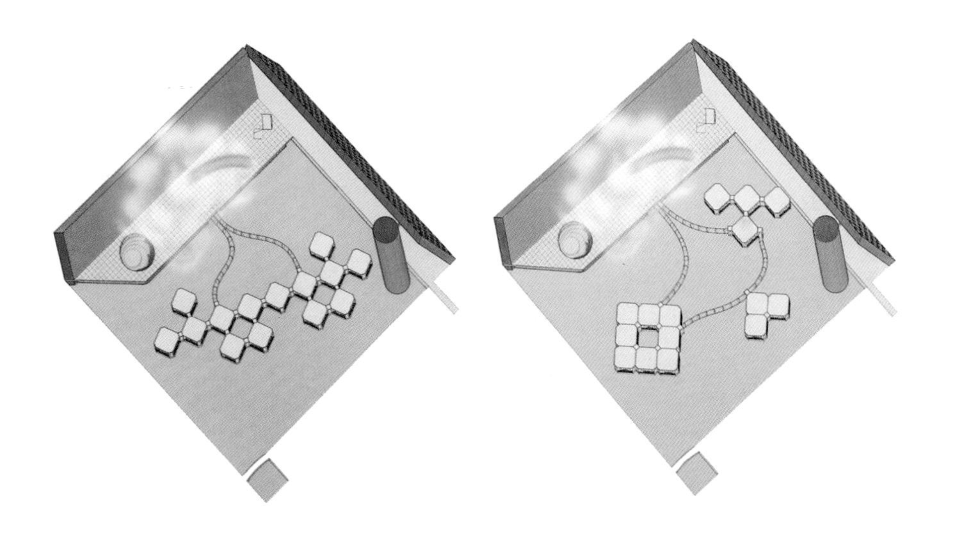

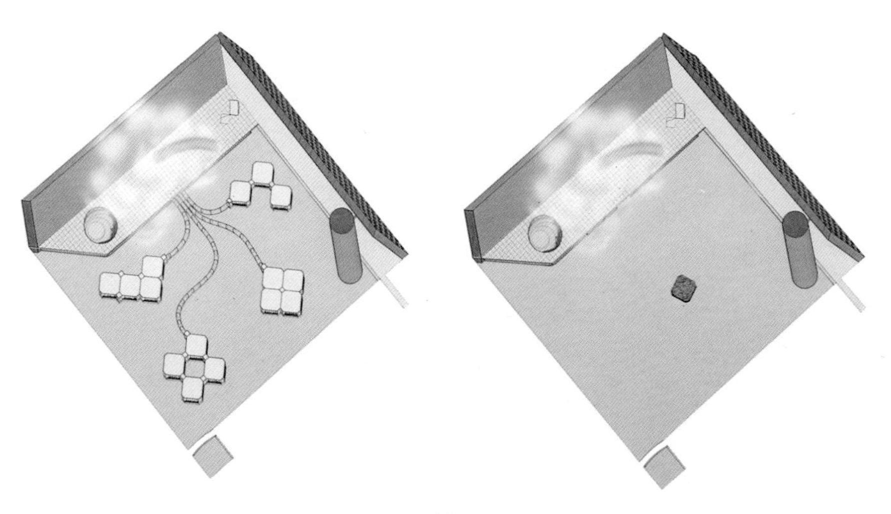

Union Station

Kansas City, Missouri

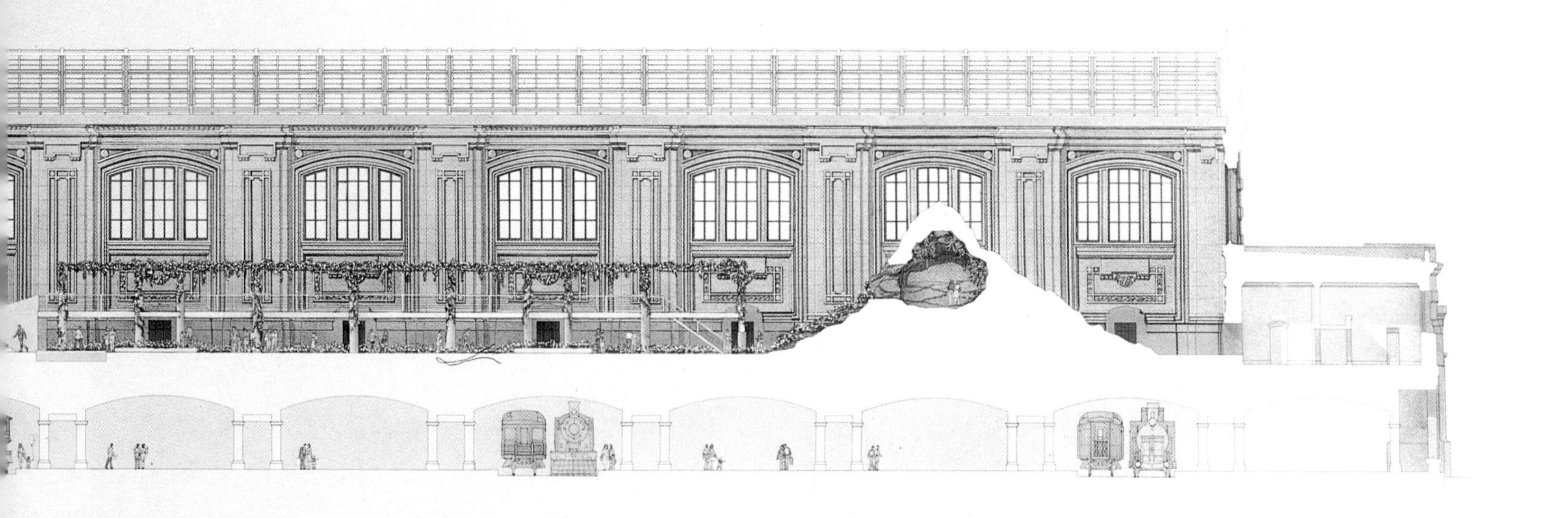

Grand Rapids Museum

Grand Rapids, Michigan

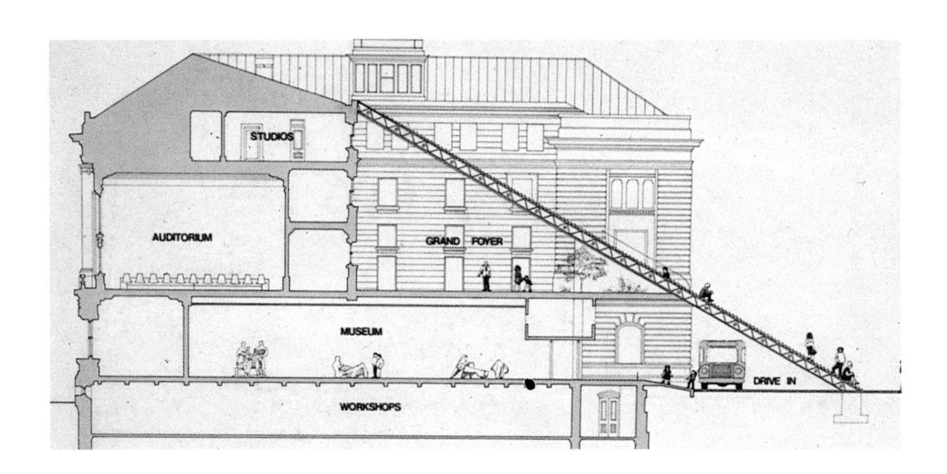

Banque Bruxelles Lambert

Lausanne, Switzerland

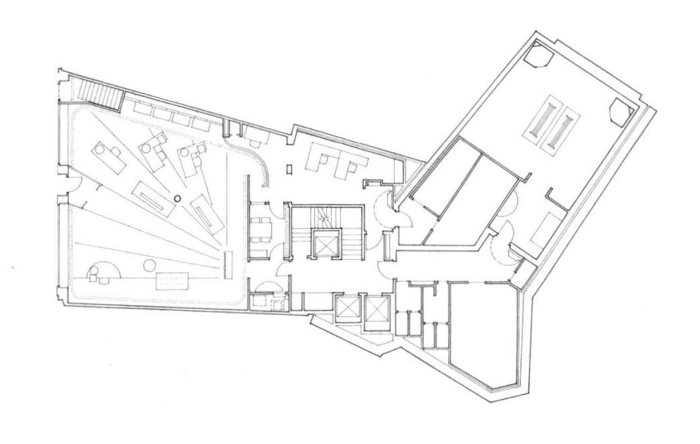

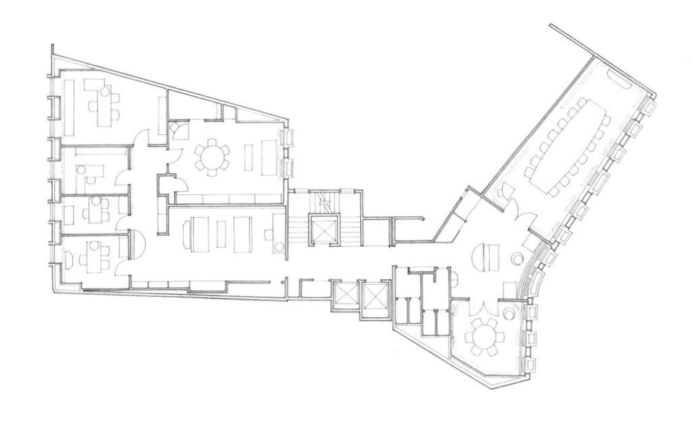

Schlumberger Research Laboratories

Austin, Texas

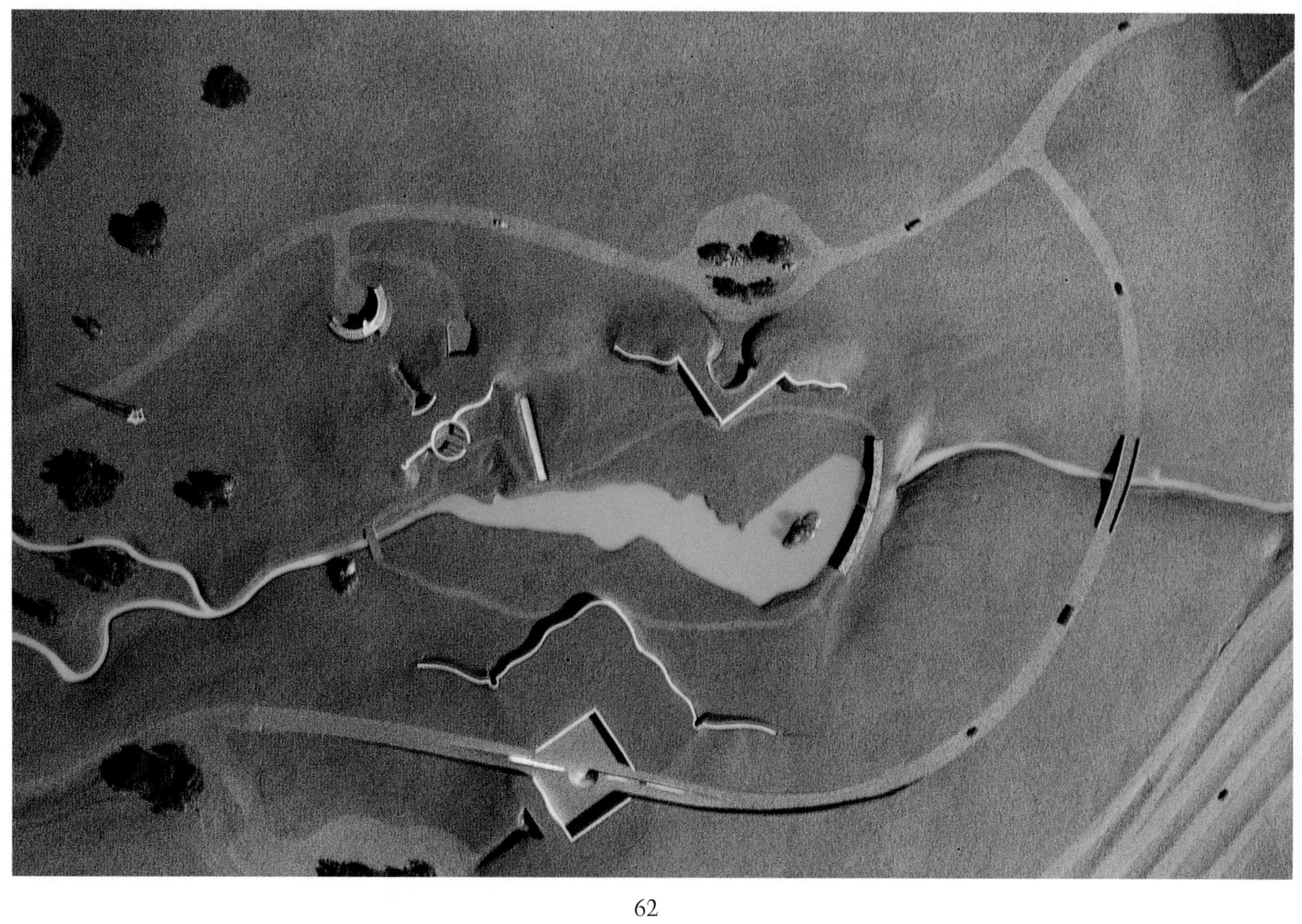

Mercedez-Benz Showroom

New Jersey, USA

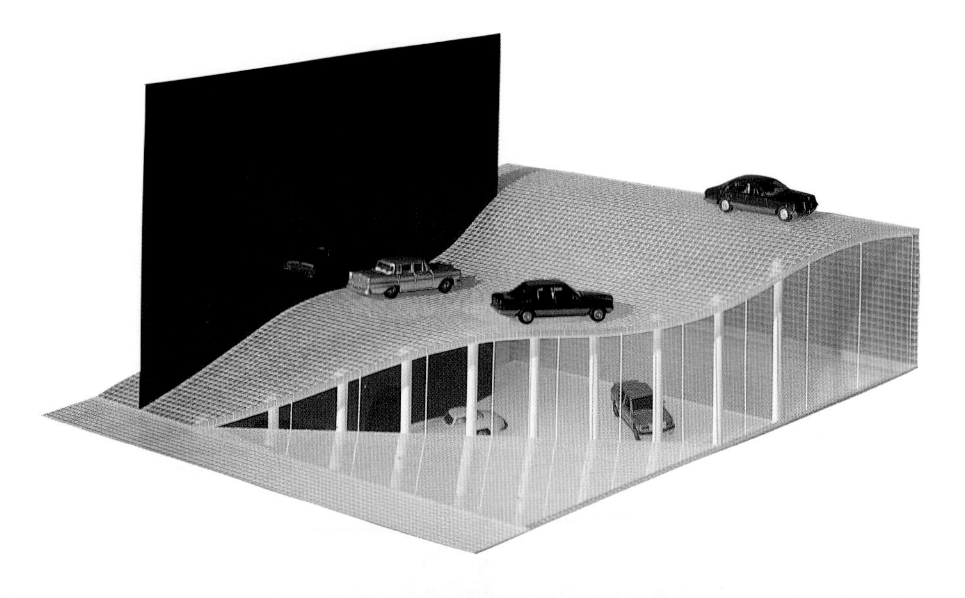

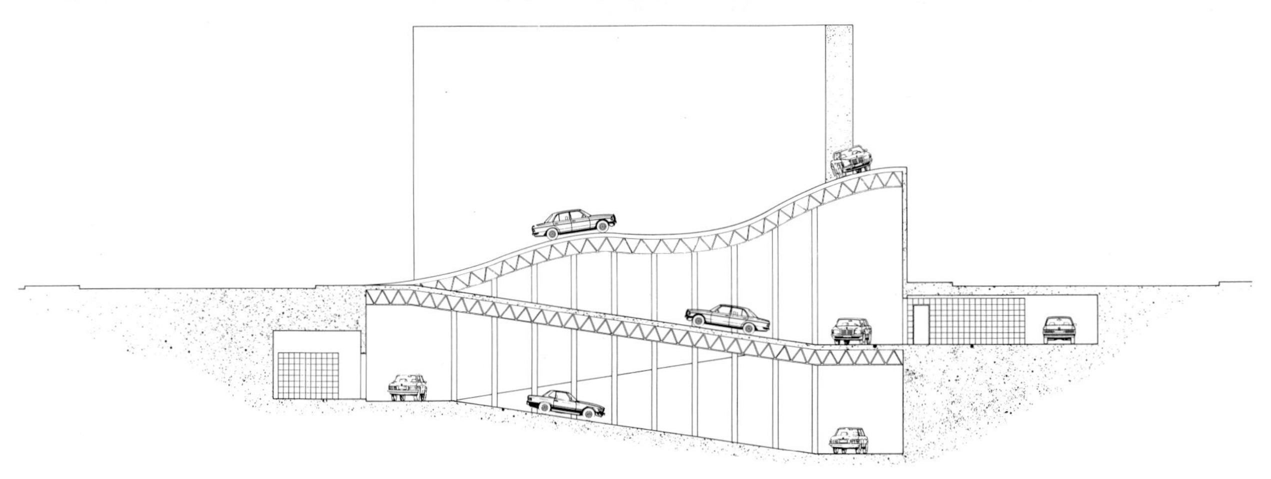

Nichii Obihiro Department Store

Obihiro, Japan

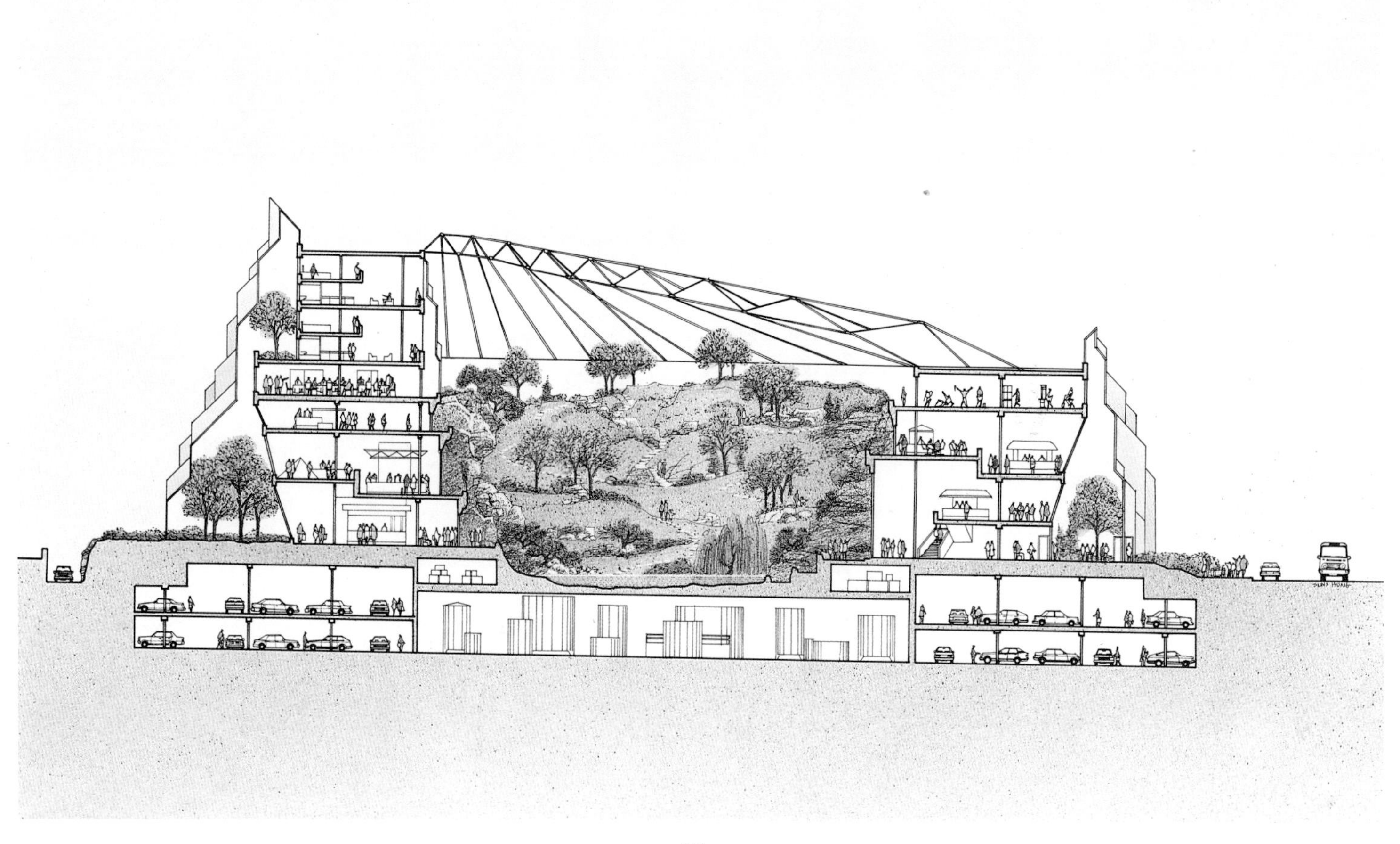

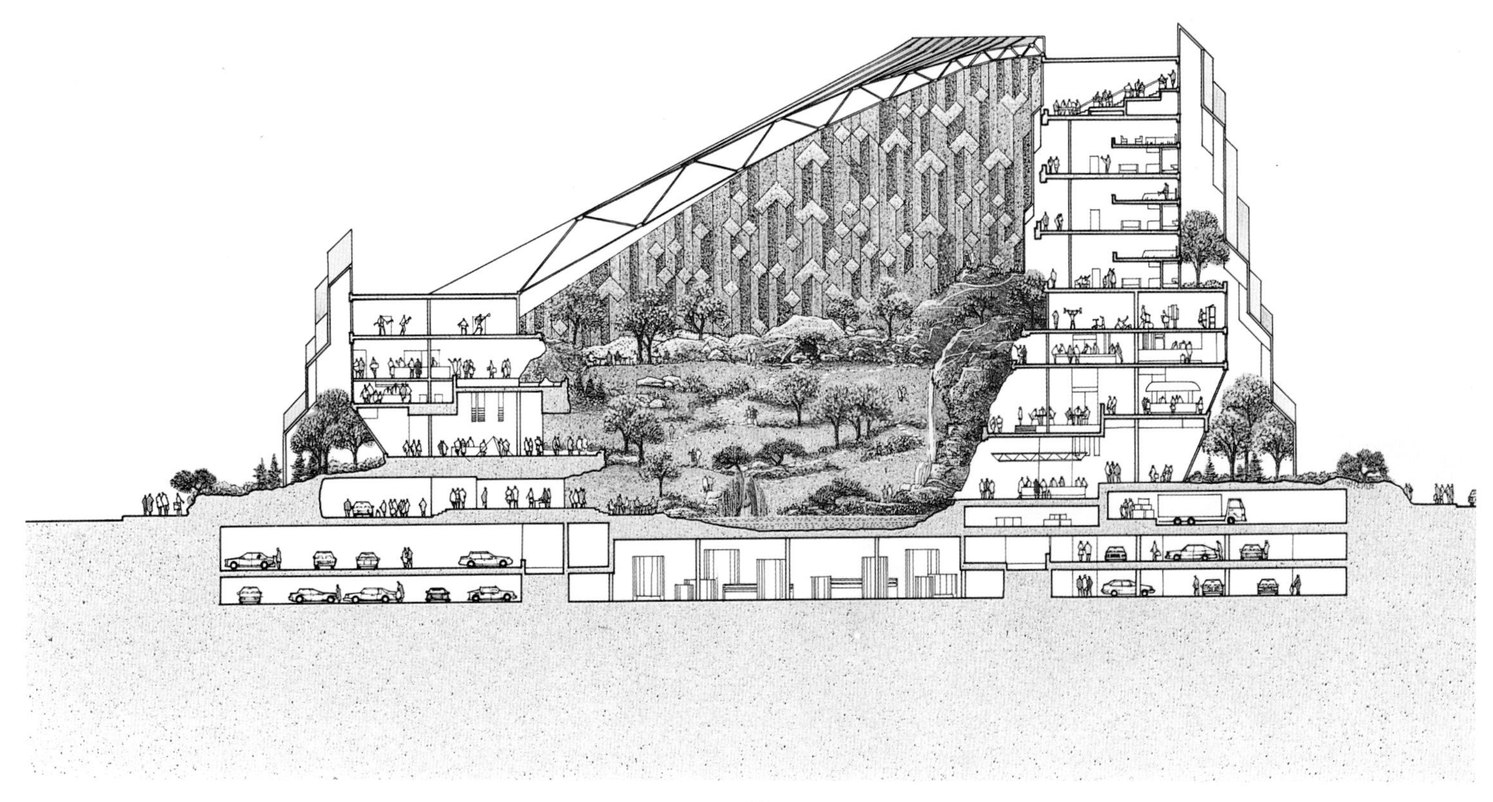

Worldbridge Trade and Investment Center

Baltimore, Maryland

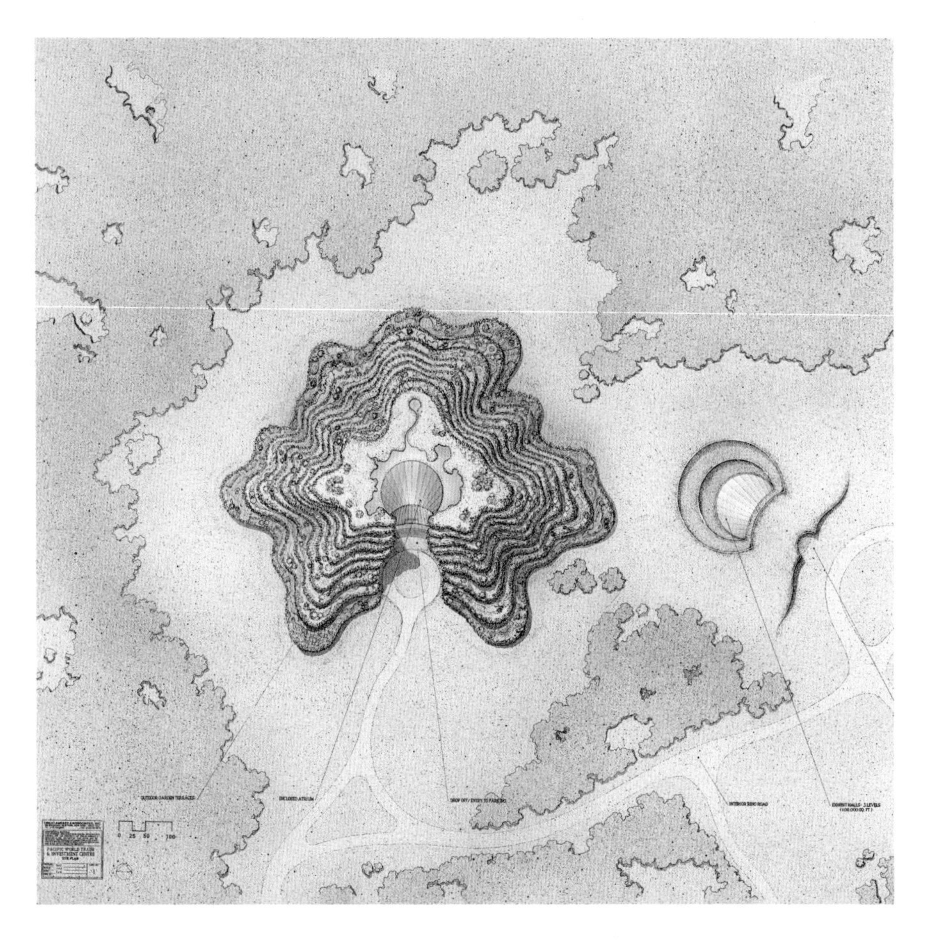

Residence-au-Lac

Lugano, Switzerland

Focchi Shopping Center

Rimini, Italy

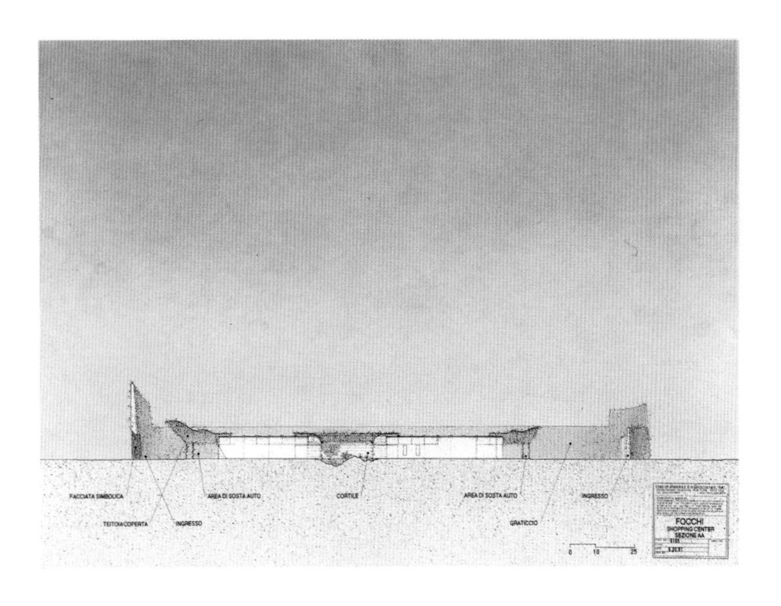
FACCIATA SIMBOLICA
TETTOIA COPERTA
INGRESSO
AREA DI SOSTA AUTO
CORTILE
AREA DI SOSTA AUTO
GRATICCIO
INGRESSO
FOCCHI
SHOPPING CENTER
SEZIONE AA
0
10
25

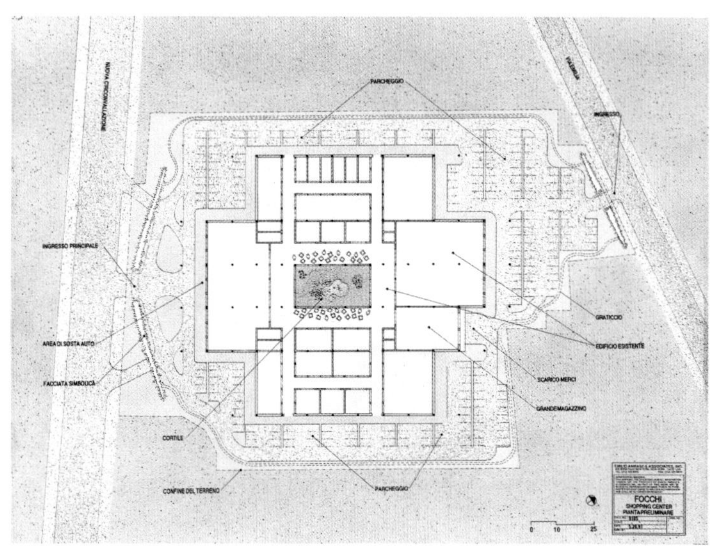
NUOVA CIRCONVALLAZIONE
PARCHEGGIO
INGRESSO
INGRESSO PRINCIPALE
AREA DI SOSTA AUTO
FACCIATA SIMBOLICA
CORTILE
CONFINE DEL TERRENO
PARCHEGGIO
GRATICCIO
EDIFICIO ESISTENTE
SCARICO MERCI
GRANDE MAGAZZINO
FOCCHI
SHOPPING CENTER
PIANTA PRELIMINARE
0
10
25

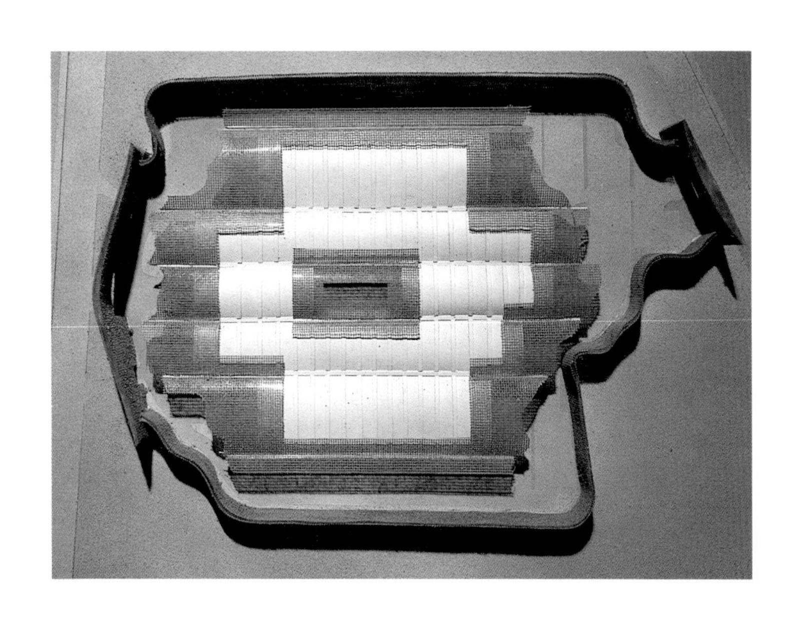

House for Leo Castelli

Northeast, USA

Casa de Retiro Espiritual

Cordoba, Spain

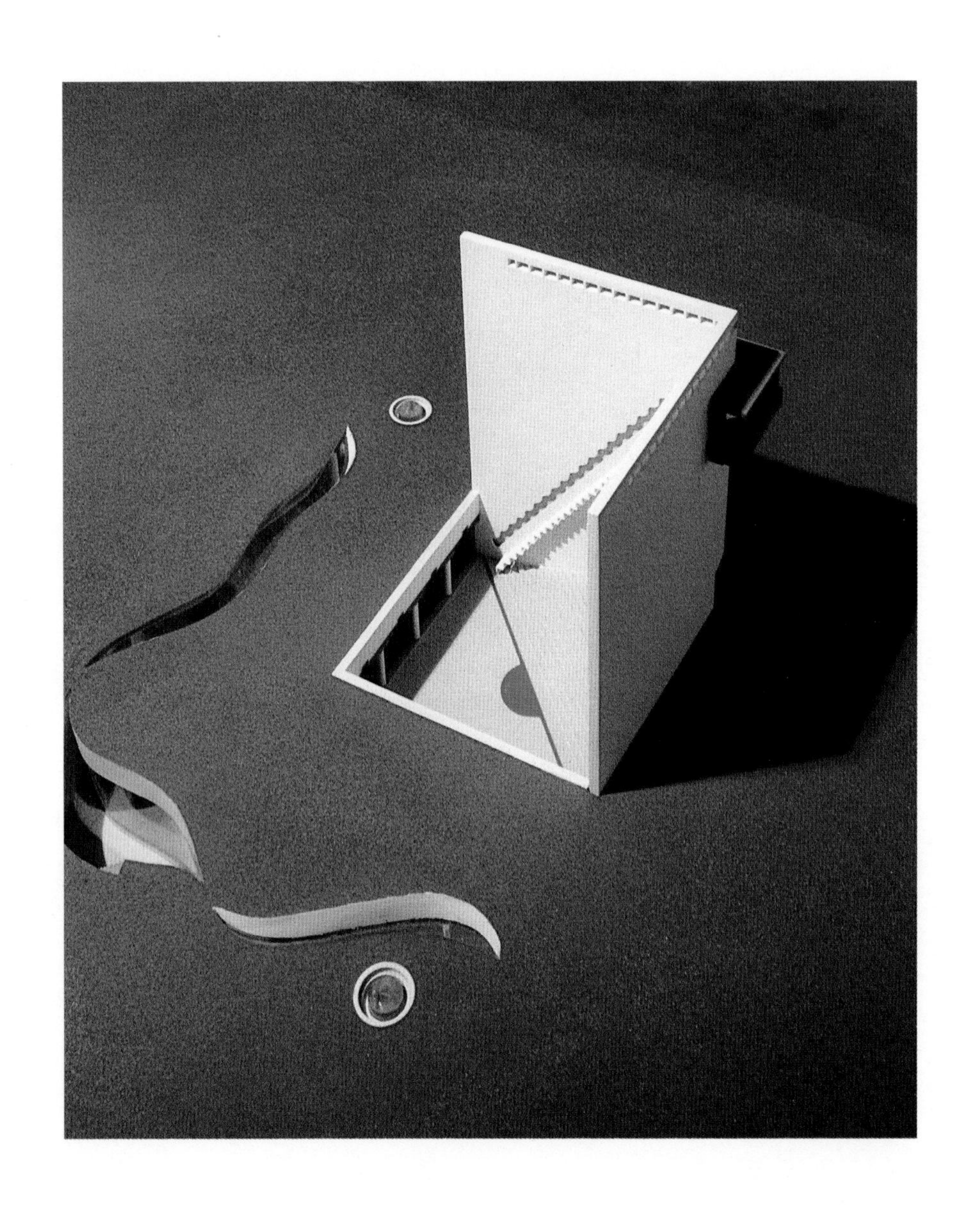

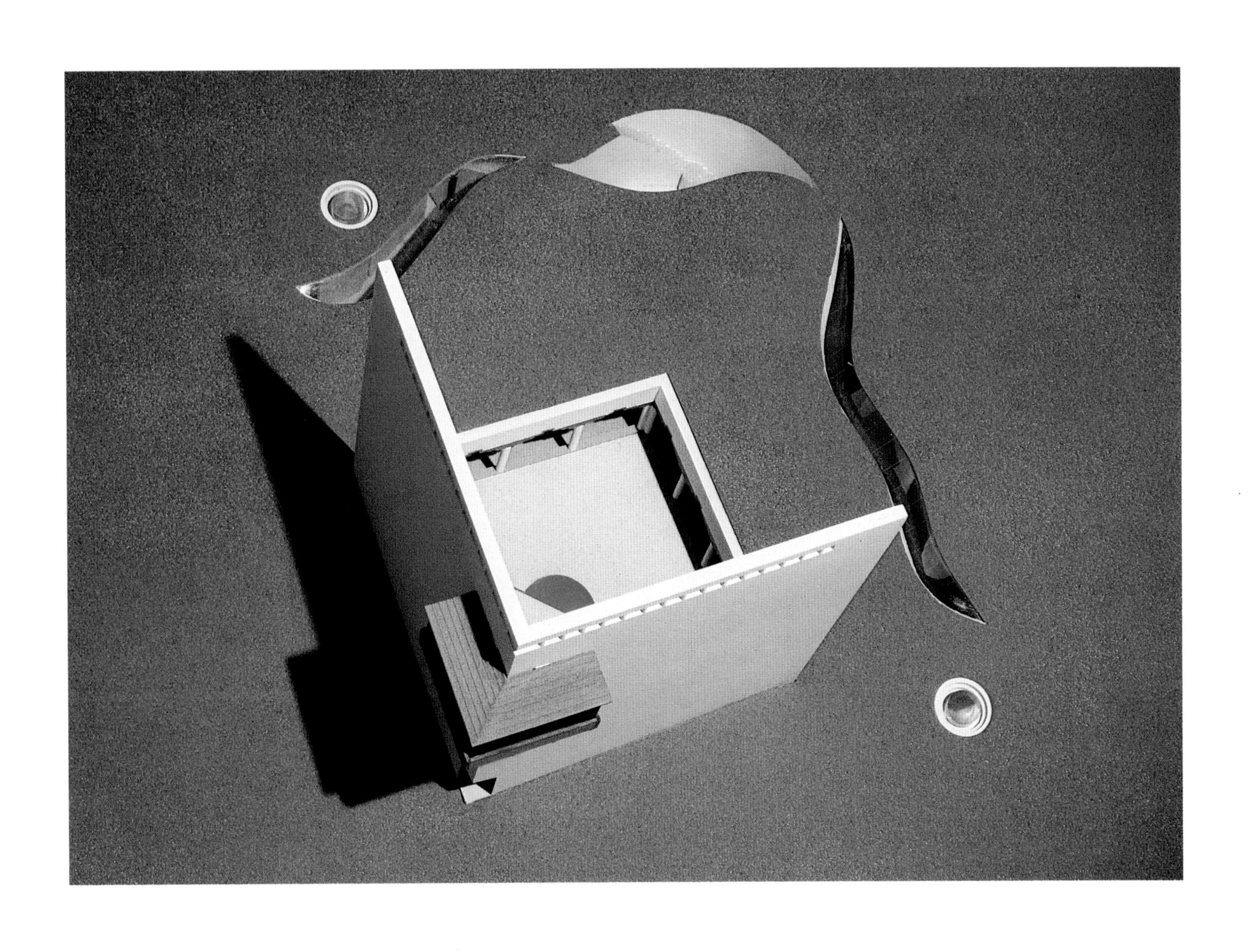

Manoir d'Angoussart

Bierges, Belgium

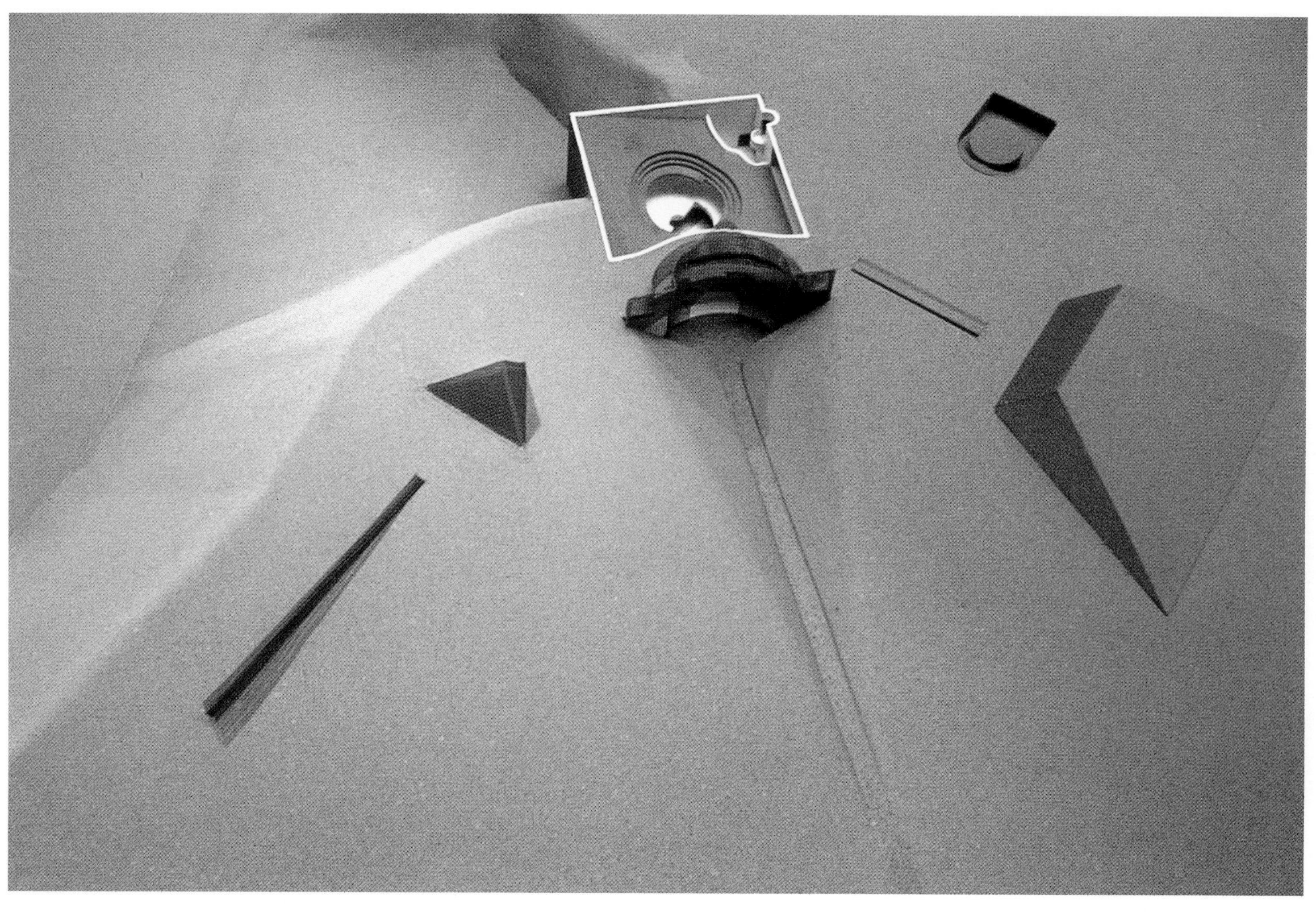

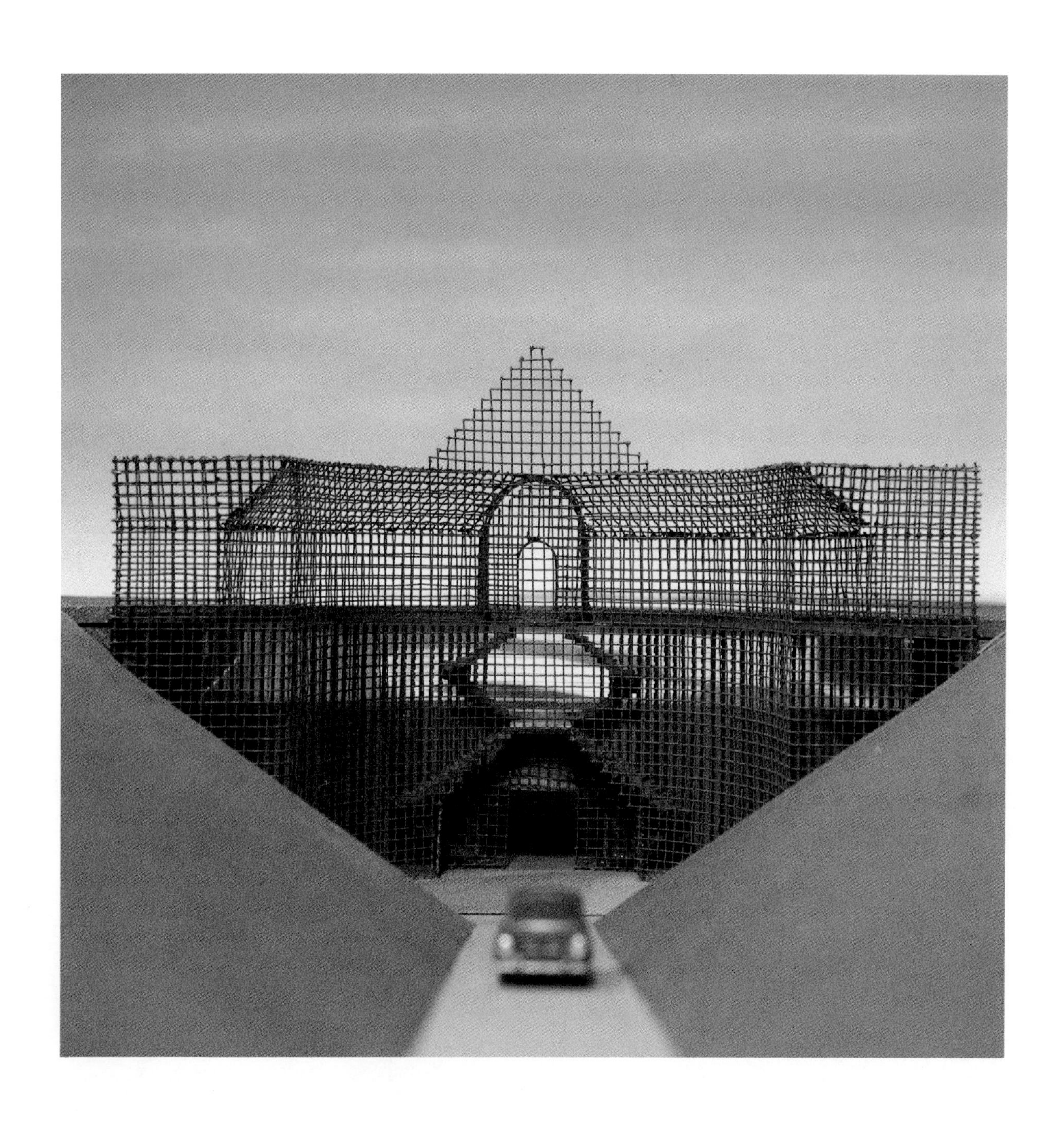

Private Estate

Montana, USA

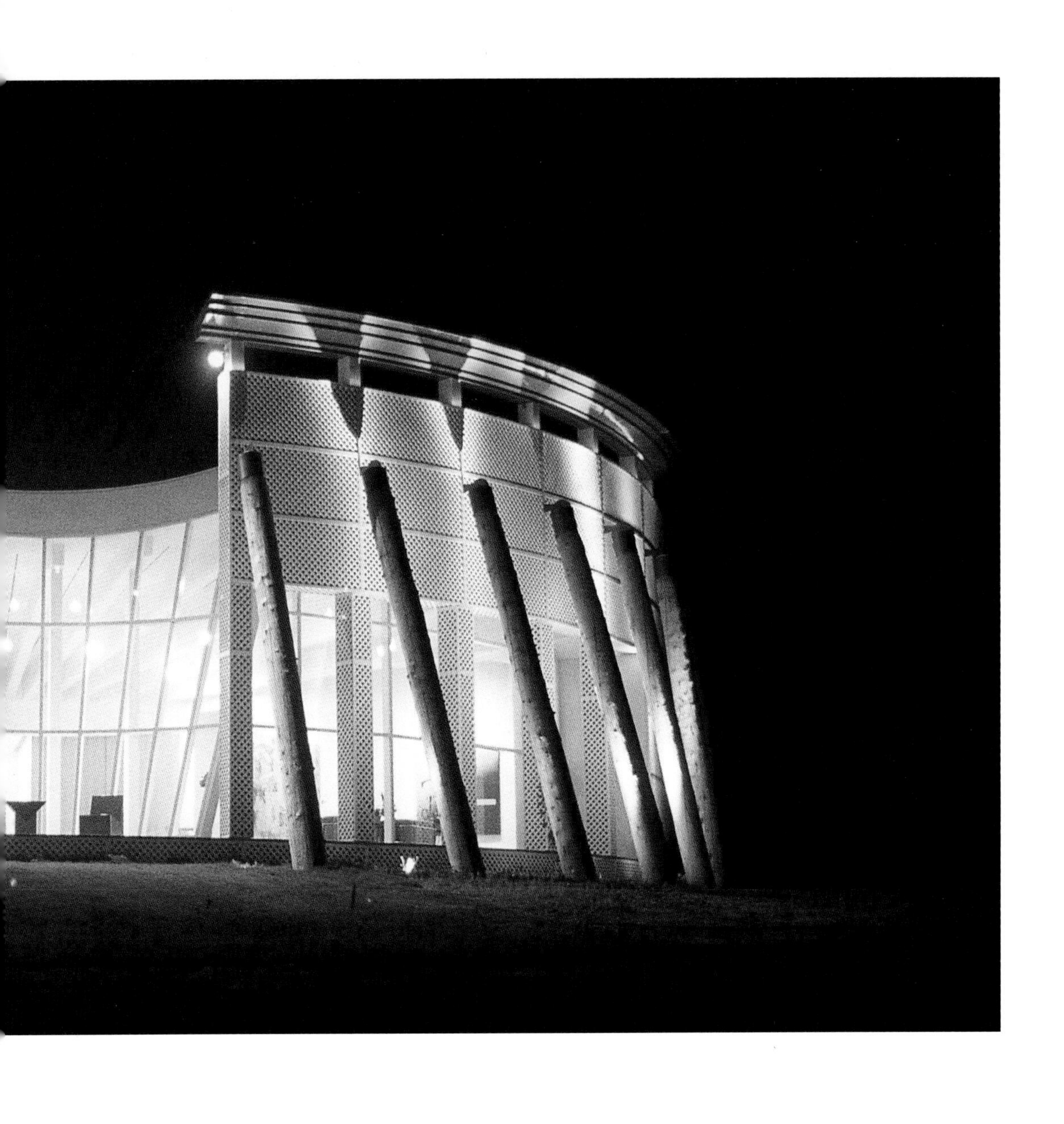

Casa Canales

Monterrey, Mexico

Financial Guaranty Insurance Company

New York, New York

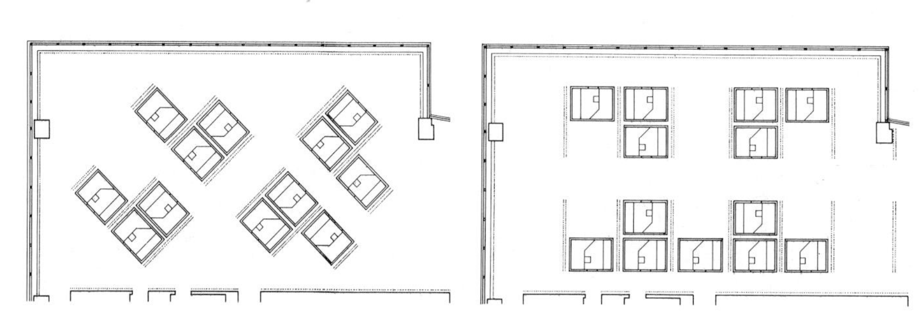

Plaza Mayor

Salamanca, Spain

Houston Center Plaza

Houston, Texas

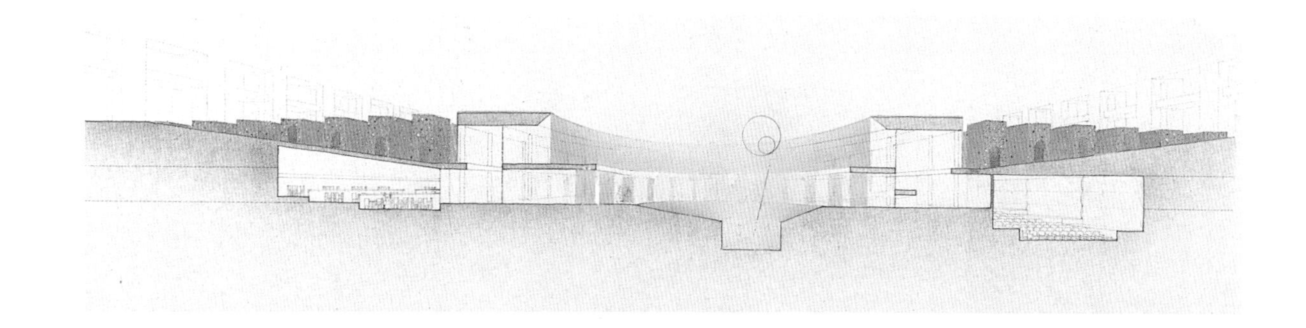

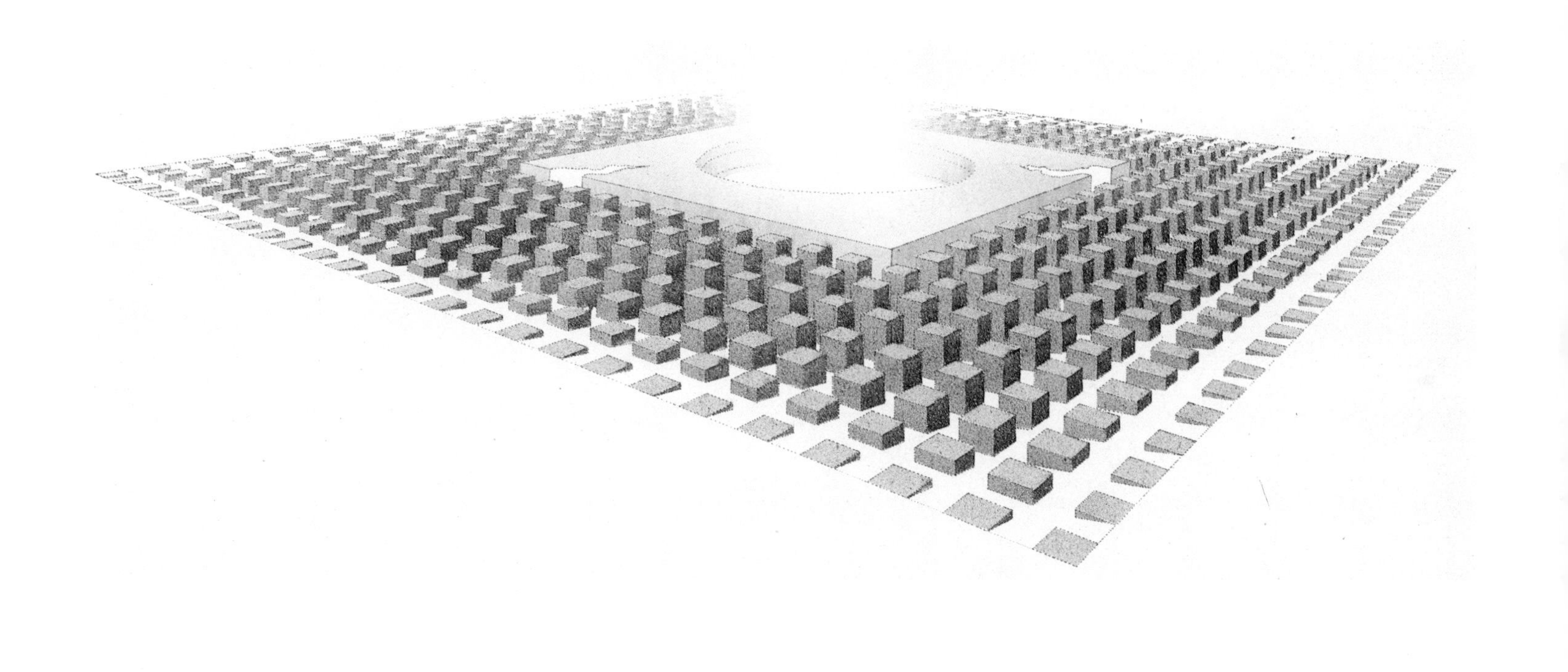

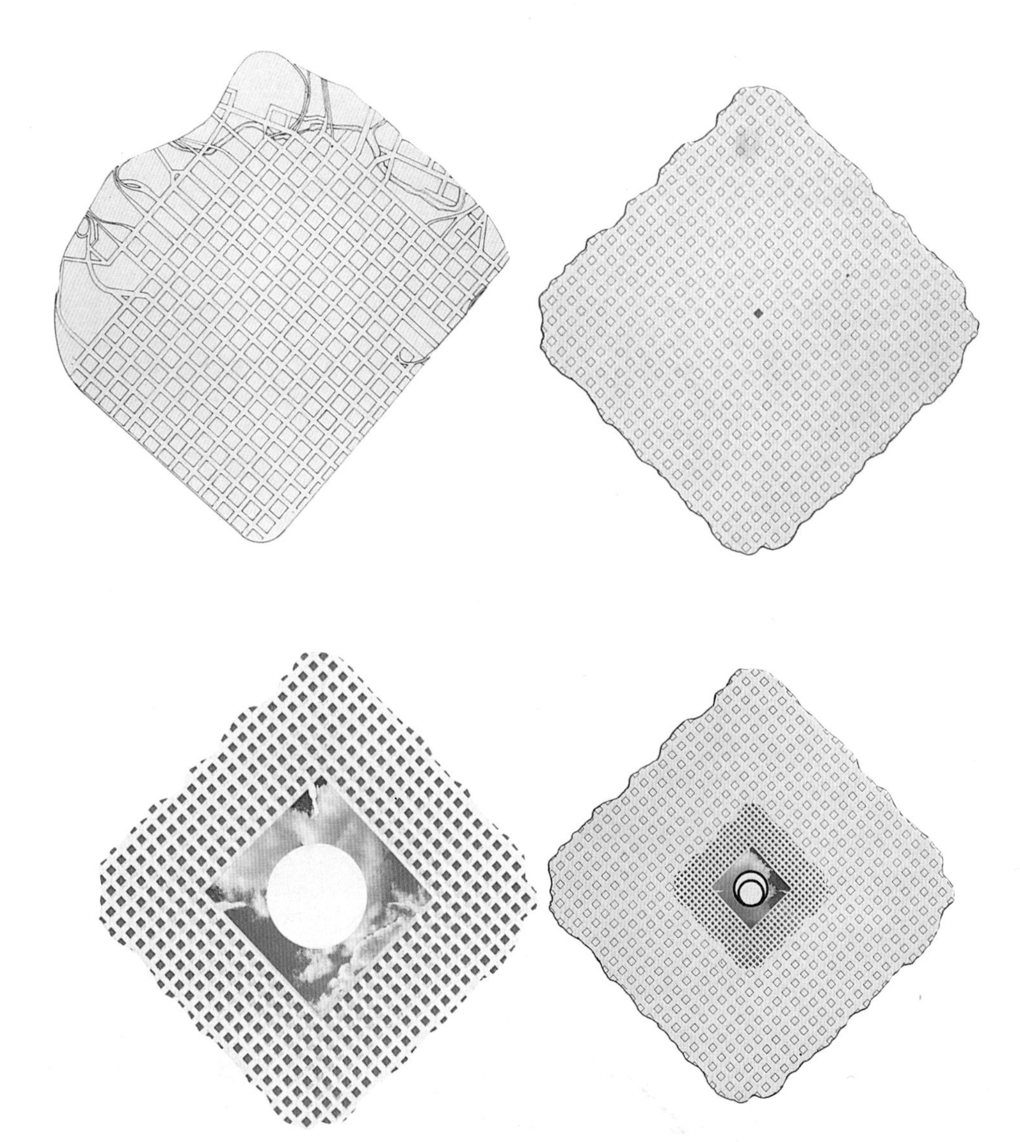

Pro Memoria Garden

Ludenhausen, Germany

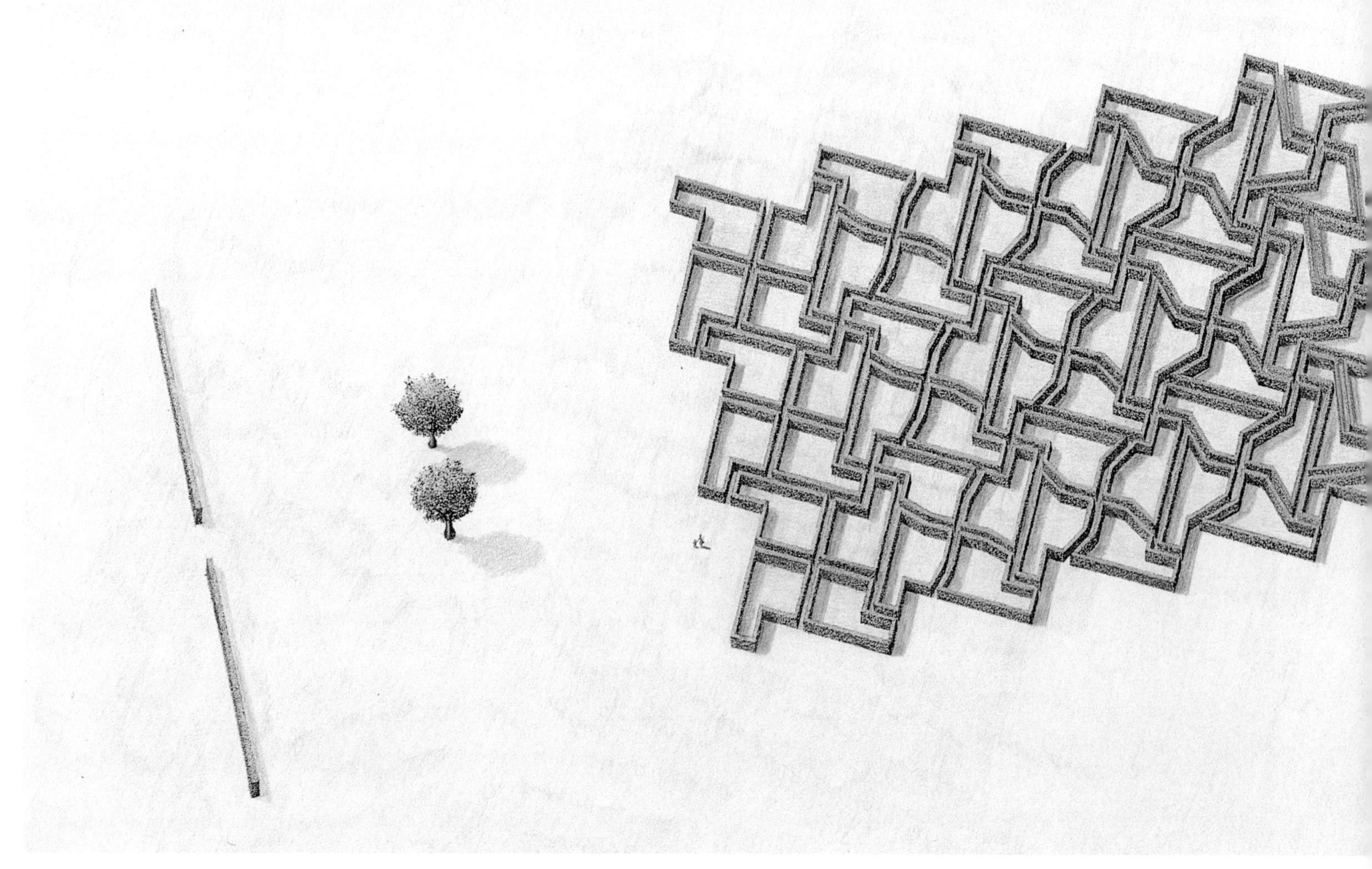

Emilio's Folly

Man is an Island

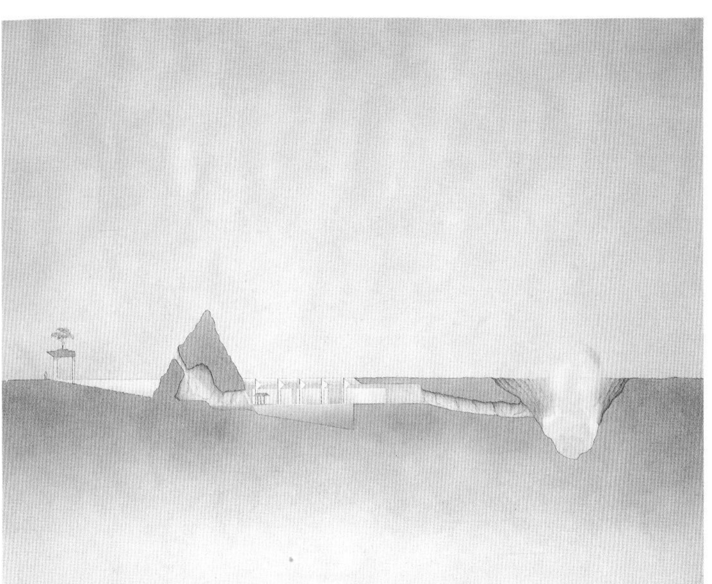

Hortus Conclusus—Centre Georges Pompidou

Paris, France

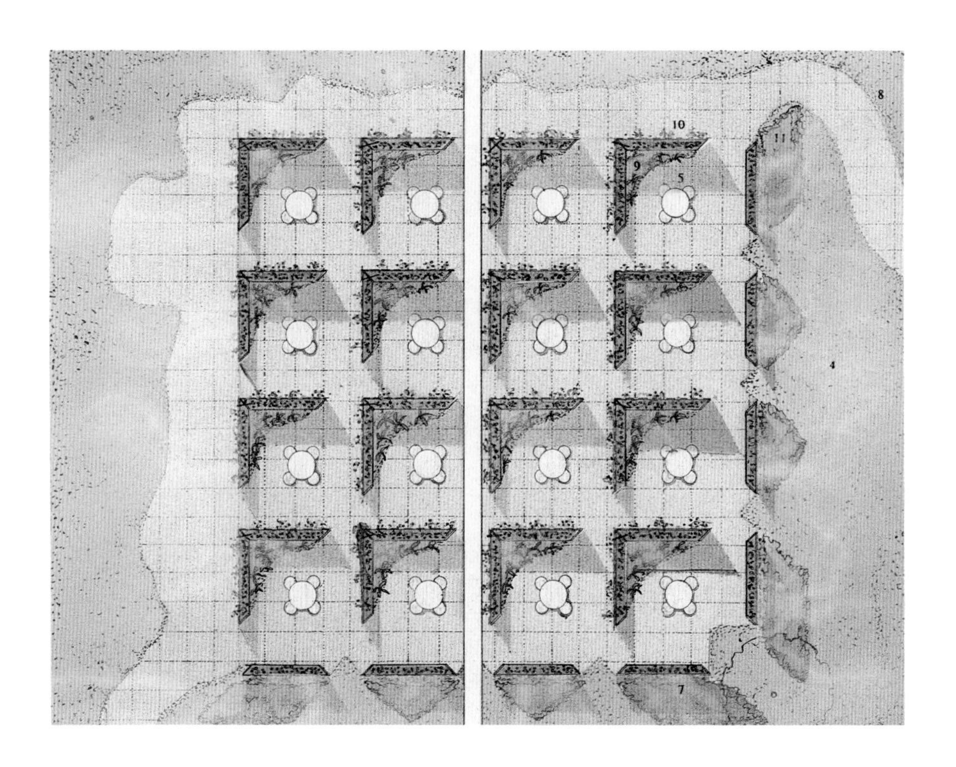

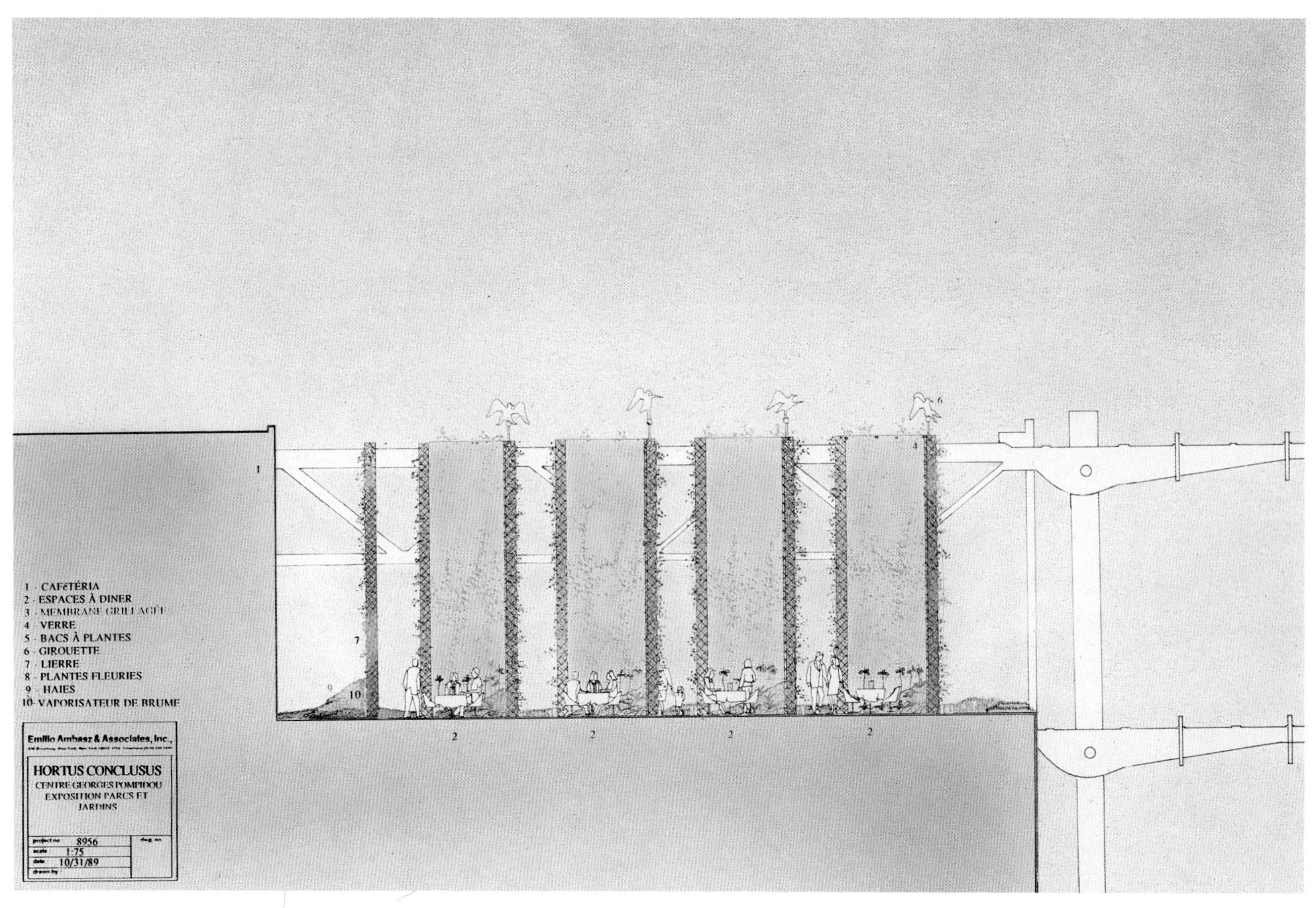

Columbus Bridge

Columbus, Indiana

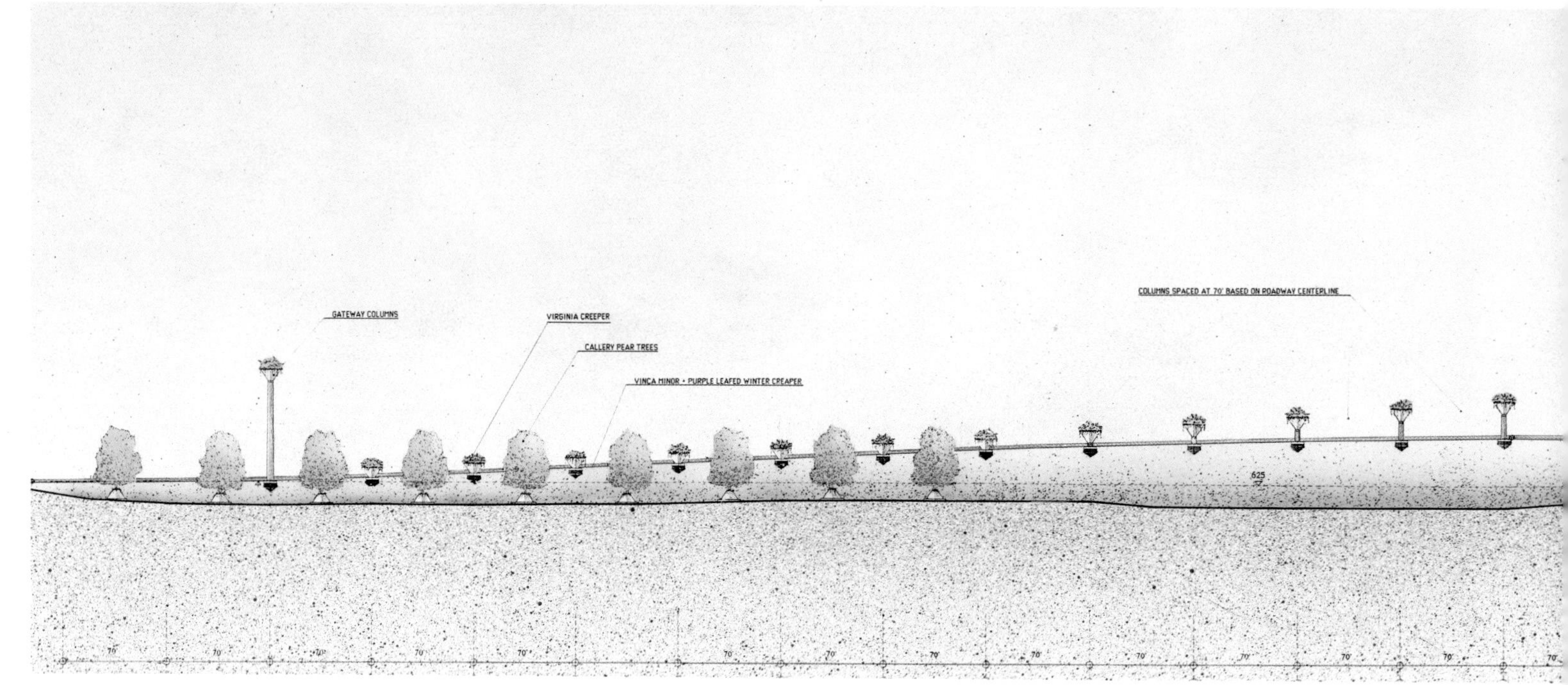
GATEWAY COLUMNS
VIRGINIA CREEPER
CALLERY PEAR TREES
VINCA MINOR + PURPLE LEAFED WINTER CREAPER
COLUMNS SPACED AT 70' BASED ON ROADWAY CENTERLINE

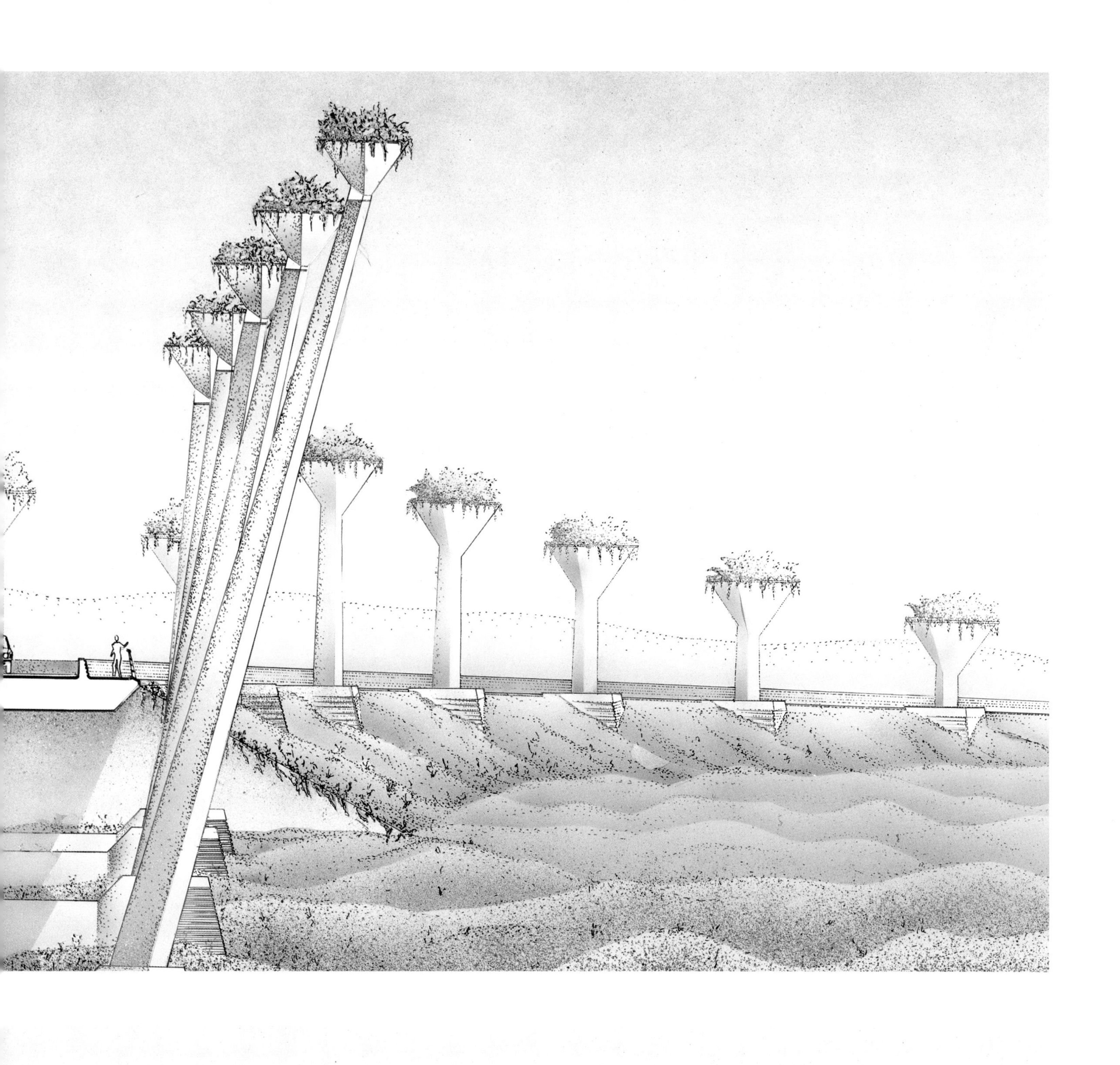

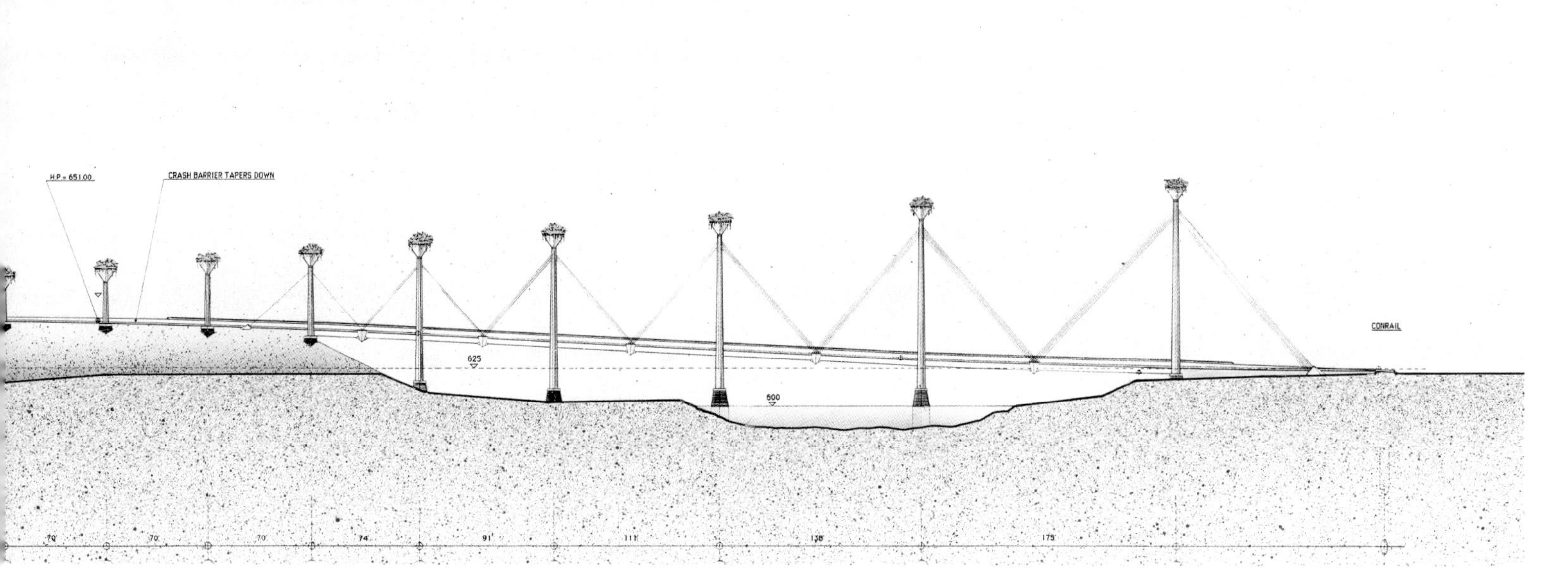
H.P. = 651.00
CRASH BARRIER TAPERS DOWN
CONRAIL
625
600
70'
70'
70'
74'
91'
111'
138'
175'

Master Plan for the Universal Exposition—Seville 1992

Seville, Spain

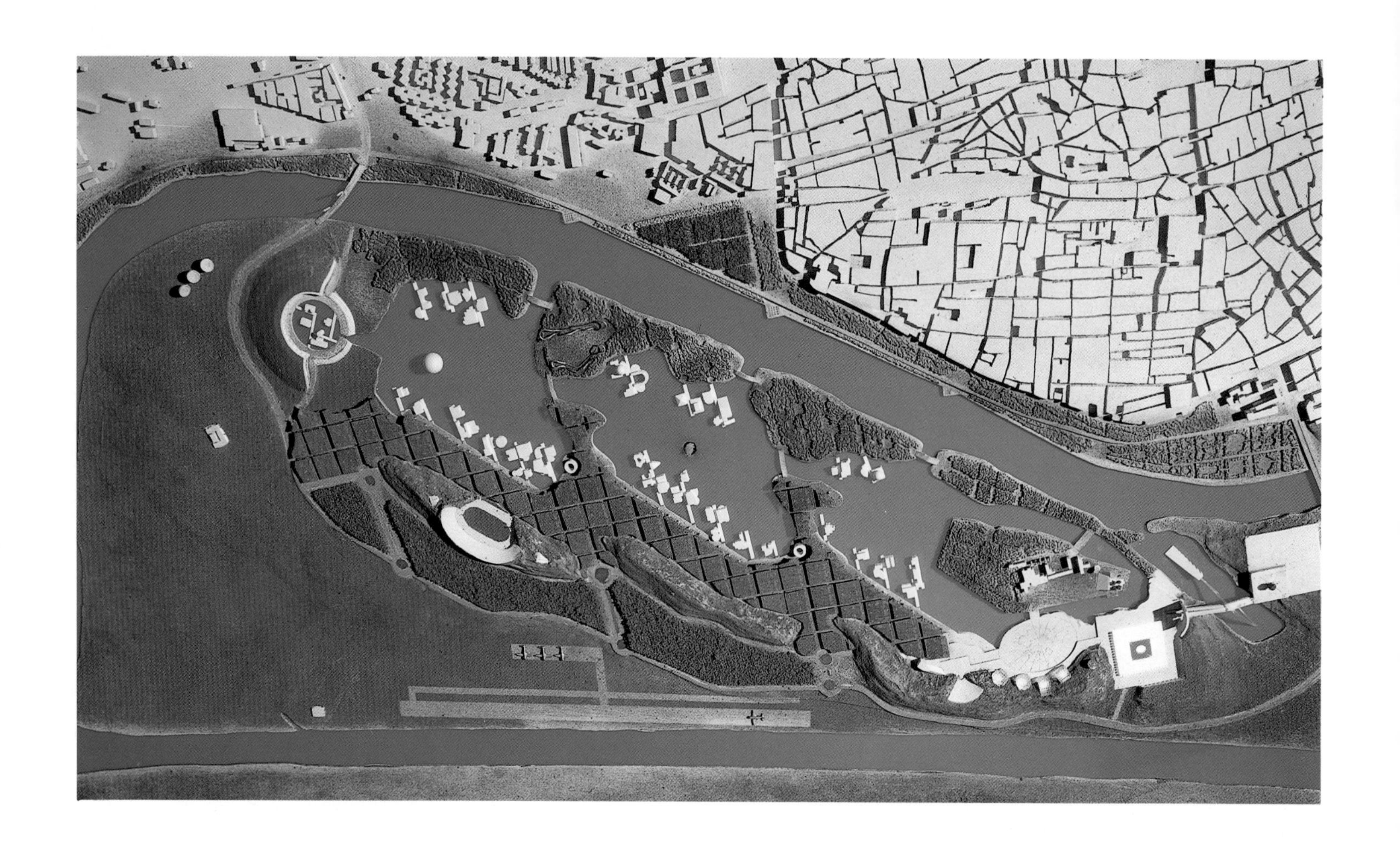

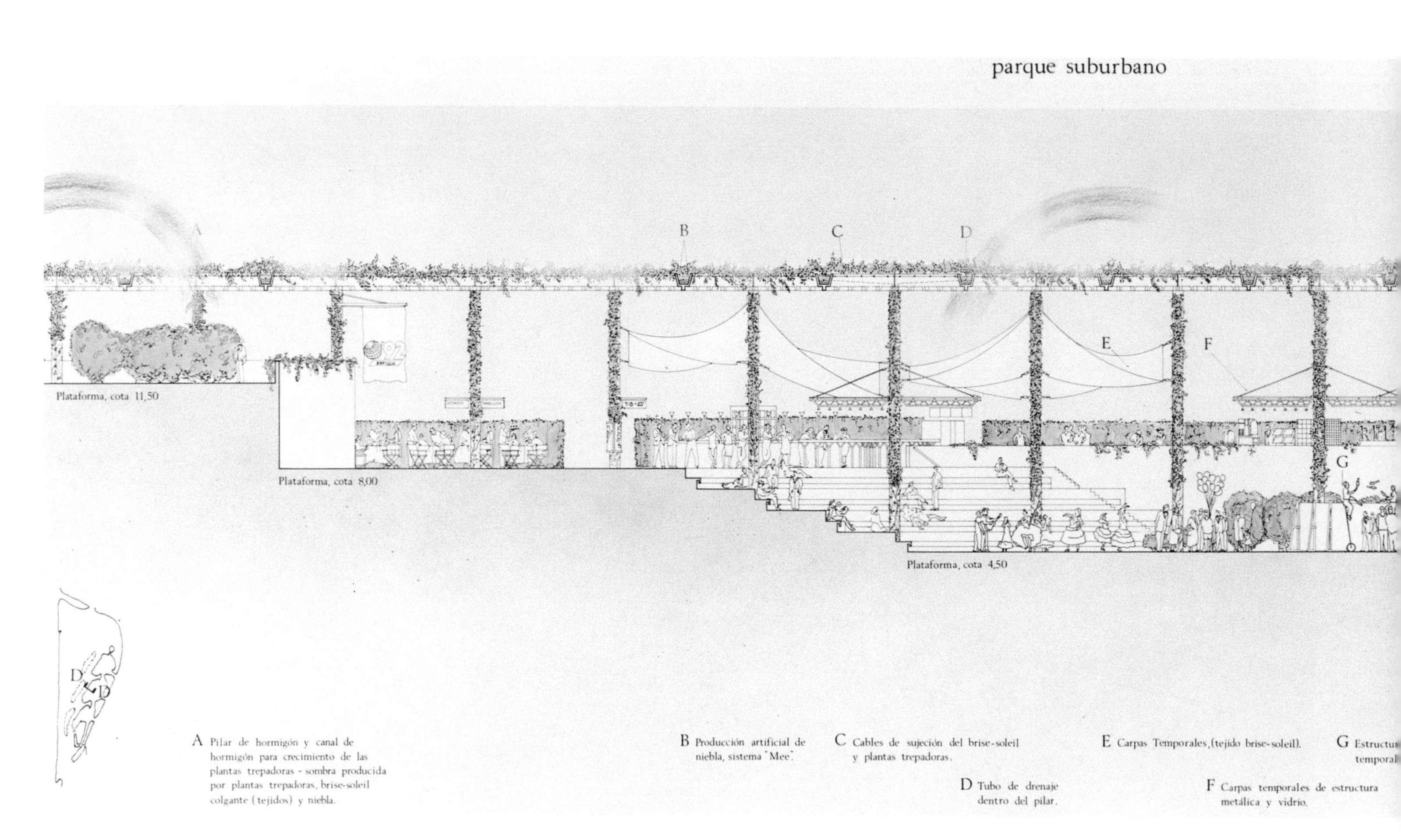

parque suburbano
A
B
C
D
E
F
G
Plataforma, cota 11,50
Plataforma, cota 8,00
Plataforma, cota 4,50
A Pilar de hormigón y canal de hormigón para crecimiento de las plantas trepadoras - sombra producida por plantas trepadoras, brise-soleil colgante (tejidos) y niebla.
B Producción artificial de niebla, sistema "Mee".
C Cables de sujeción del brise-soleil y plantas trepadoras.
D Tubo de drenaje dentro del pilar.
E Carpas Temporales, (tejido brise-soleil).
F Carpas temporales de estructura metálica y vidrio.
G Estructu
temporal

olímpico al fondo

sistema de riego

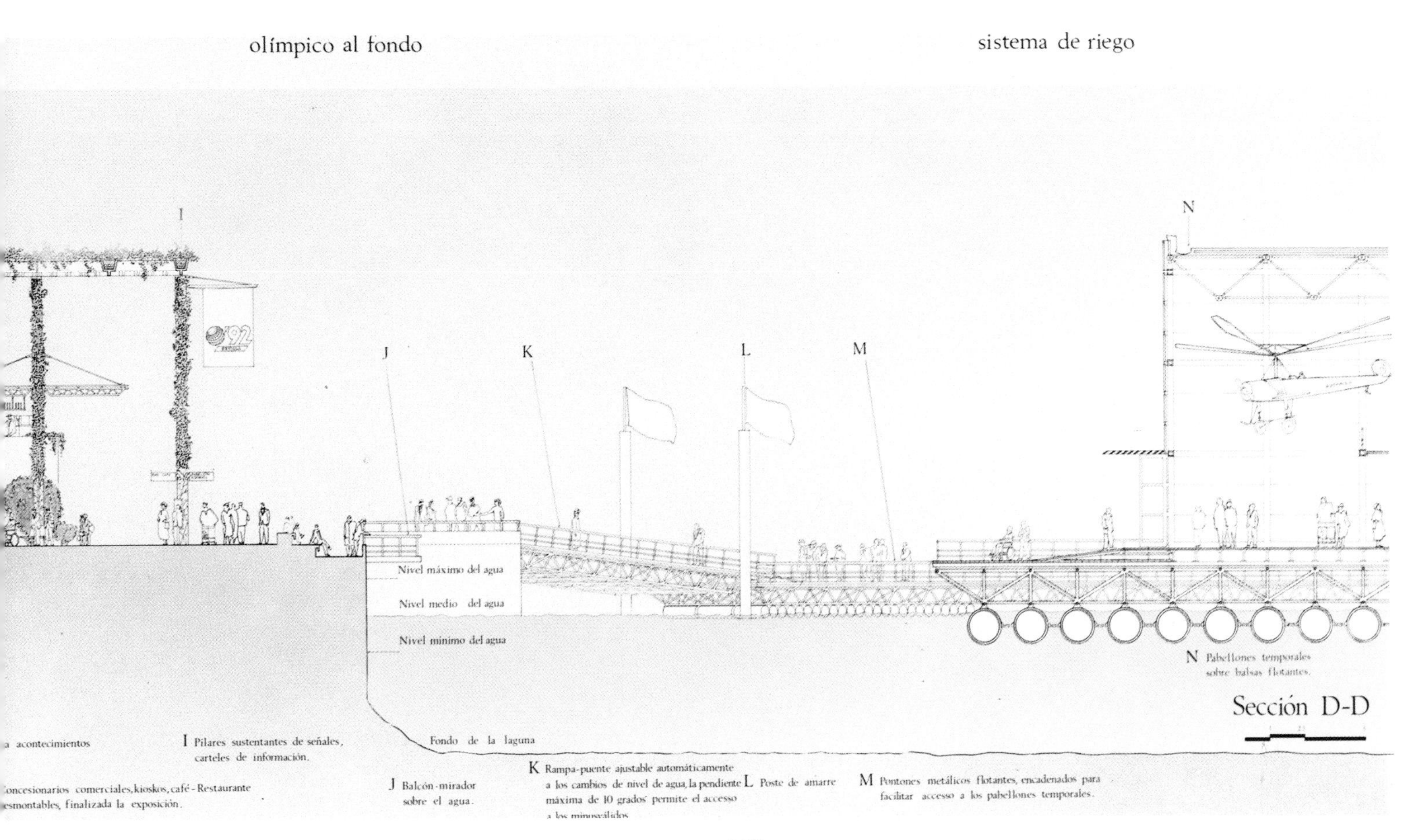

I
J
K
L
M
N
Nivel máximo del agua
Nivel medio del agua
Nivel mínimo del agua
Fondo de la laguna
N Pabellones temporales sobre balsas flotantes.
Sección D-D
a acontecimientos
I Pilares sustentantes de señales, carteles de información.
oncesionarios comerciales, kioskos, café - Restaurante esmontables, finalizada la exposición.
J Balcón-mirador sobre el agua.
K Rampa-puente ajustable automáticamente a los cambios de nivel de agua, la pendiente máxima de 10 grados permite el acceso a los minusválidos
L Poste de amarre
M Pontones metálicos flotantes, encadenados para facilitar acceso a los pabellones temporales.

Eschenheimer Tower

Frankfurt, Germany

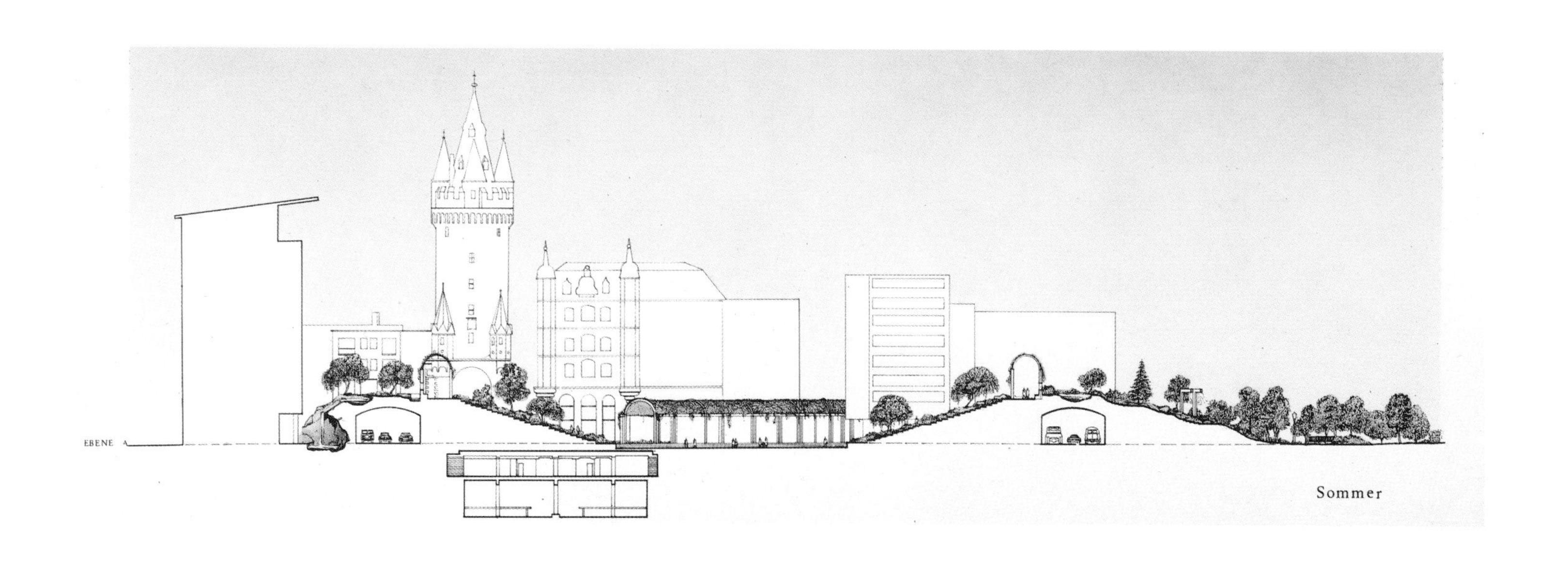
EBENE A
Sommer

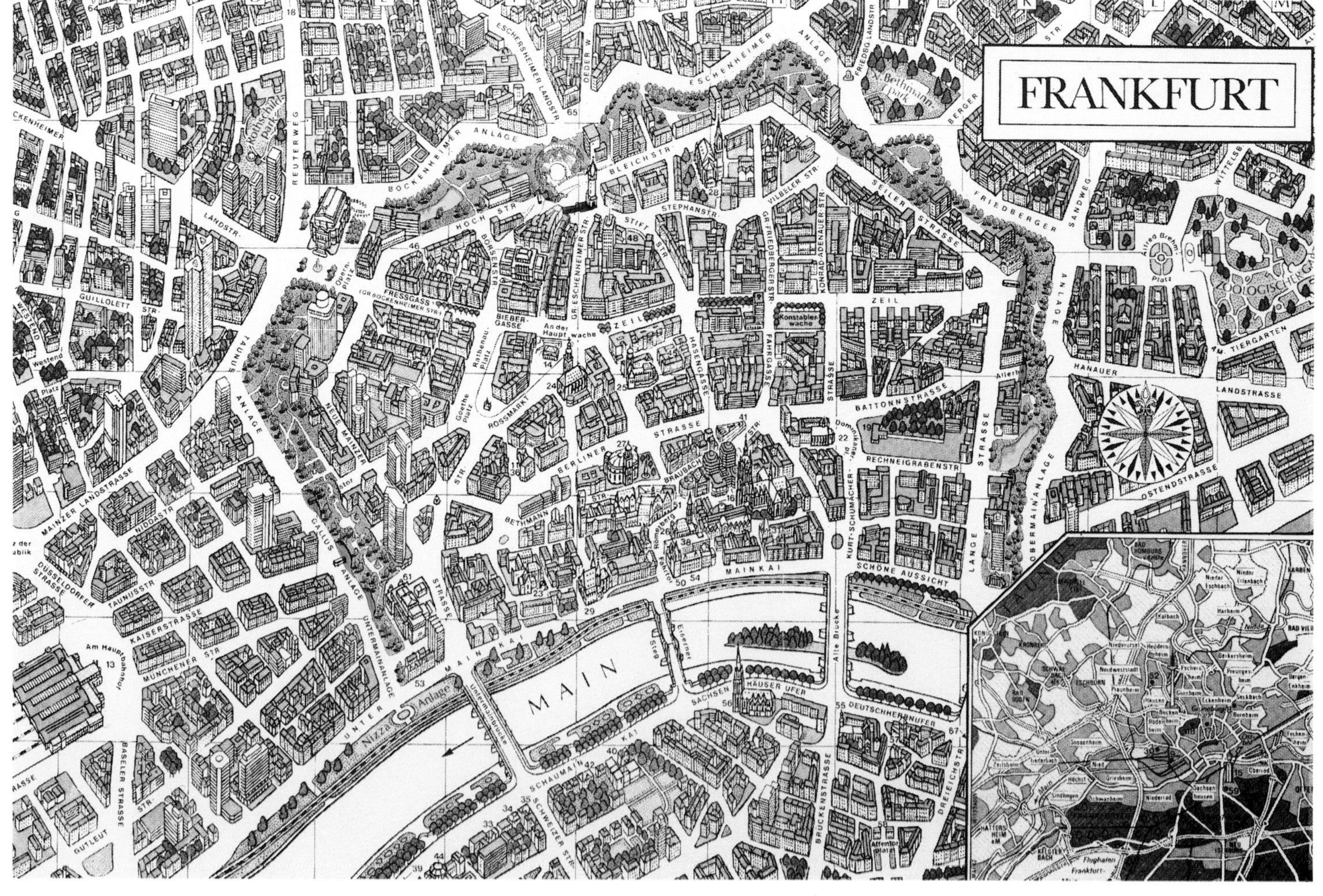
FRANKFURT
MAIN
ESCHENHEIMER ANLAGE
BOCKENHEIMER ANLAGE
BLEICHSTR.
STEPHANSTR.
ZEIL
KONRAD-ADENAUER STR.
BATTONNSTRASSE
RECHNEIGRABENSTR.
MAINKAI
SCHÖNE AUSSICHT
HANAUER LANDSTRASSE
OSTENDSTRASSE
LANGE STRASSE
SEILER STRASSE
FRIEDBERGER ANLAGE
TAUNUS ANLAGE
GALLUS ANLAGE
KAISERSTRASSE
MÜNCHENER STR.
NIDDASTR.
DEUTSCHHERRNUFER
SCHAUMAIN KAI
BRÜCKENSTRASSE

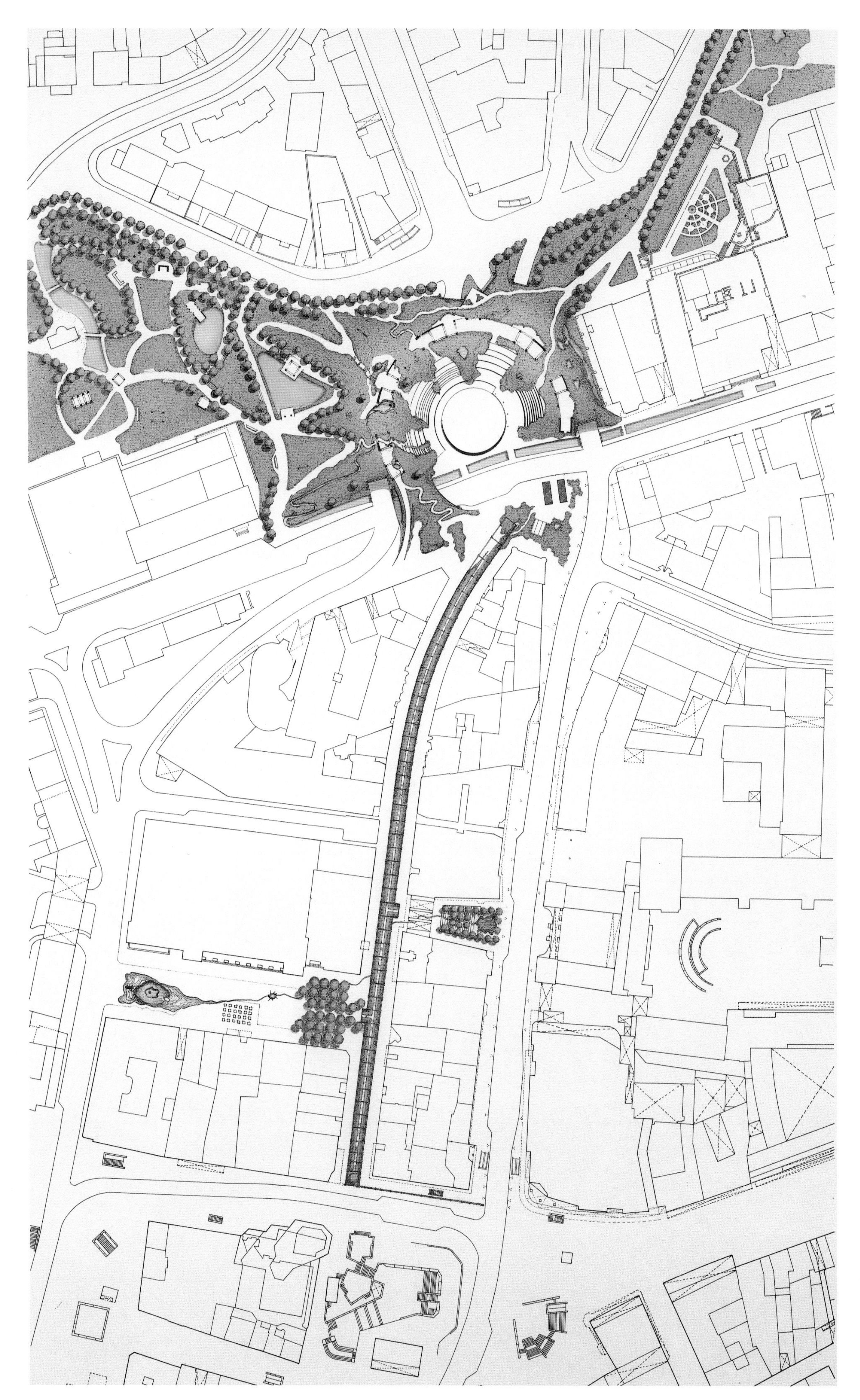

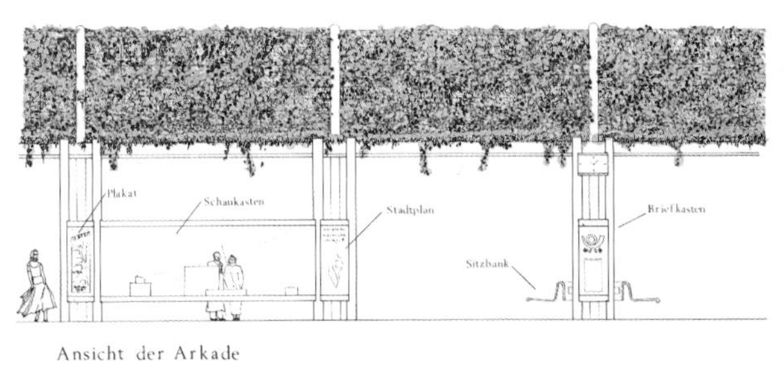

Ansicht der Arkade

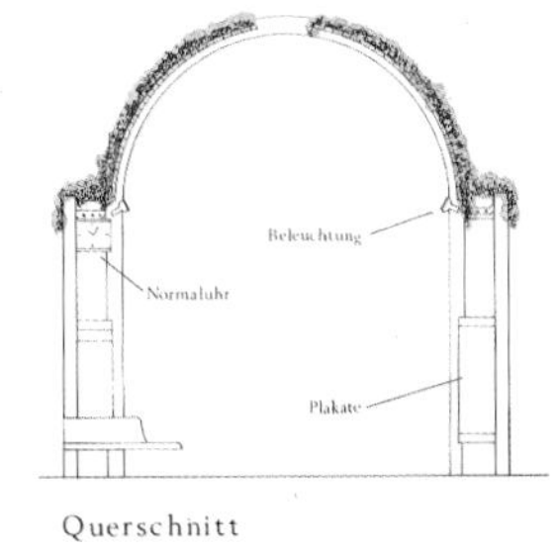

Querschnitt

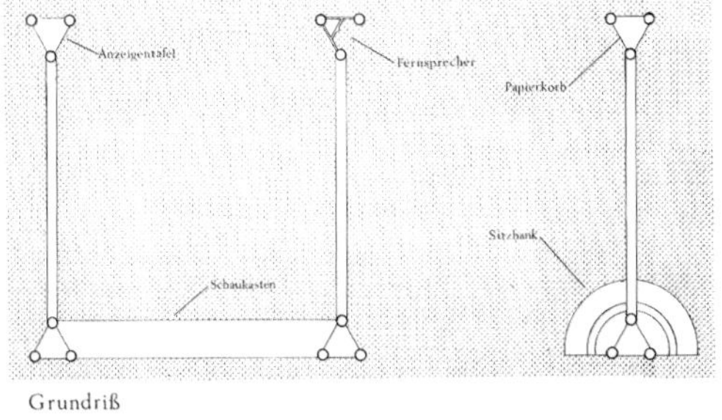

Grundriß

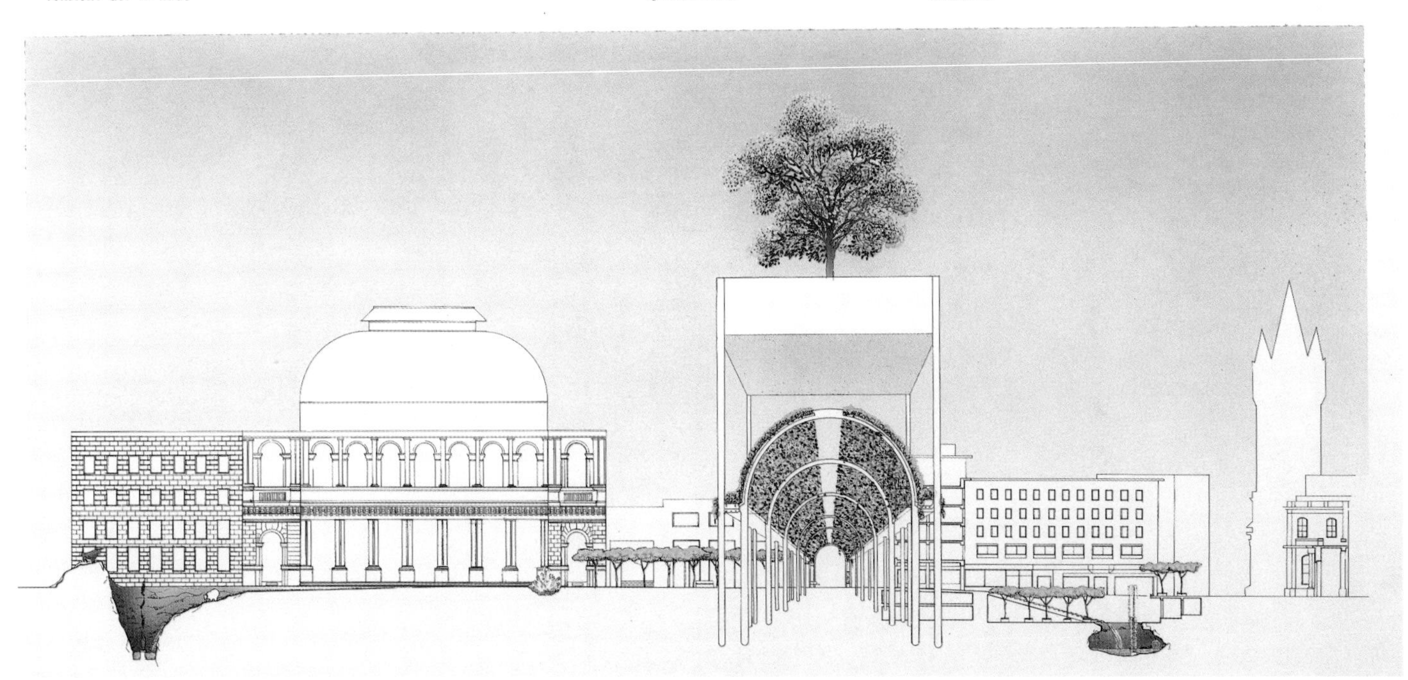

Banque Bruxelles Lambert

Milan, Italy

Frankfurt Zoo

Frankfurt, Germany

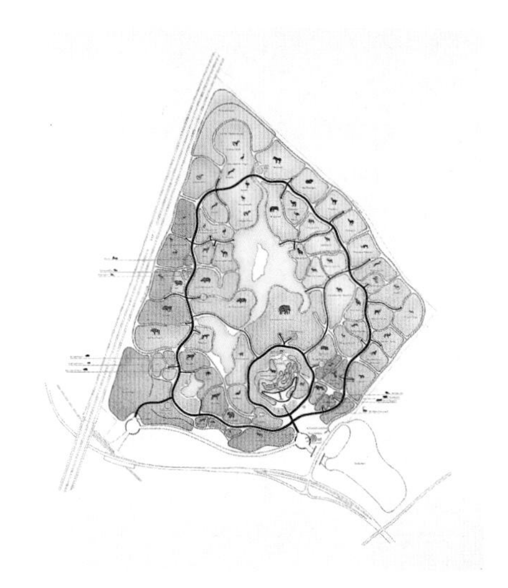

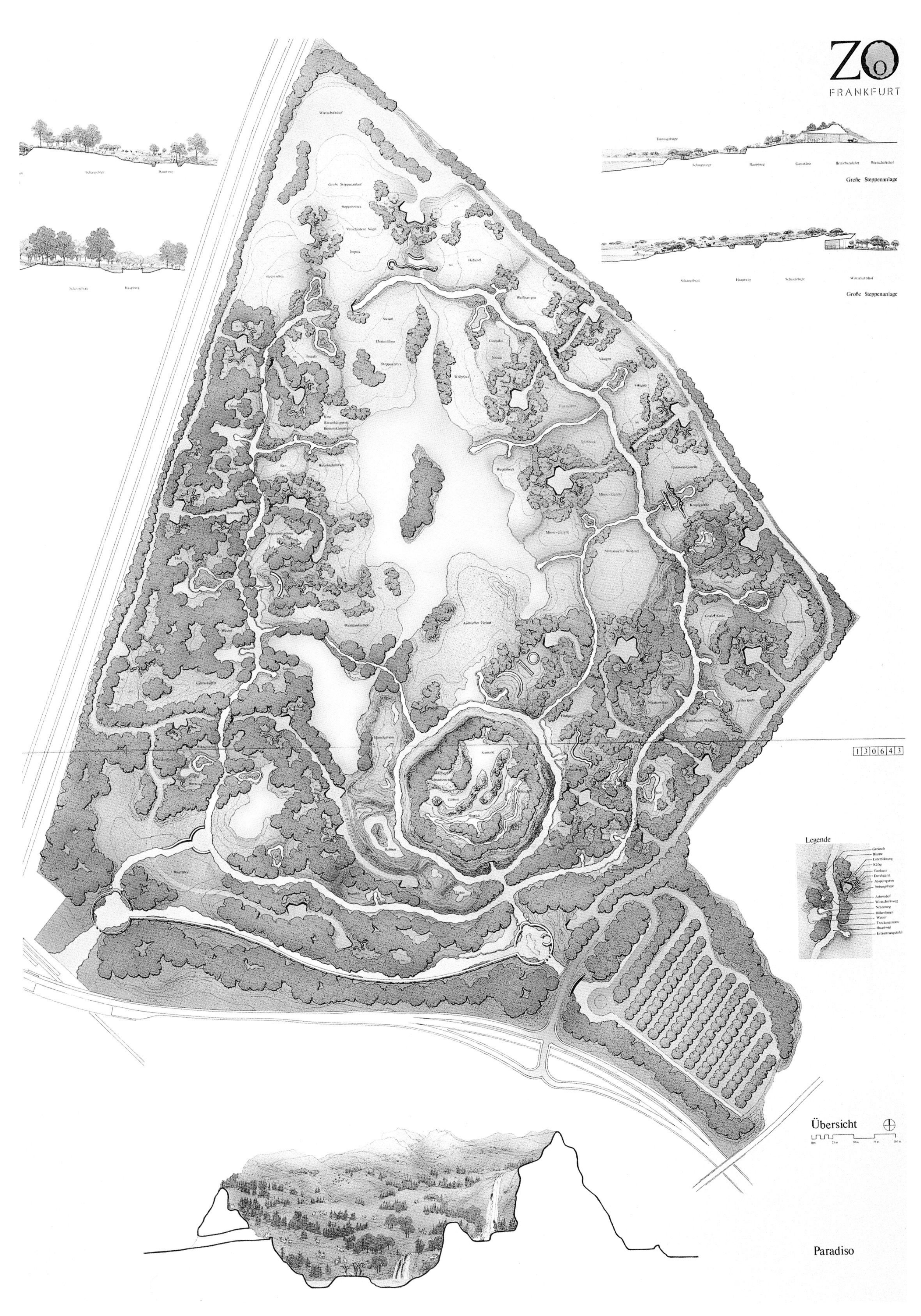
ZOO
FRANKFURT
Große Steppenanlage
Große Steppenanlage
Legende
Übersicht
Paradiso

Cooperative of Mexican-American Grapegrowers

Borrego Springs, California

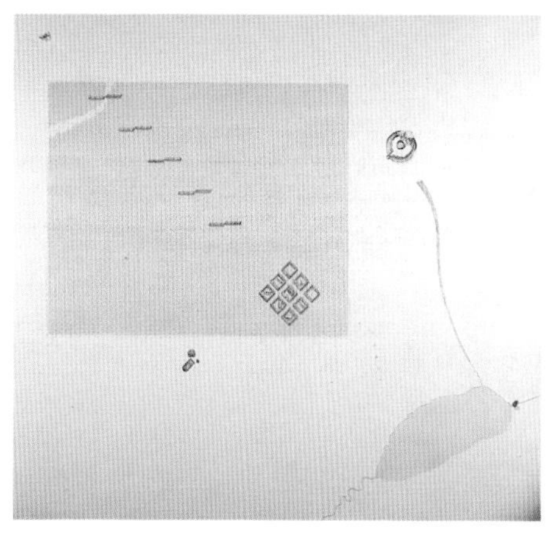

New Town Center

Chiba Prefecture, Japan

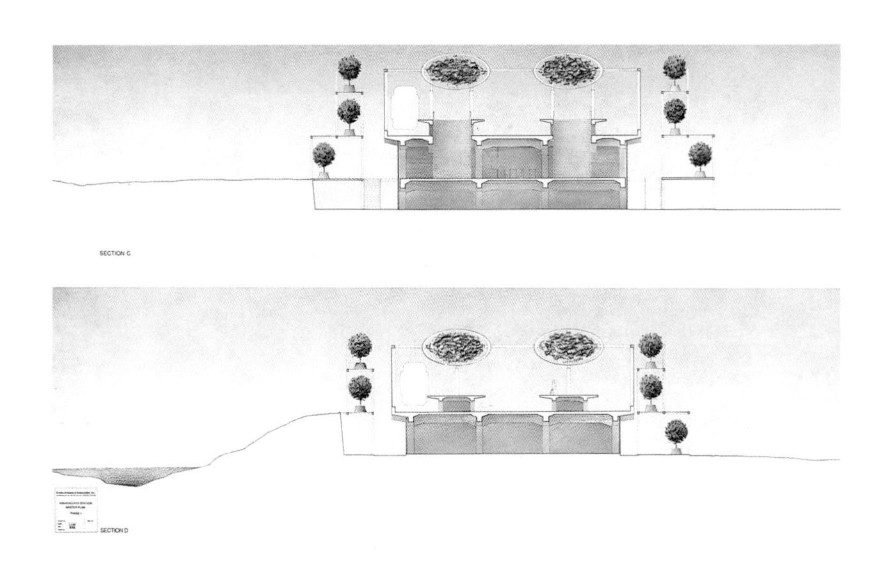
SECTION C
SECTION D

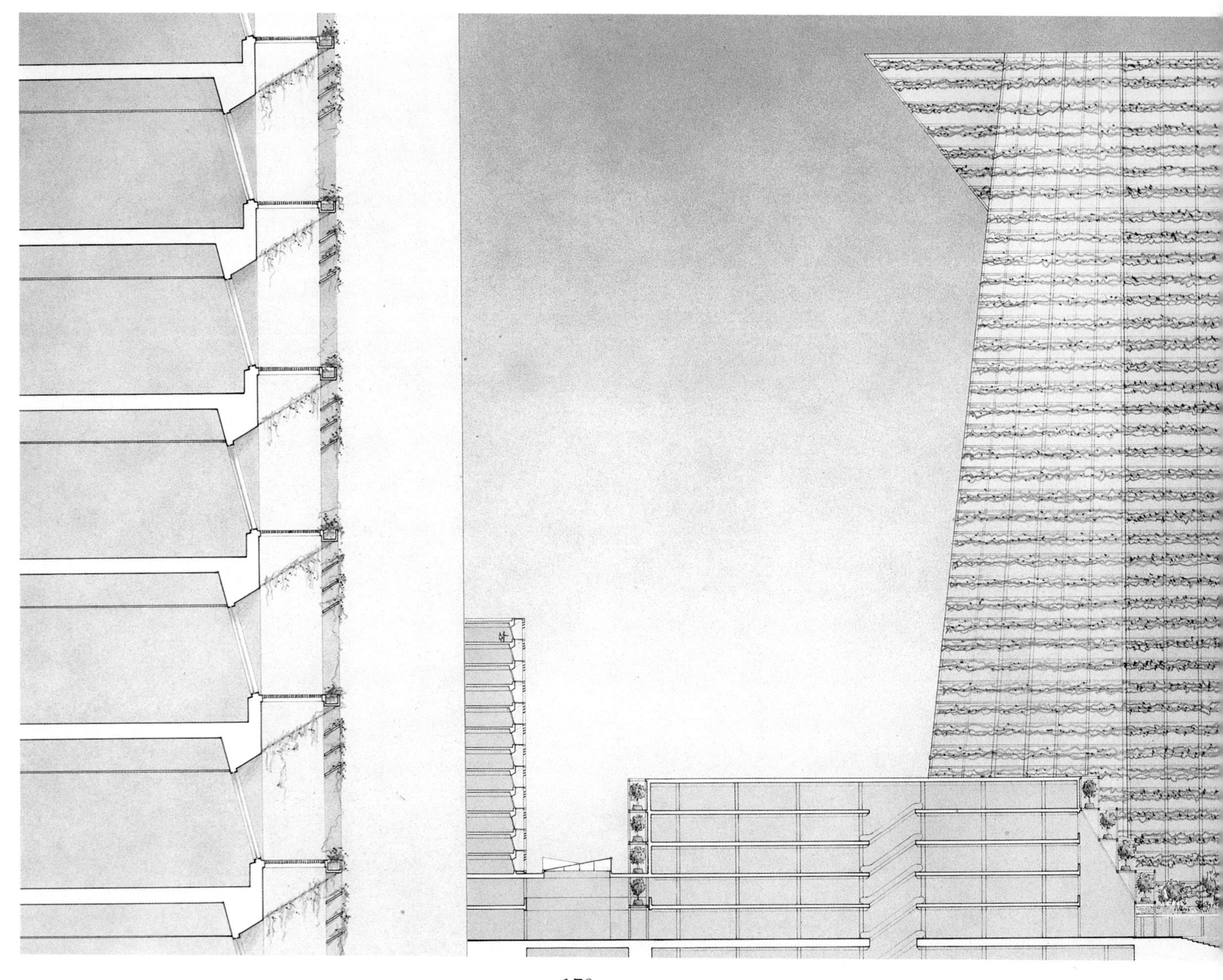

Realworld Theme Park

Barcelona, Spain

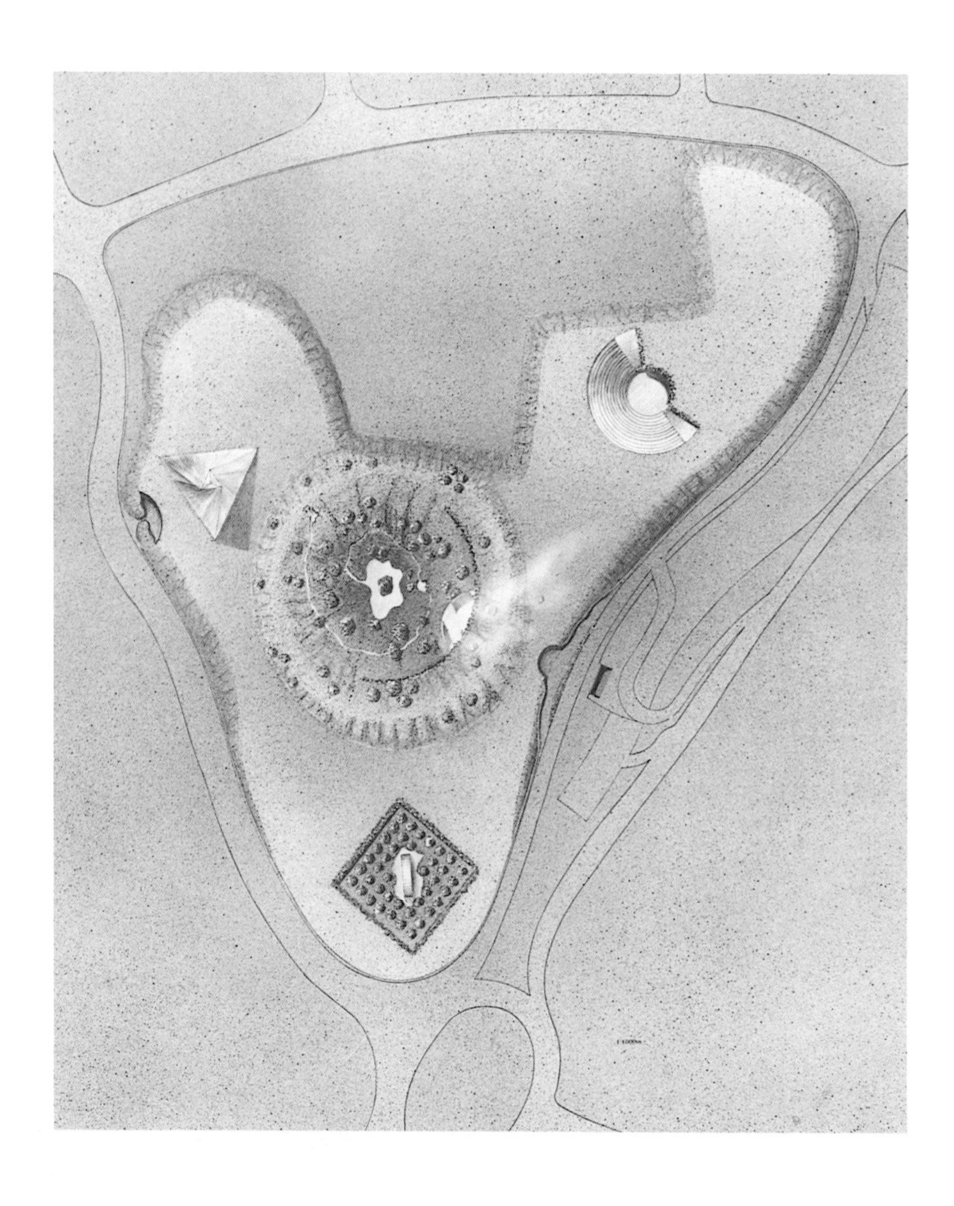

EMILIO AMBASZ & ASSOCIATES, INC.
REAL WORLD
Barcelona Spain
Floor Plan
21 January 1992
2

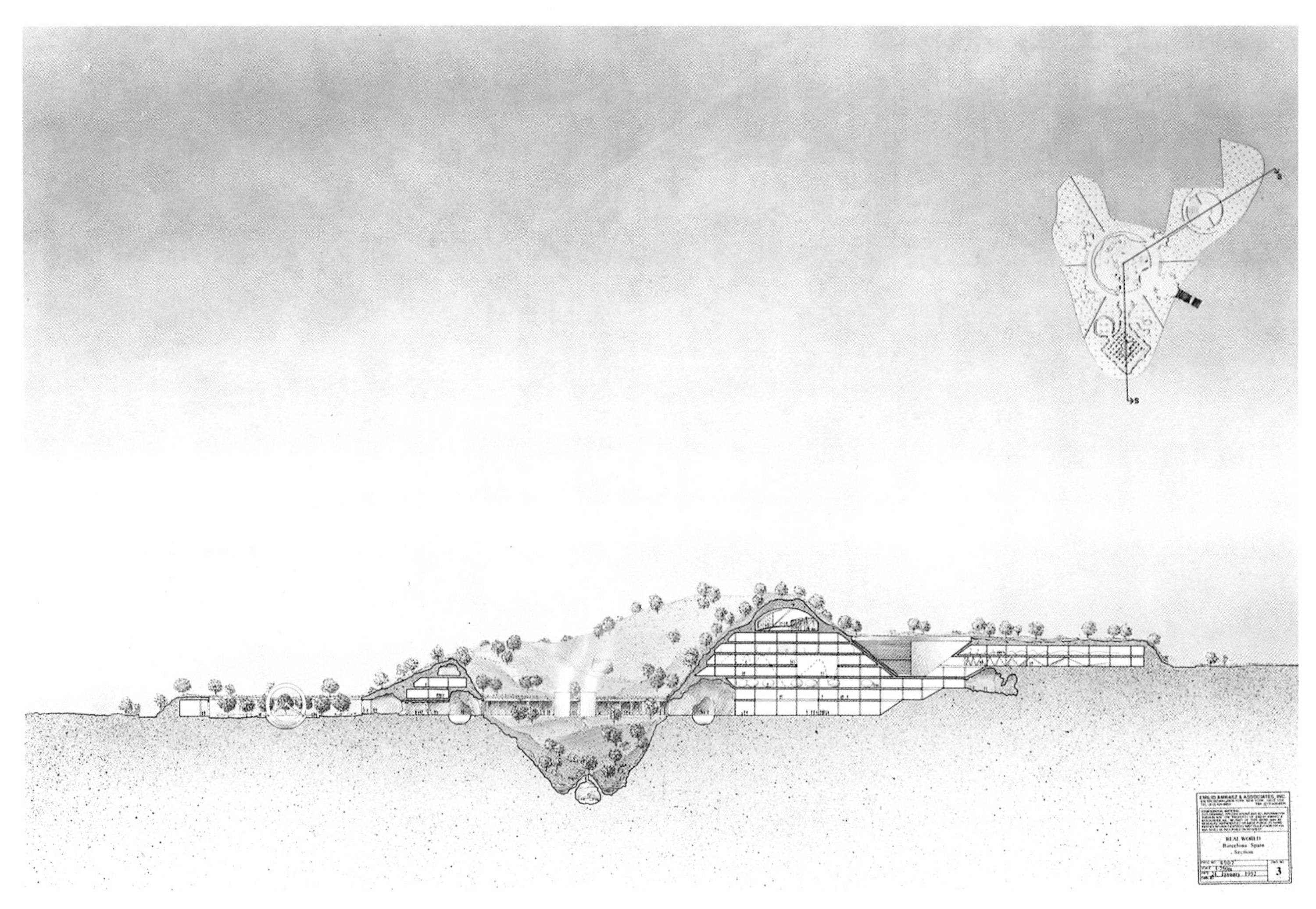
EMILIO AMBASZ & ASSOCIATES, INC.
REAL WORLD
Barcelona Spain
Section
1:750m
21 January 1992
3

Marine City Waterfront Development

Otaru, Hokkaido Island, Japan

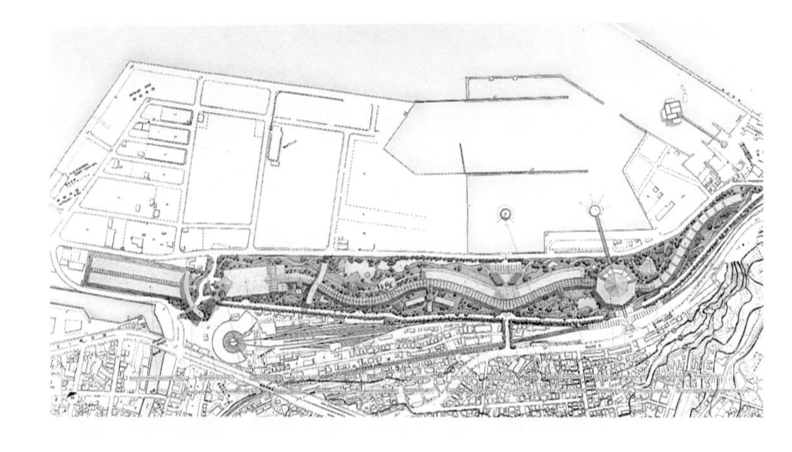
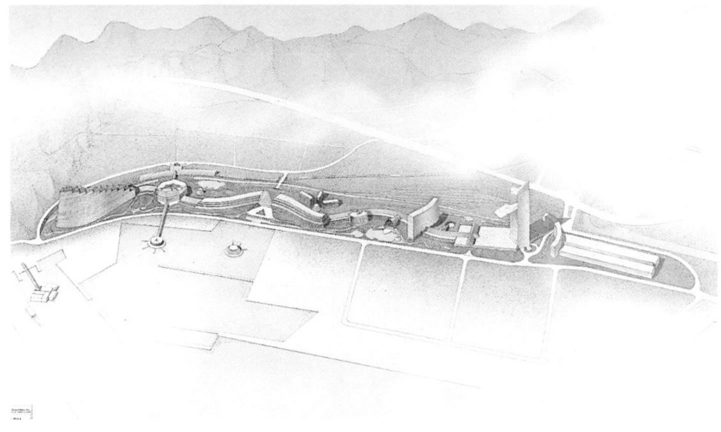

Rimini Beachfront

Rimini, Italy

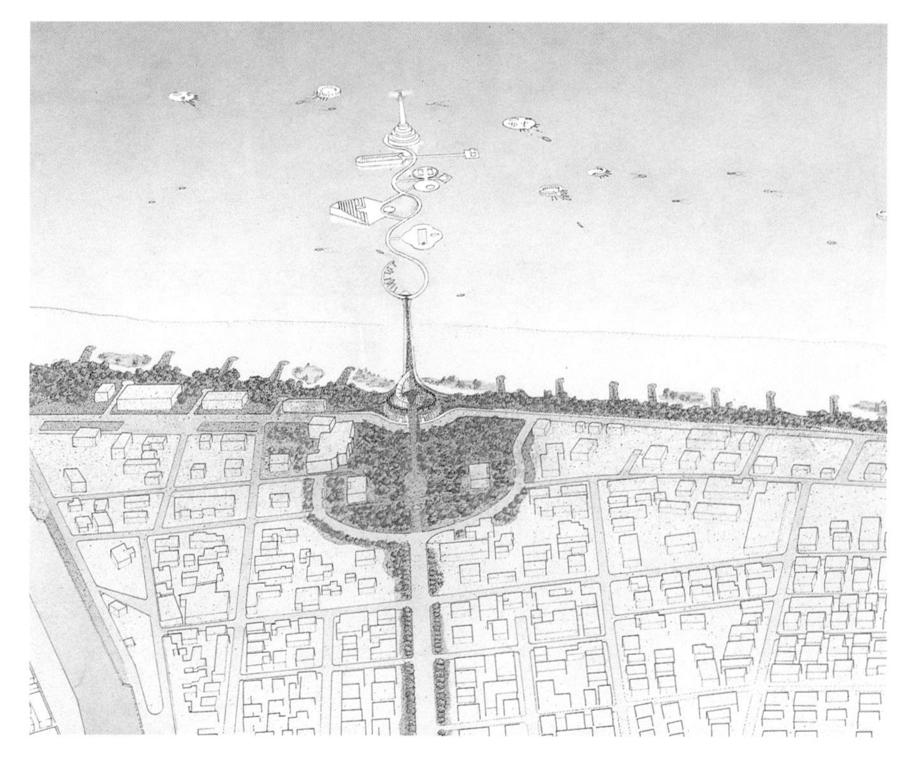

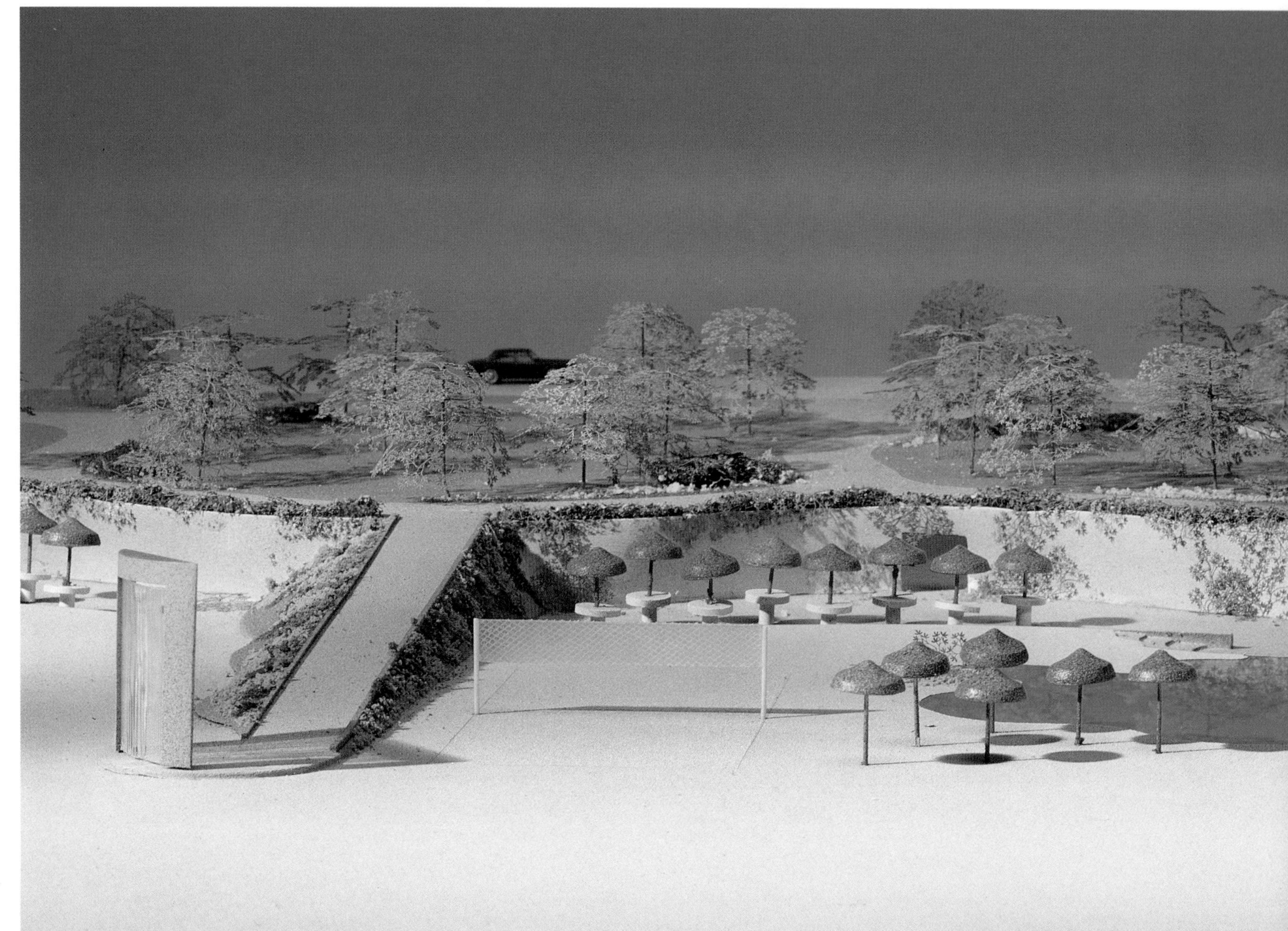

CONDIZIONE ESISTENTE DELLA SPIAGGIA
SCHEMA DELLA PROPOSTA
RIMINI

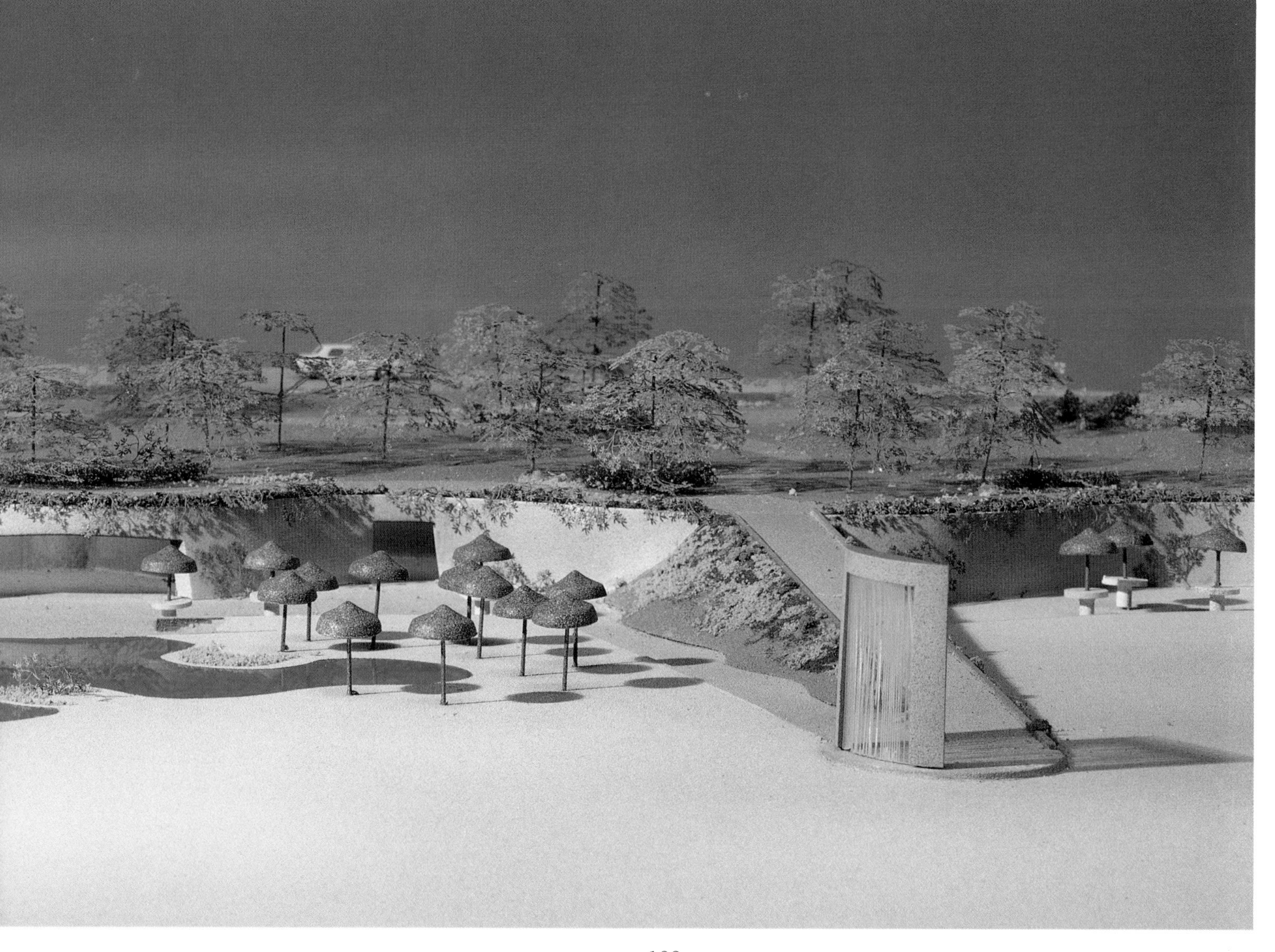

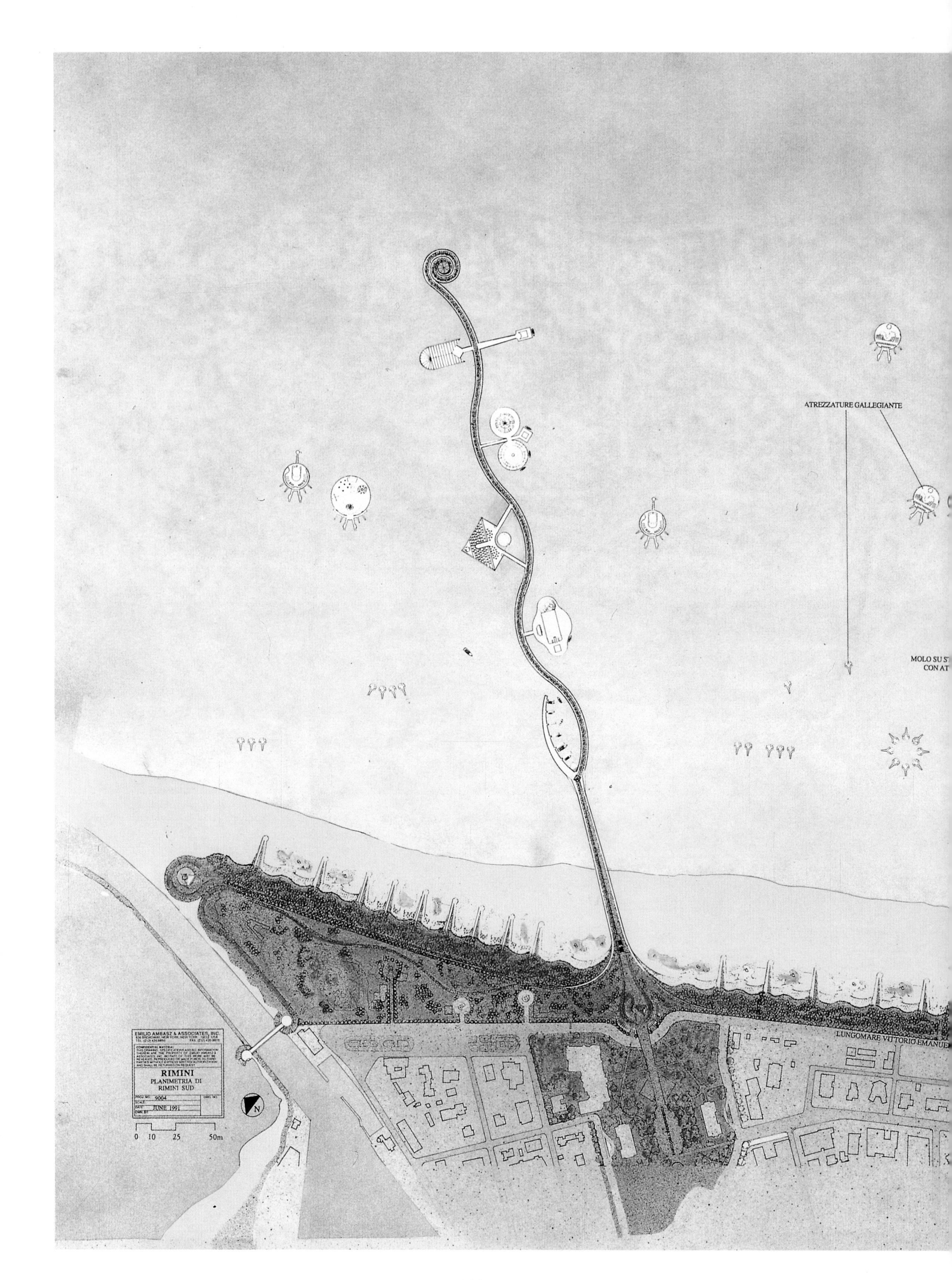
ATREZZATURE GALLEGIANTE
MOLO SU ST
CON AT
LUNGOMARE VITTORIO EMANUE
EMILIO AMBASZ & ASSOCIATES, INC.
CONFIDENTIAL MATERIAL
RIMINI
PLANIMETRIA DI
RIMINI SUD
9004
JUNE 1991
N
0 10 25 50m

RA GALLEGGIANTE
TURE DI SVAGO
PASSEGGIATA ALBERATA
TRA L'ARCO DI AUGUSTO
ED IL LUNGOMARE

LUNGOMARE
MARE ADRIATICO
35m
40m
40m
25m
VOLLEYBALL
BAR
FASCIA A
FASCIA B
FASCIA C
FASCIA D
RIMINI

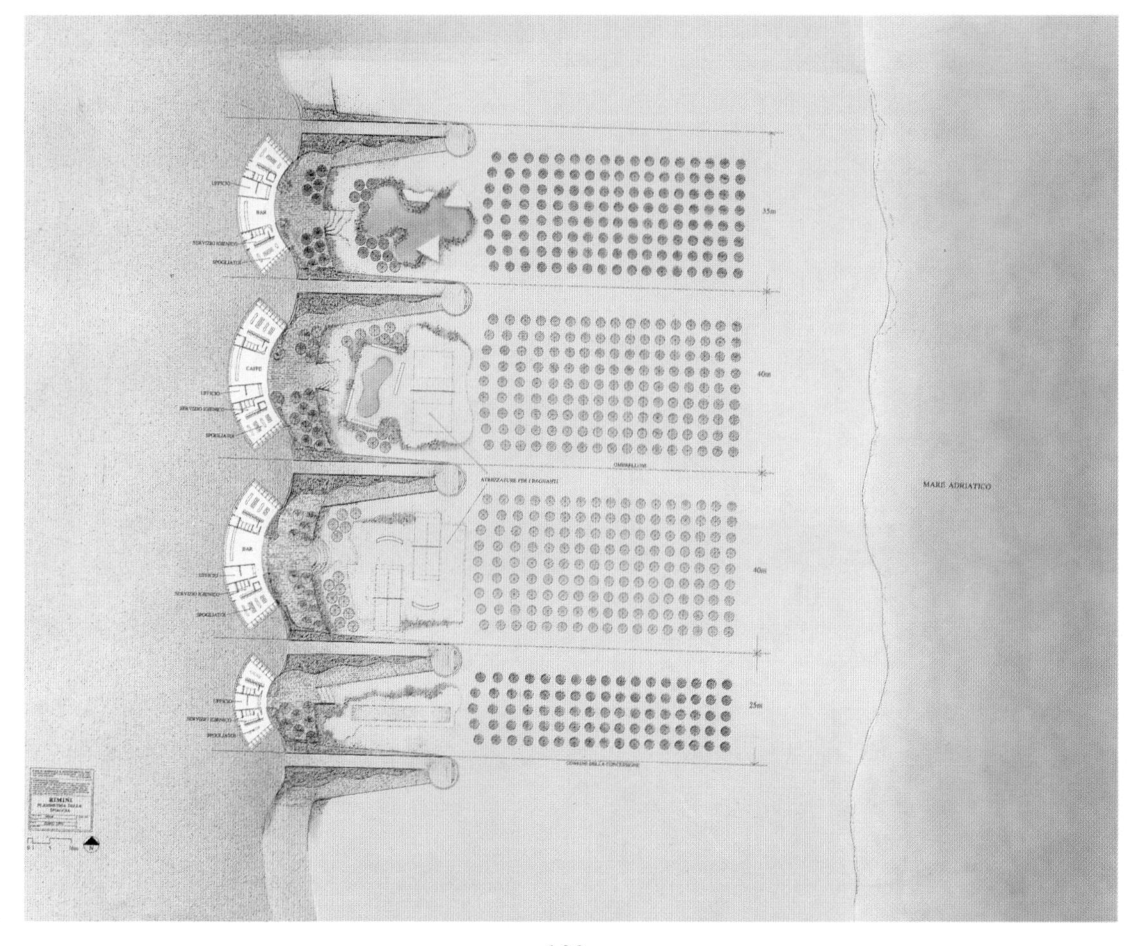

MARE ADRIATICO
35m
40m
40m
25m
BAR
UFFICIO
SPOGLIATOI
RIMINI

European School of Hotel Management

Omegna, Italy

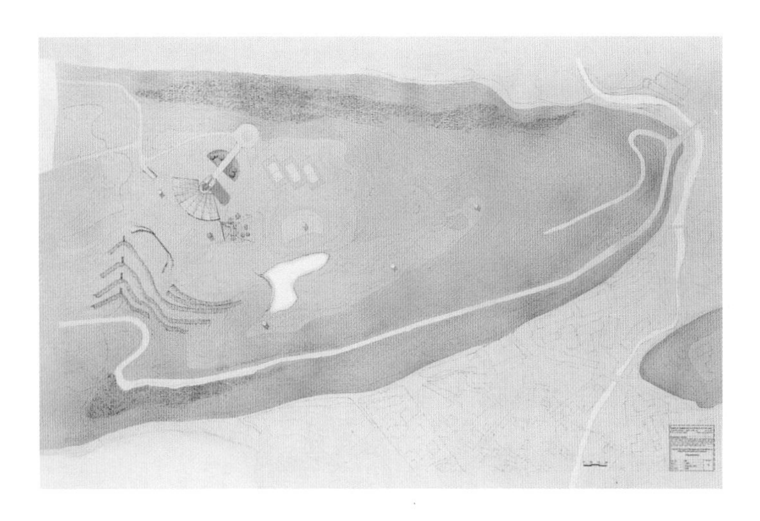

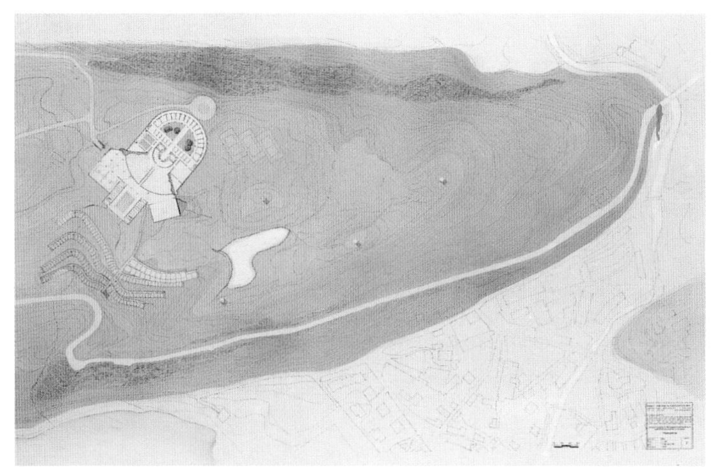

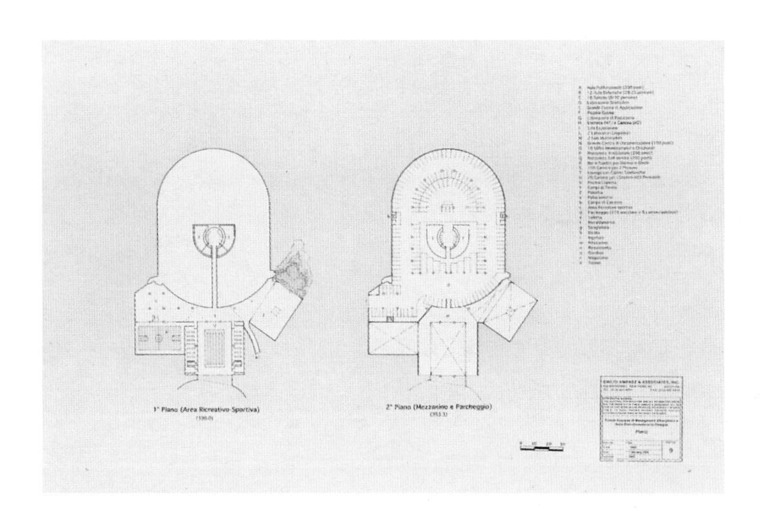
1° Piano (Area Ricreativo Sportiva)
2° Piano (Mezzanino e Parcheggio)

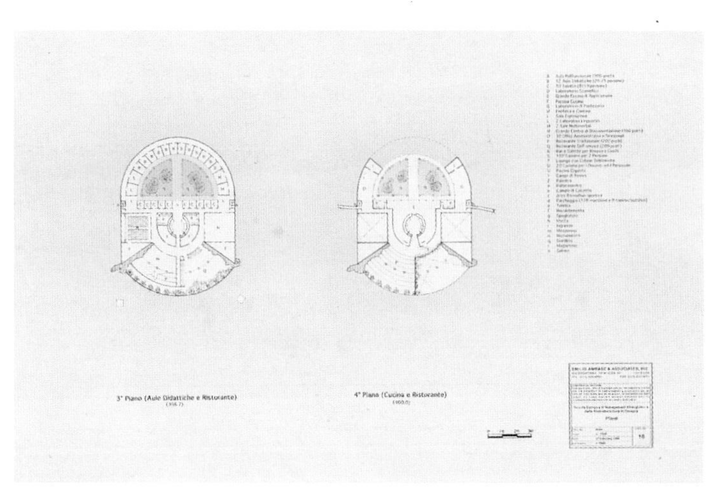
3° Piano (Aule Didattiche e Ristorante)
4° Piano (Cucina e Ristorante)

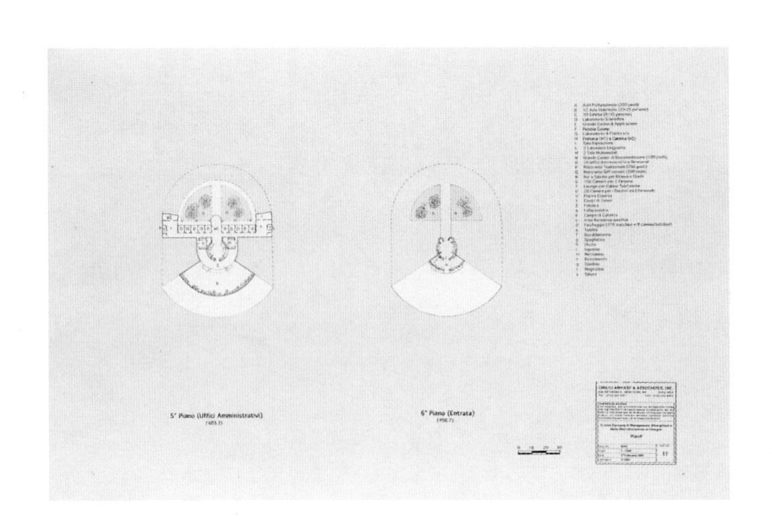
5° Piano (Uffici Amministrativi)
6° Piano (Entrata)

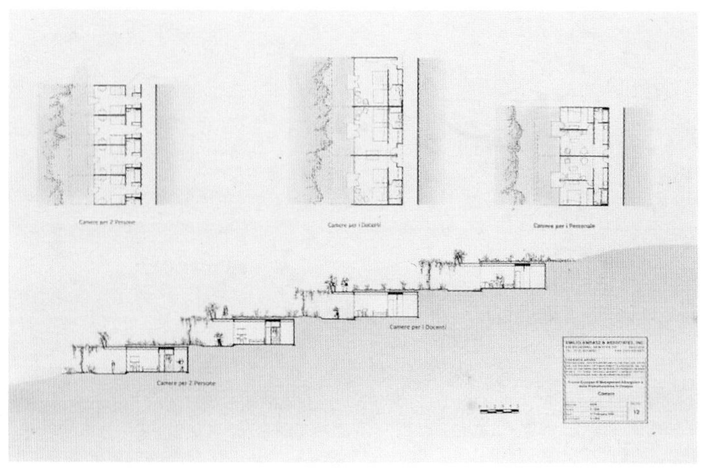
Camere per 2 Persone
Camere per i Docenti
Camere per i Personale
Camere per i Docenti
Camere per 2 Persone

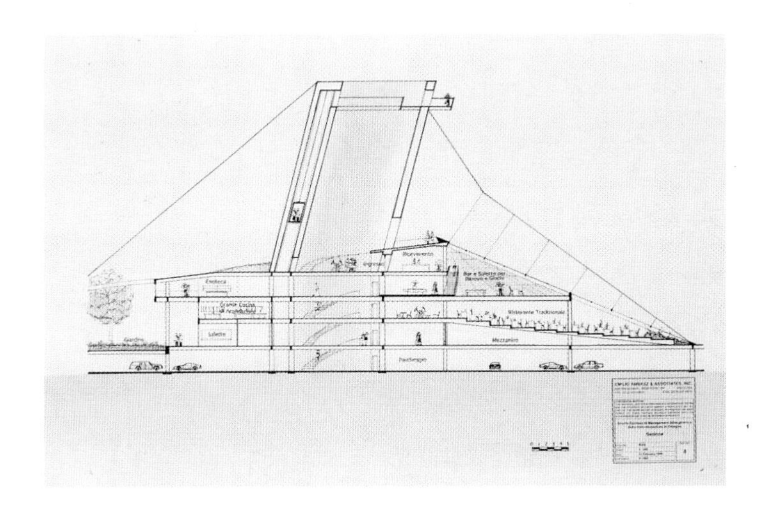
Ricevimento
Enoteca
Ristorante Tradizionale
Mezzanino
Parcheggio
Salette
Giardino

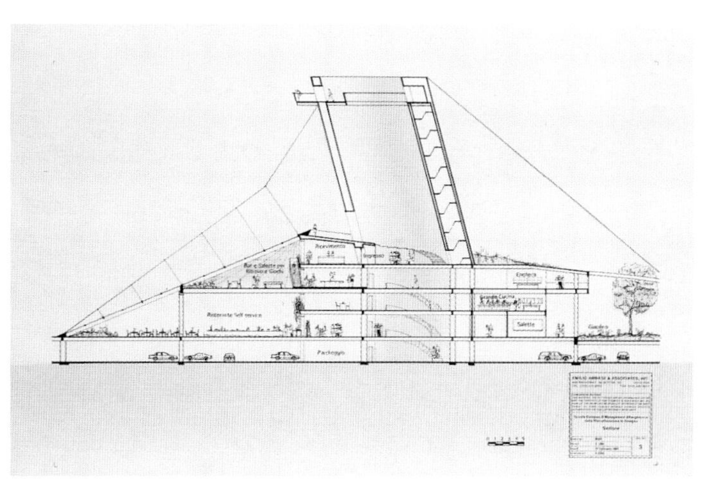
Ricevimento
Enoteca
Ristorante Self service
Salette
Parcheggio
Giardino

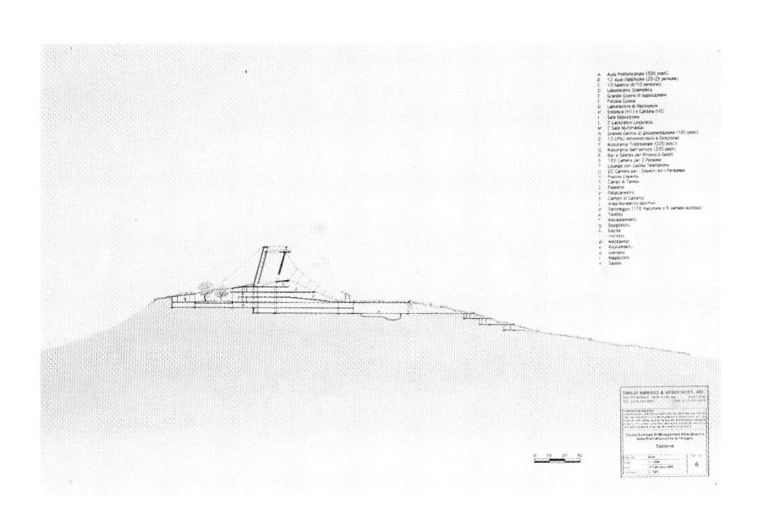

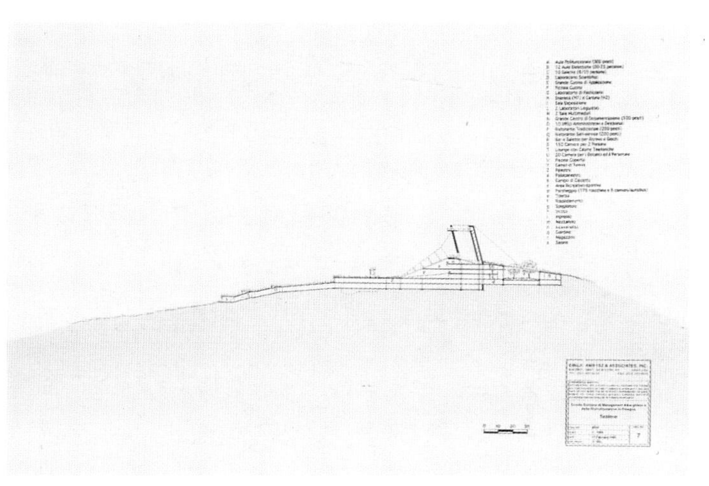

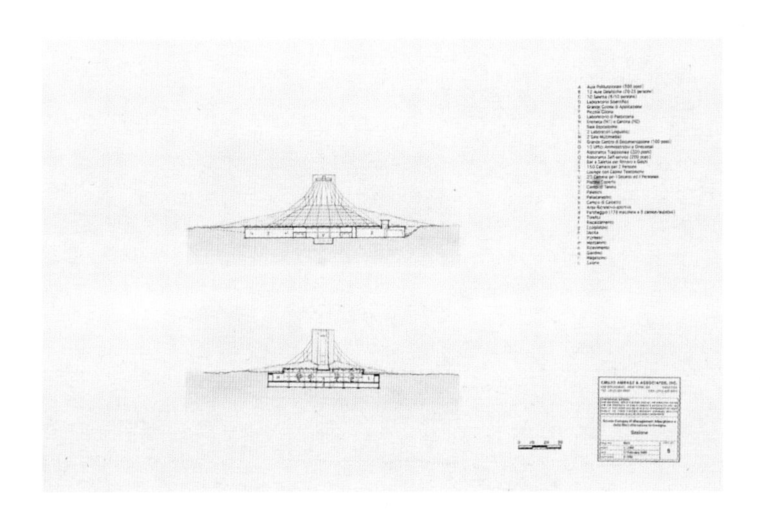

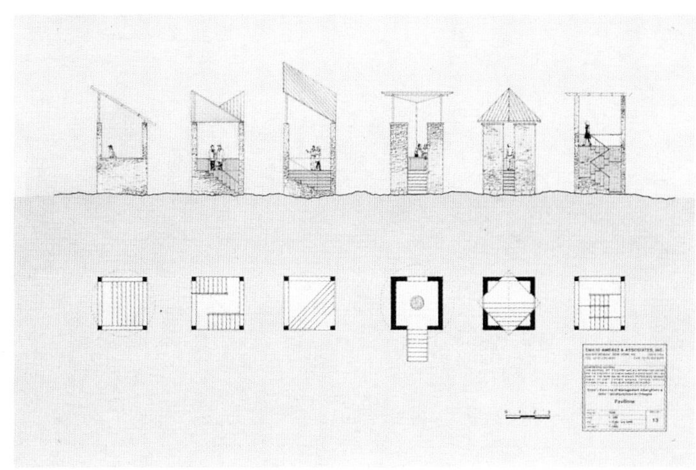

Nuova Concordia Resort Housing Development

Castellaneta, Italy

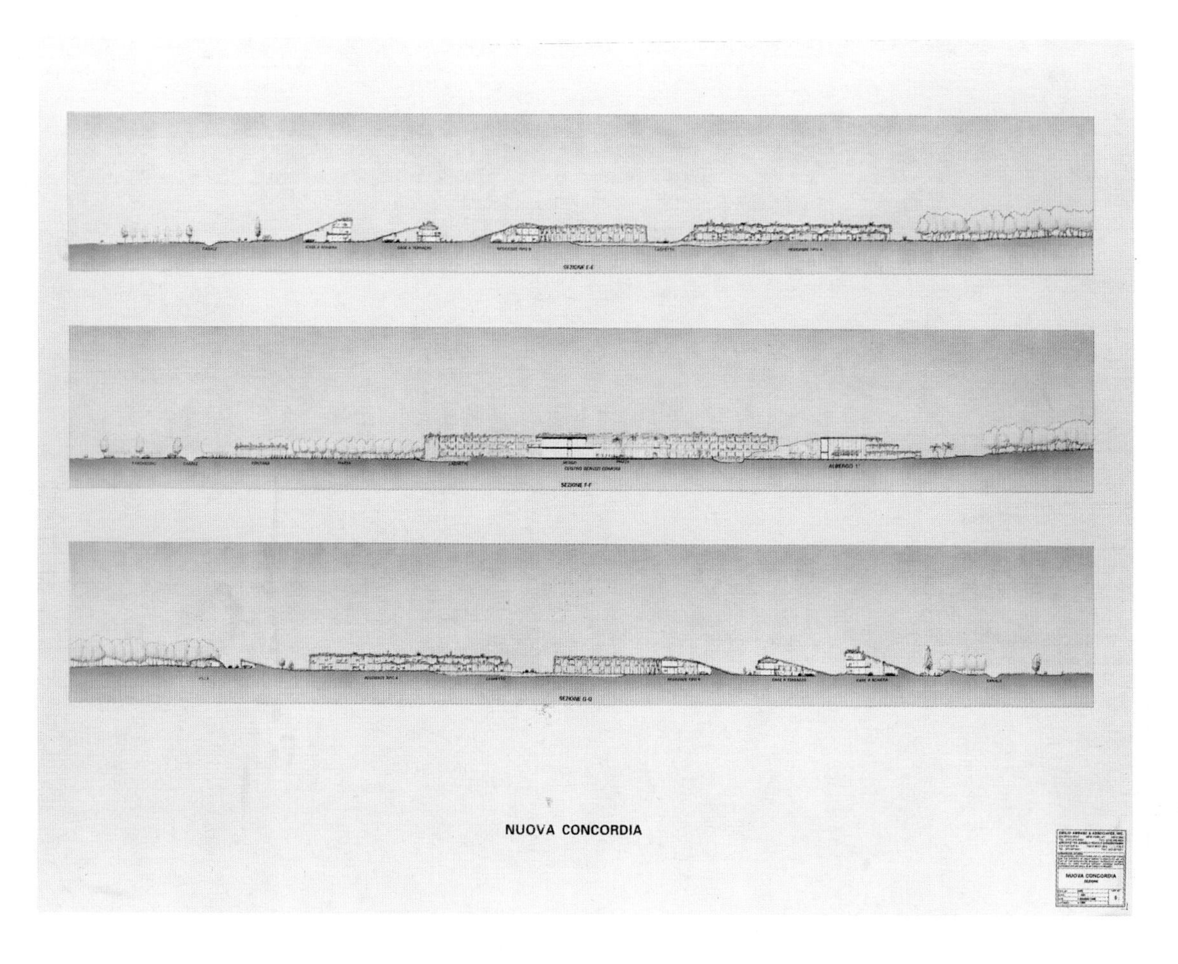
NUOVA CONCORDIA

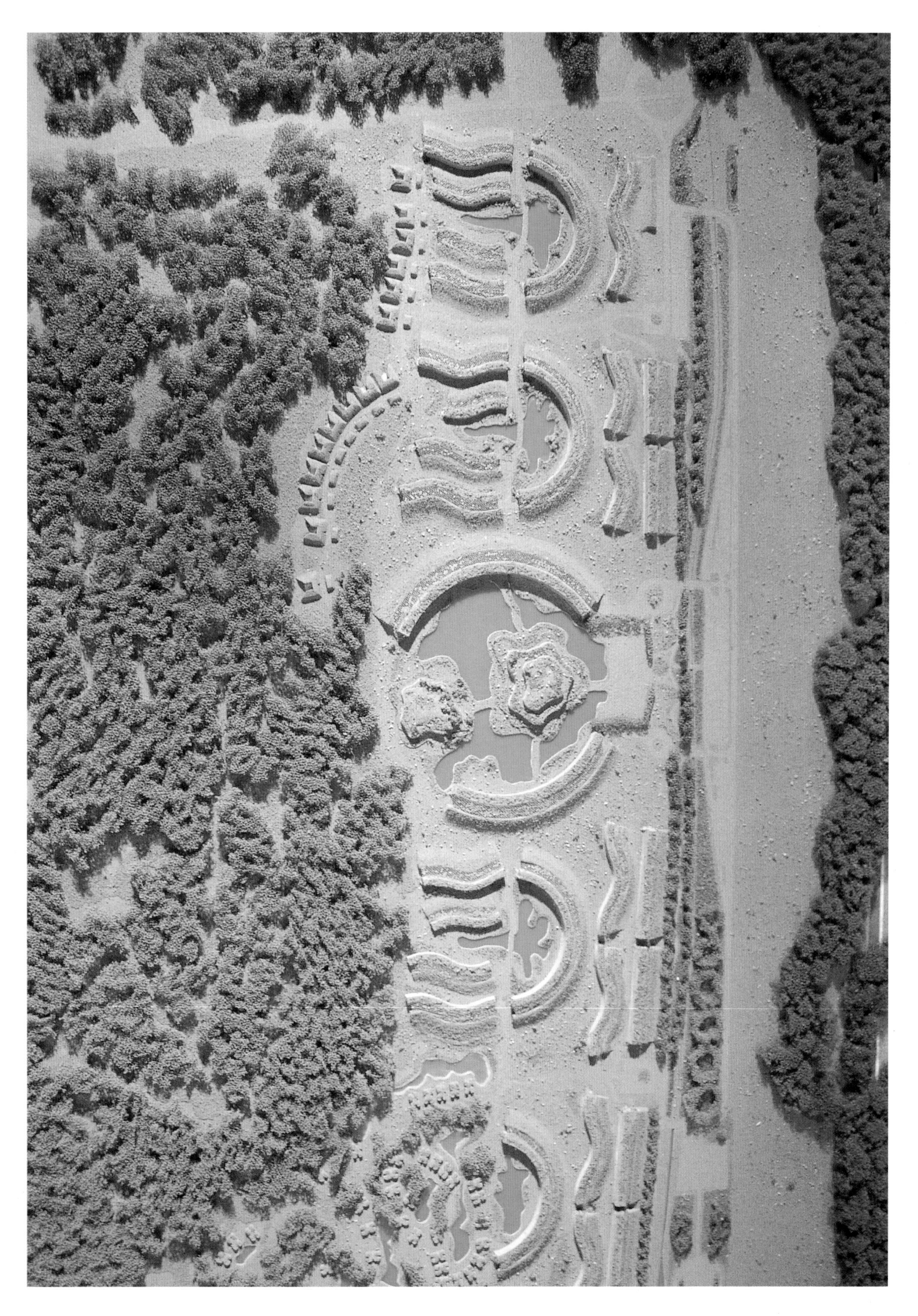

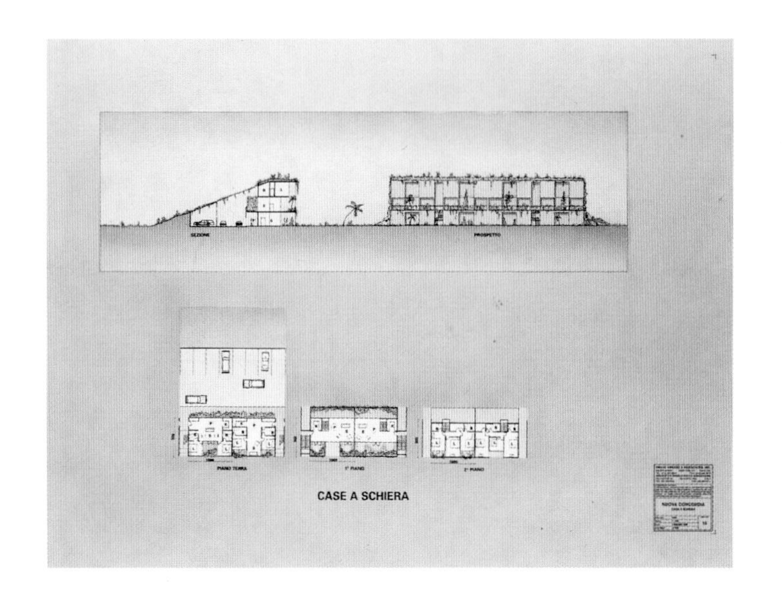
CASE A SCHIERA

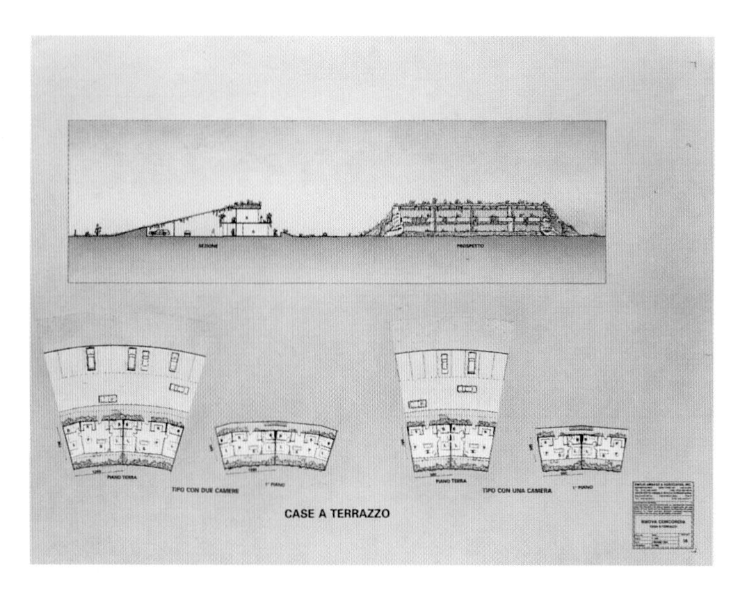
TIPO CON DUE CAMERE
TIPO CON UNA CAMERA
CASE A TERRAZZO

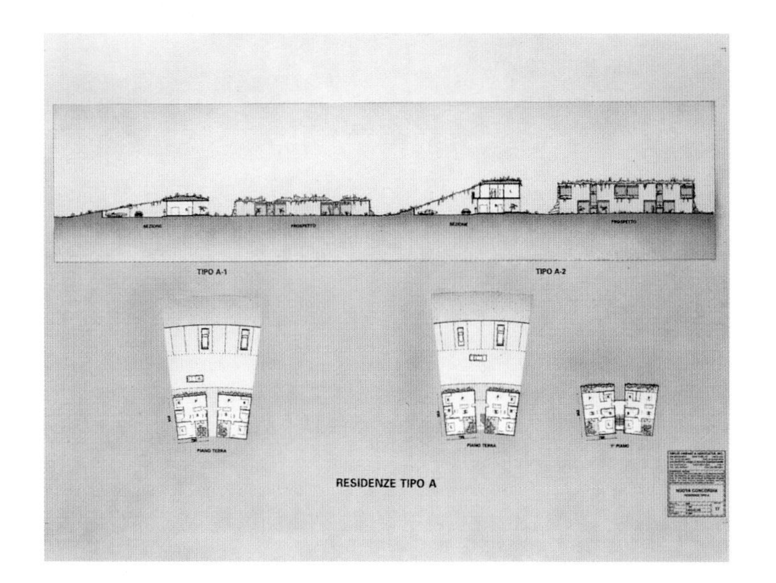
TIPO A-1
TIPO A-2
RESIDENZE TIPO A

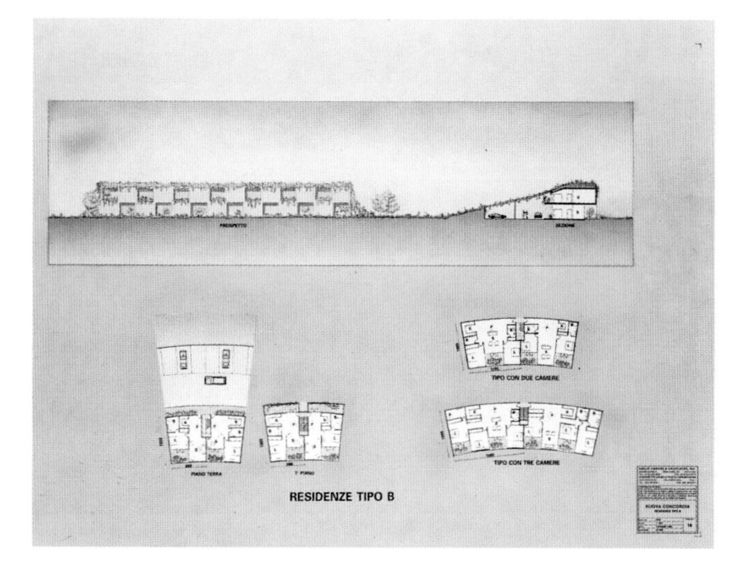
TIPO CON DUE CAMERE
TIPO CON TRE CAMERE
RESIDENZE TIPO B

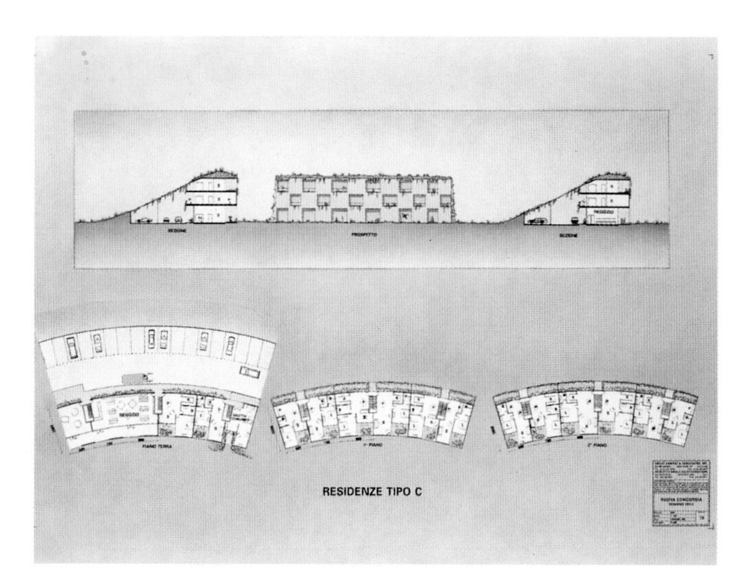
RESIDENZE TIPO C

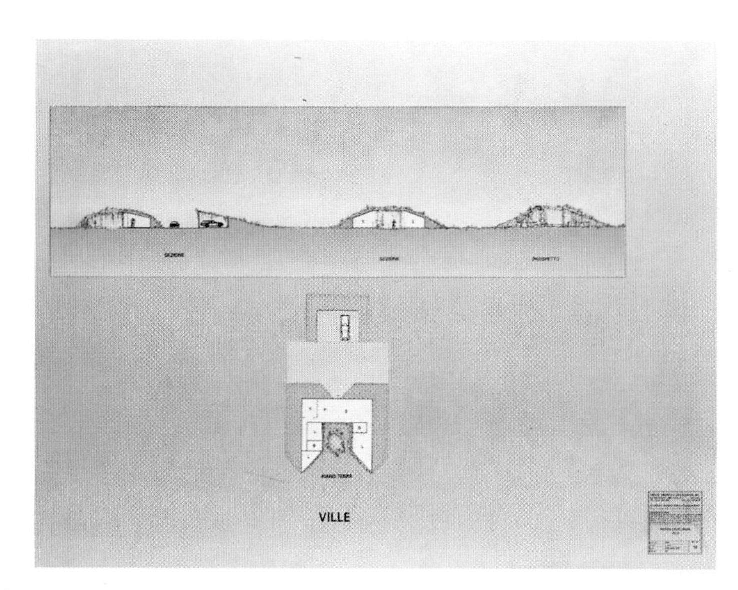
PIANO TERRA
VILLE

Baron Edmond de Rothschild Memorial Museum

Ramat Hanadiv, Israel

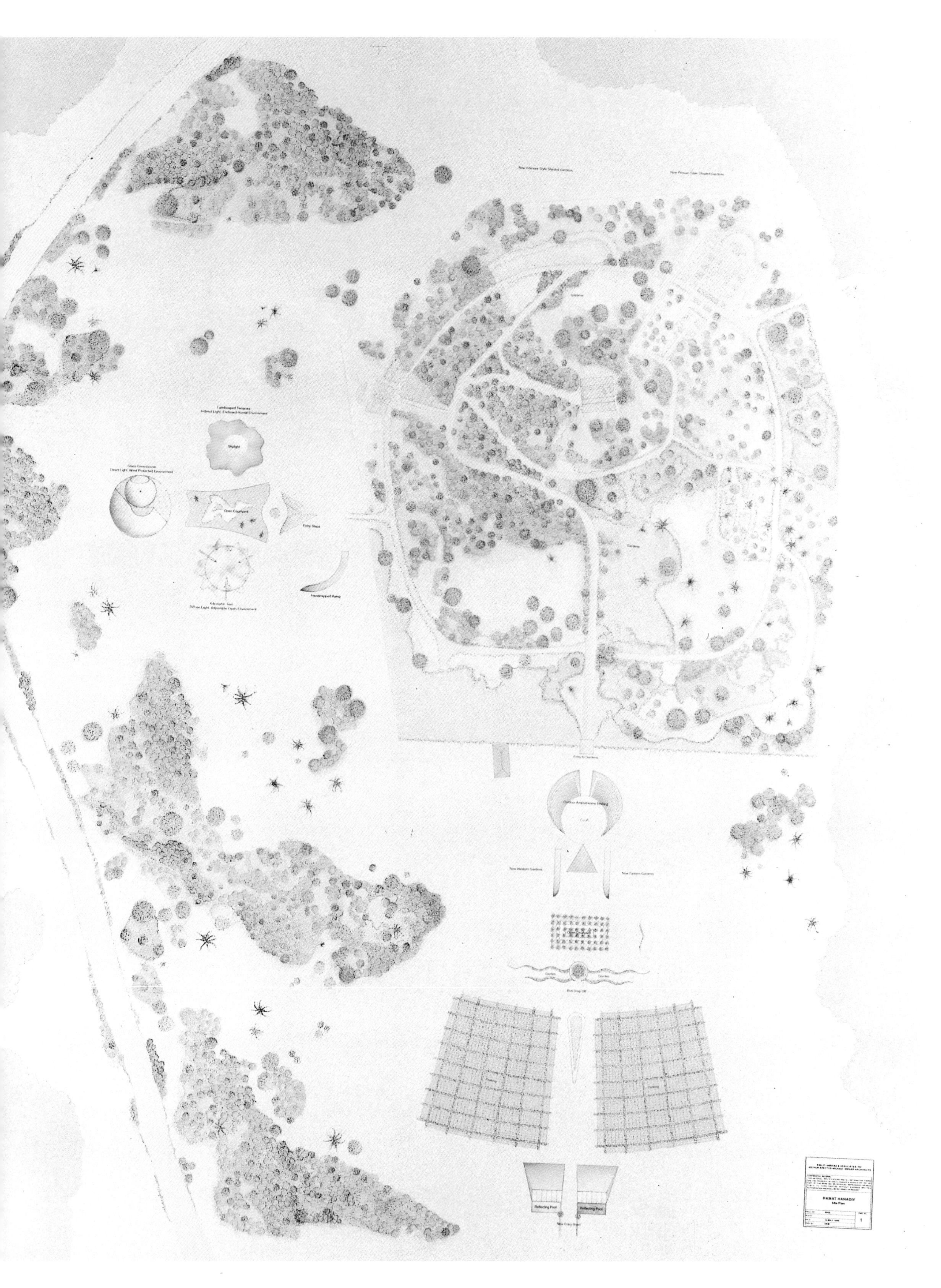
New Chinese Style Shaded Gardens
New Persian Style Shaded Gardens
Landscaped Terraces
Indirect Light, Enclosed Humid Environment
Skylight
Glass Greenhouse
Direct Light, Wind Protected Environment
Open Courtyard
Entry Steps
Adjustable Tent
Diffuse Light, Adjustable Open Environment
Handicapped Ramp
Entry to Gardens
Court
New Western Gardens
New Eastern Gardens
Garden
Bus Drop Off
Parking
Reflecting Pool
RAMAT HANADIV
Site Plan
1

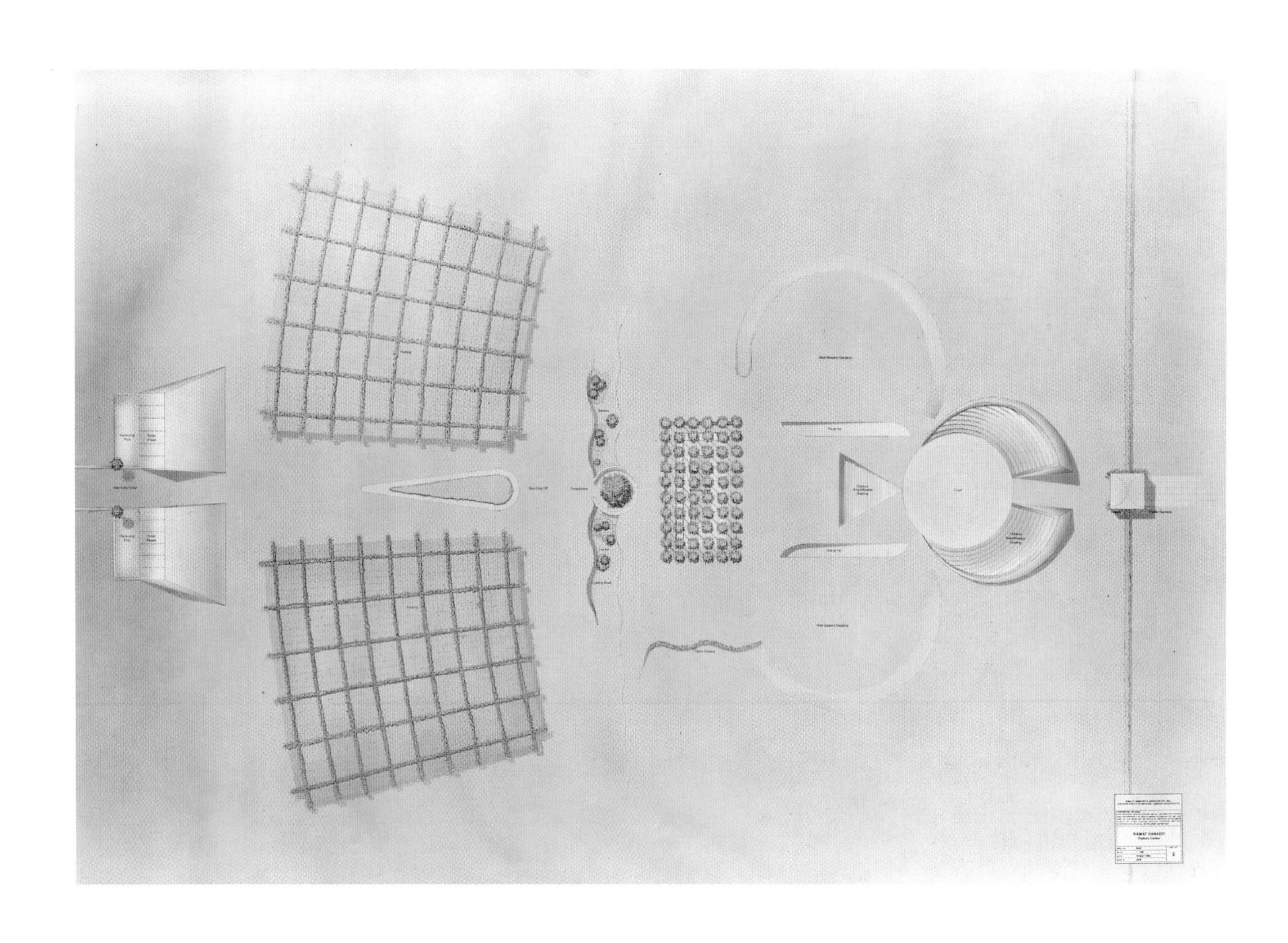

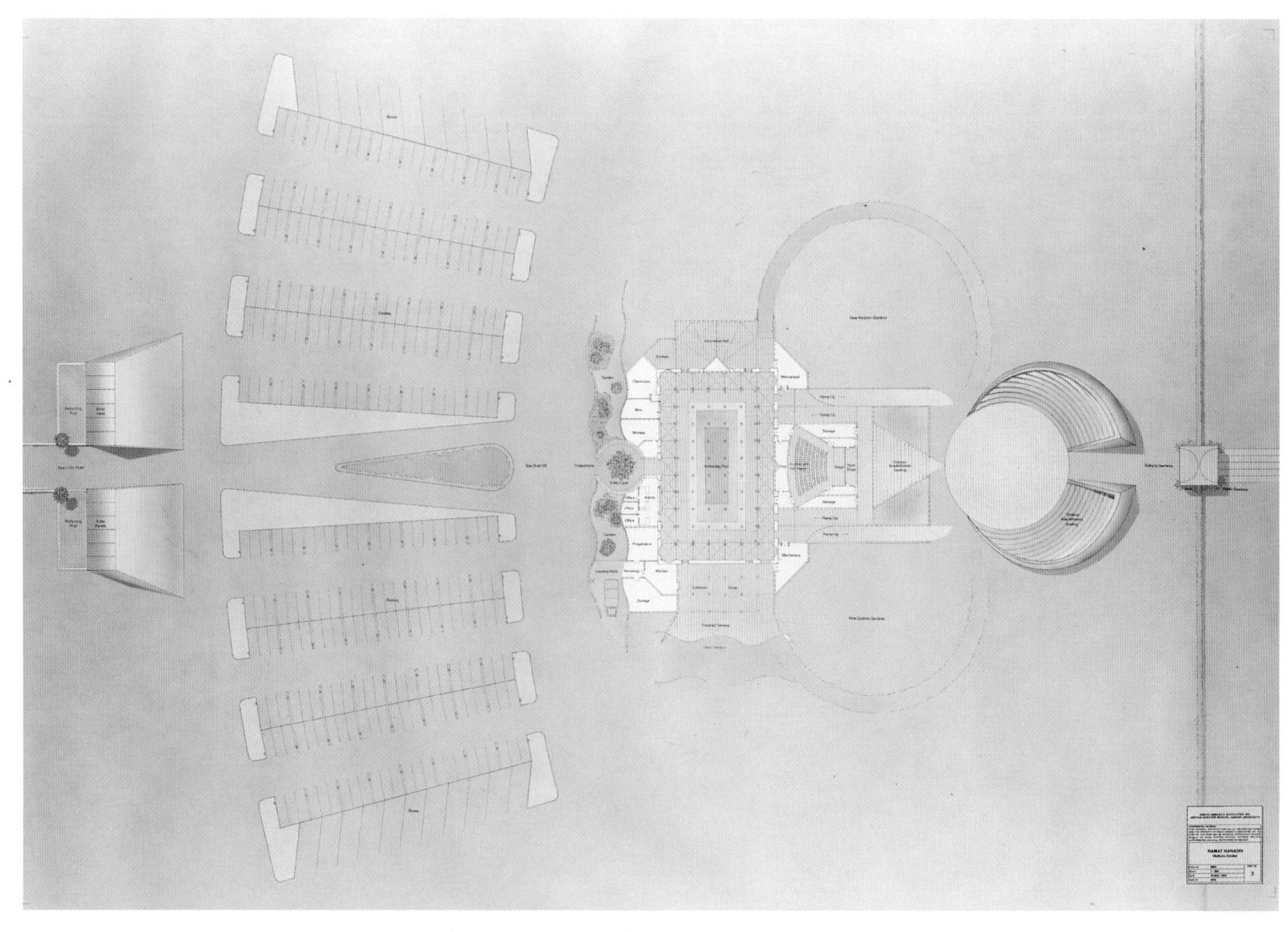

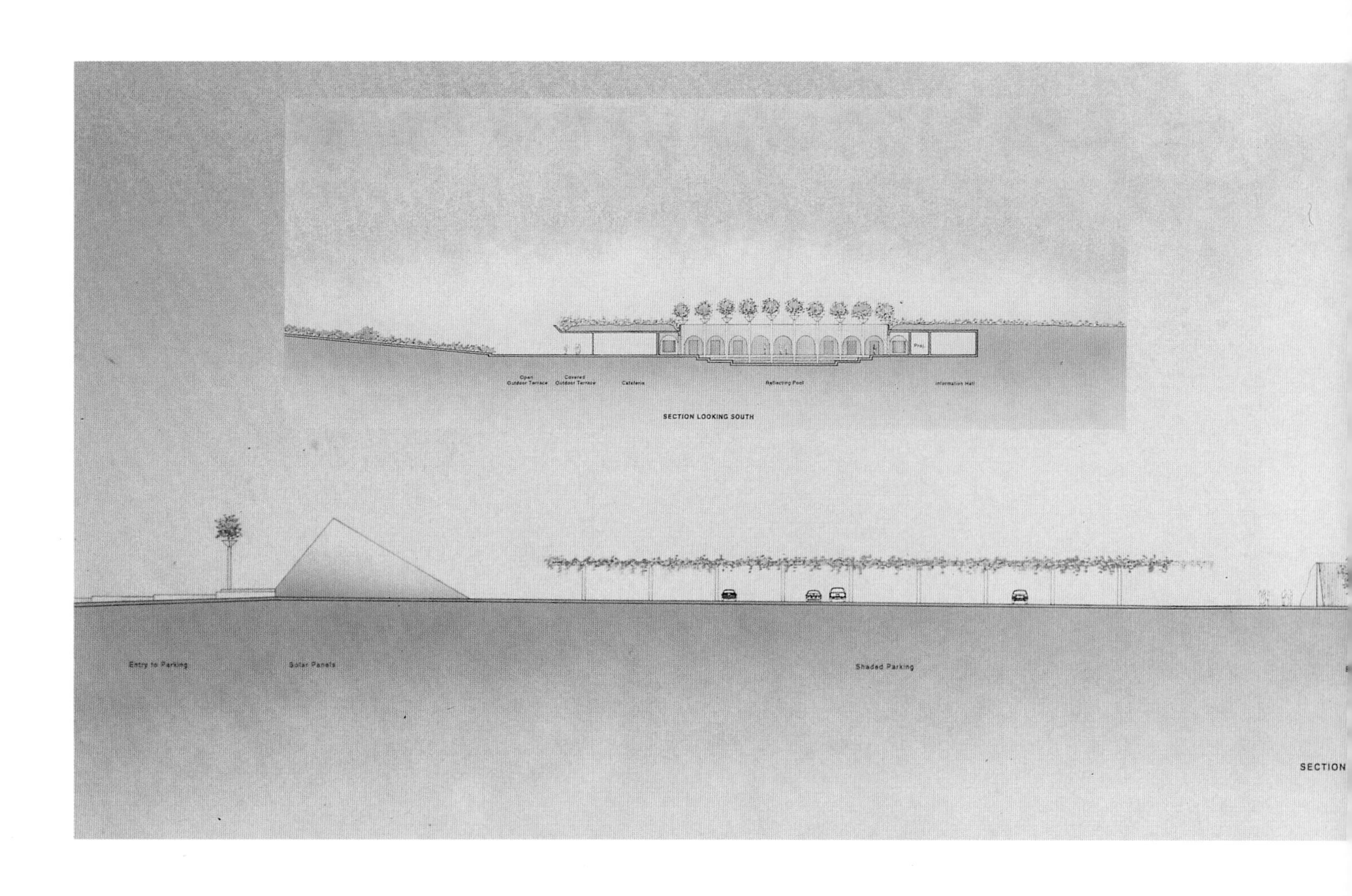
Open Outdoor Terrace
Covered Outdoor Terrace
Cafeteria
Reflecting Pool
Proj.
Information Hall
SECTION LOOKING SOUTH
Entry to Parking
Solar Panels
Shaded Parking
SECTION

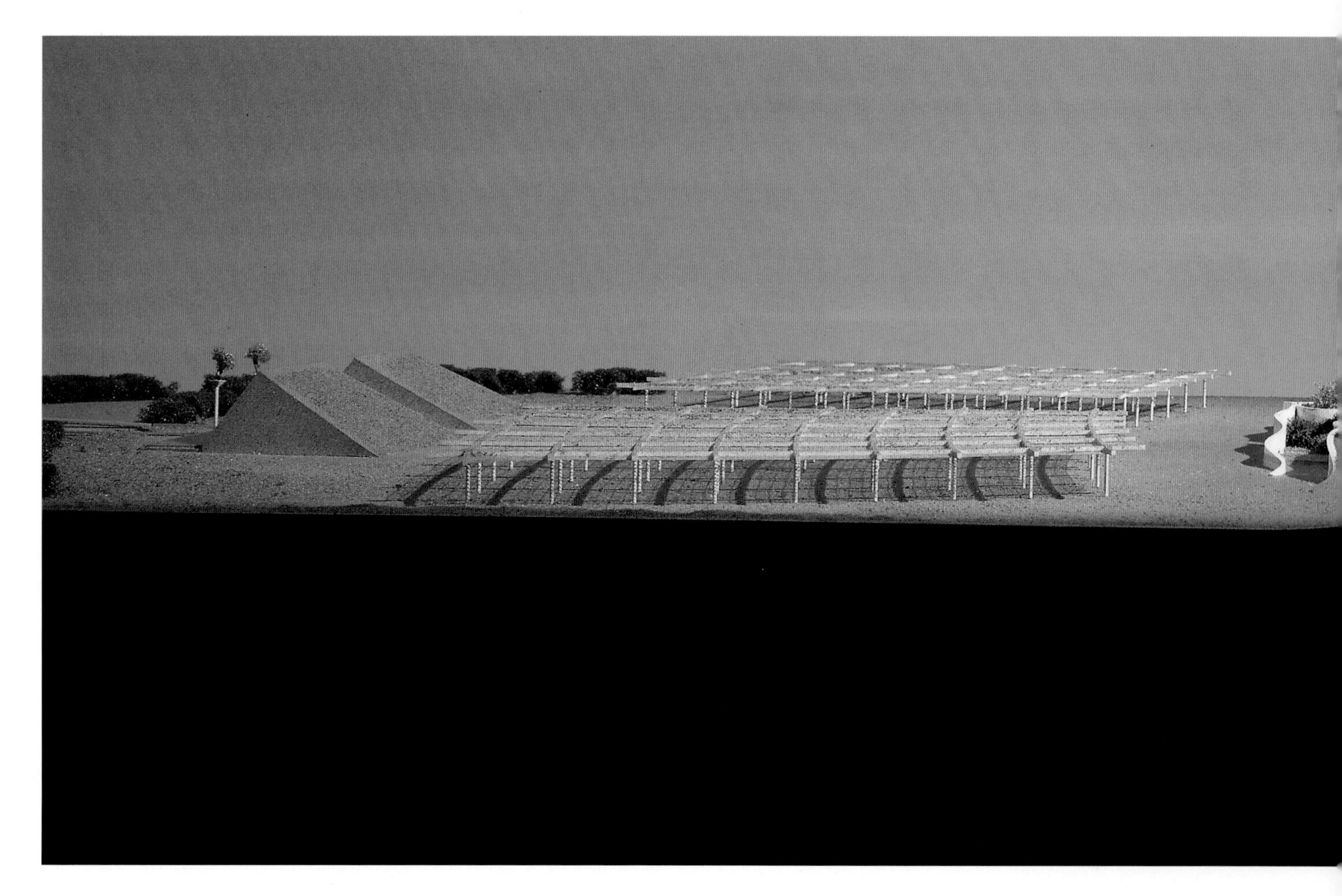

Proj.
Auditorium
urt
Reflecting Pool
Berm
Entry to Gardens
ING WEST

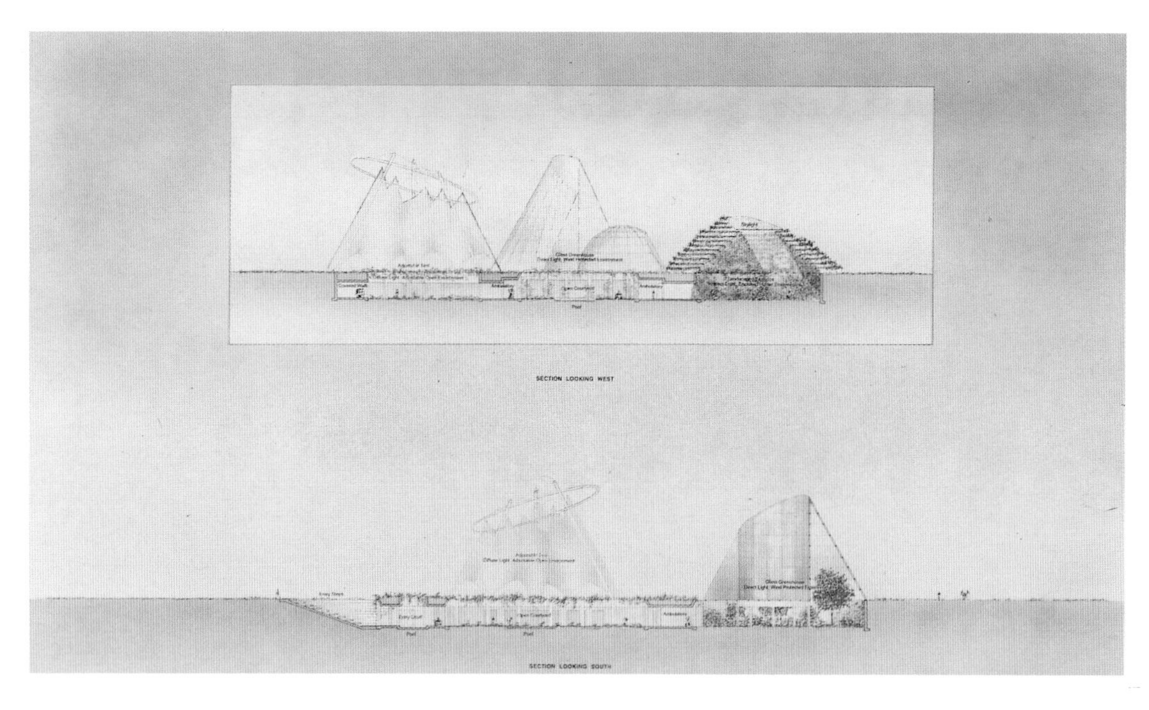
SECTION LOOKING WEST
SECTION LOOKING SOUTH

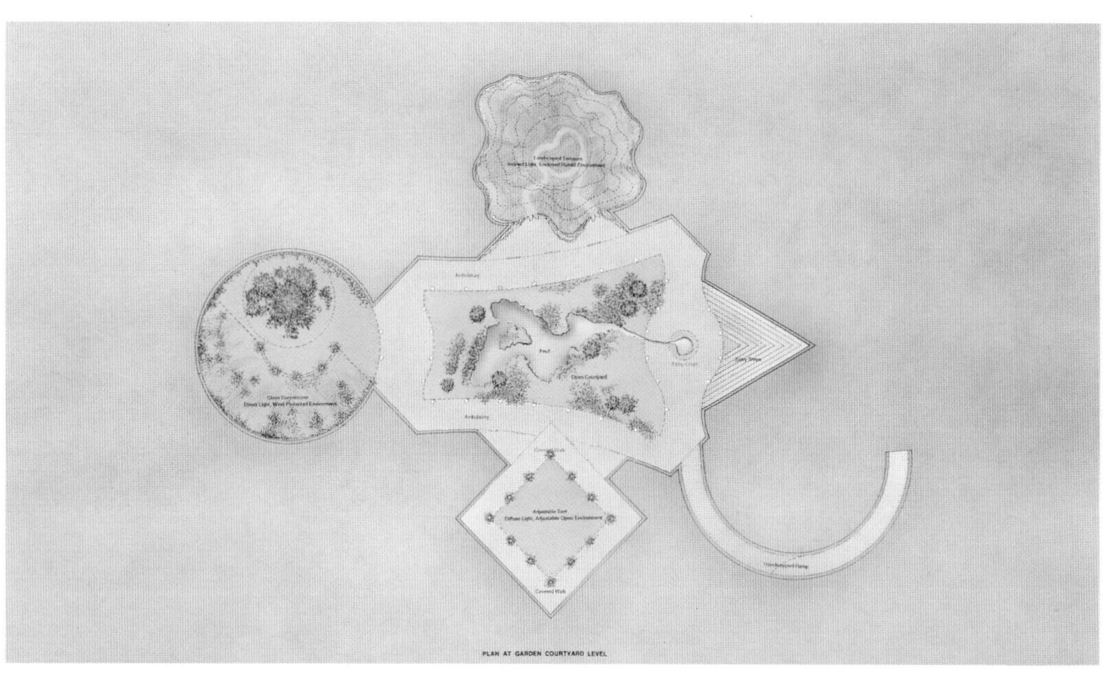
PLAN AT GARDEN COURTYARD LEVEL

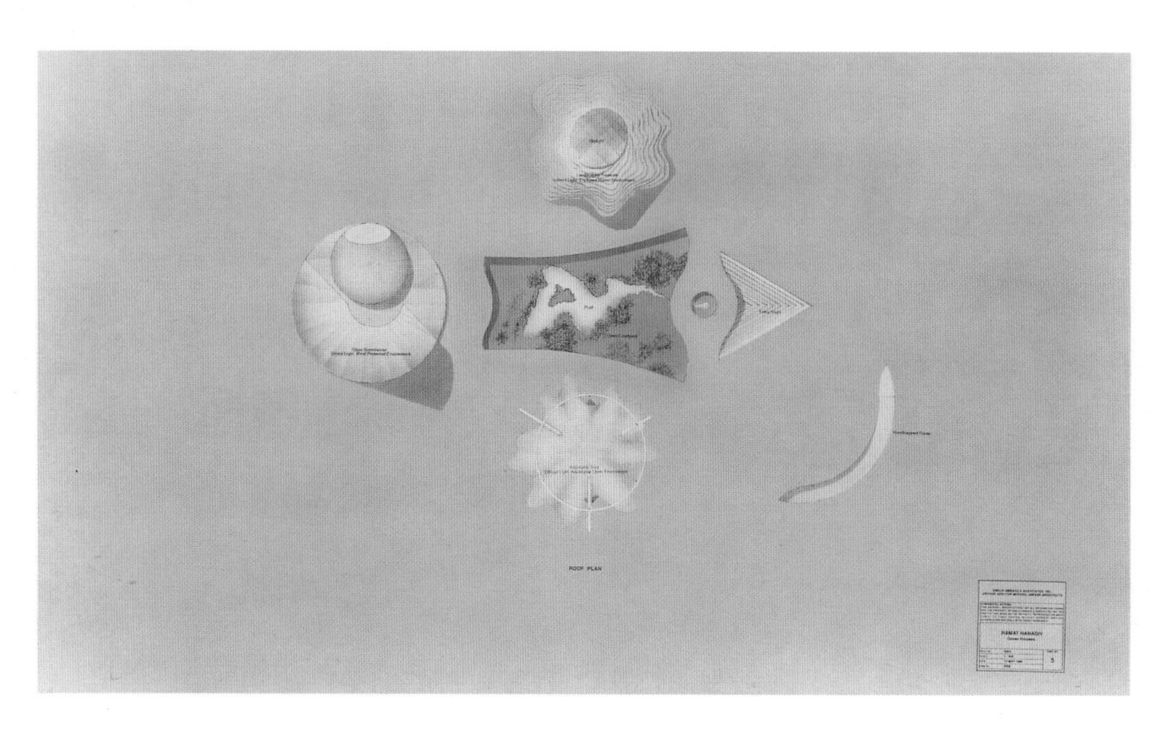
ROOF PLAN

Jardins de la France

Chaumont, France

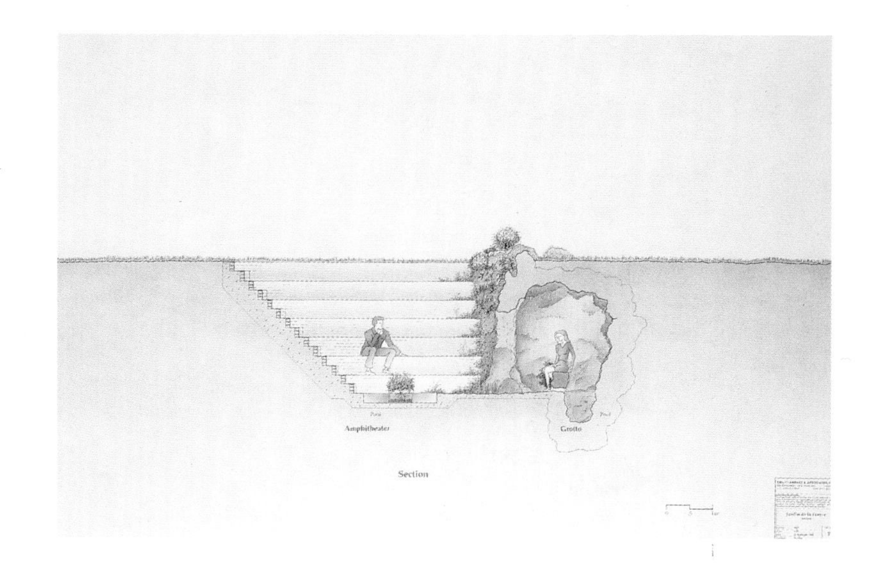

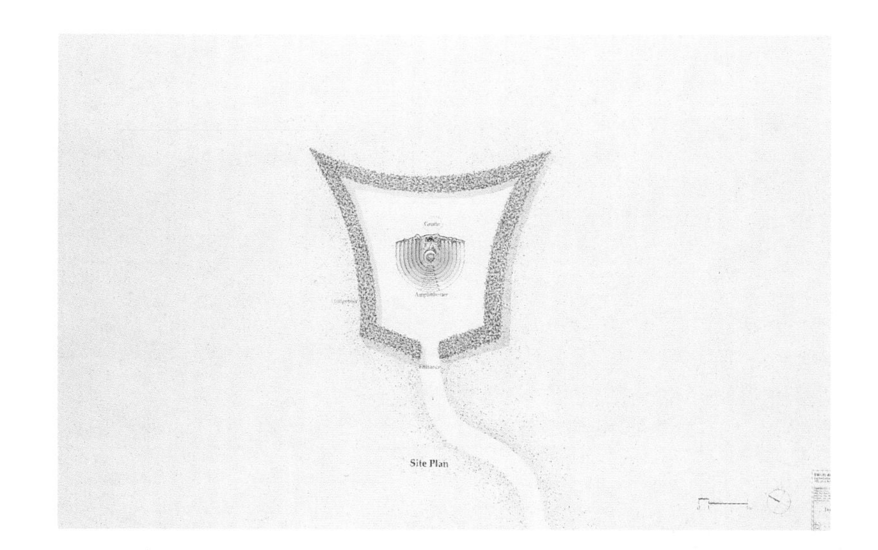

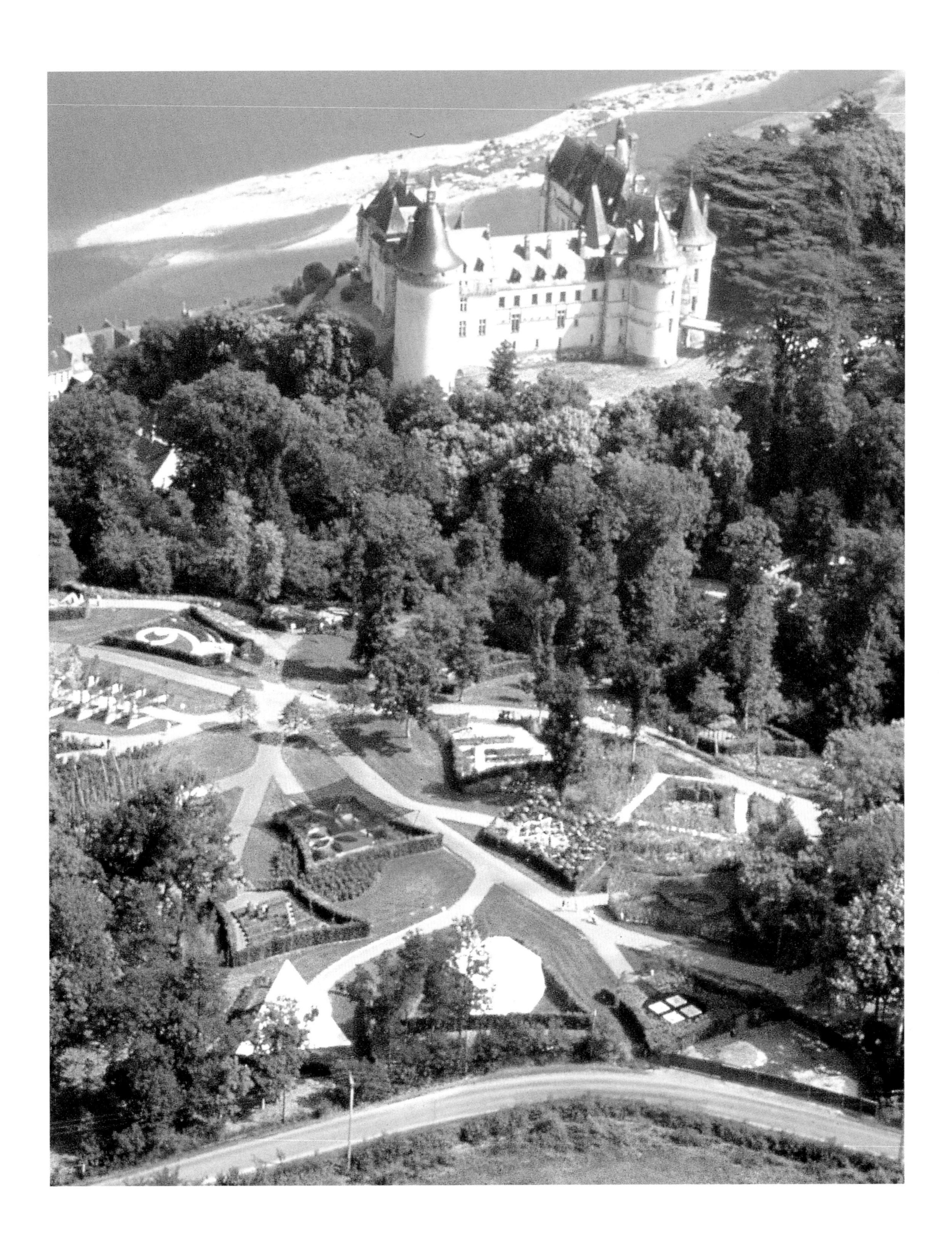

La Venta

Mexico City, Mexico

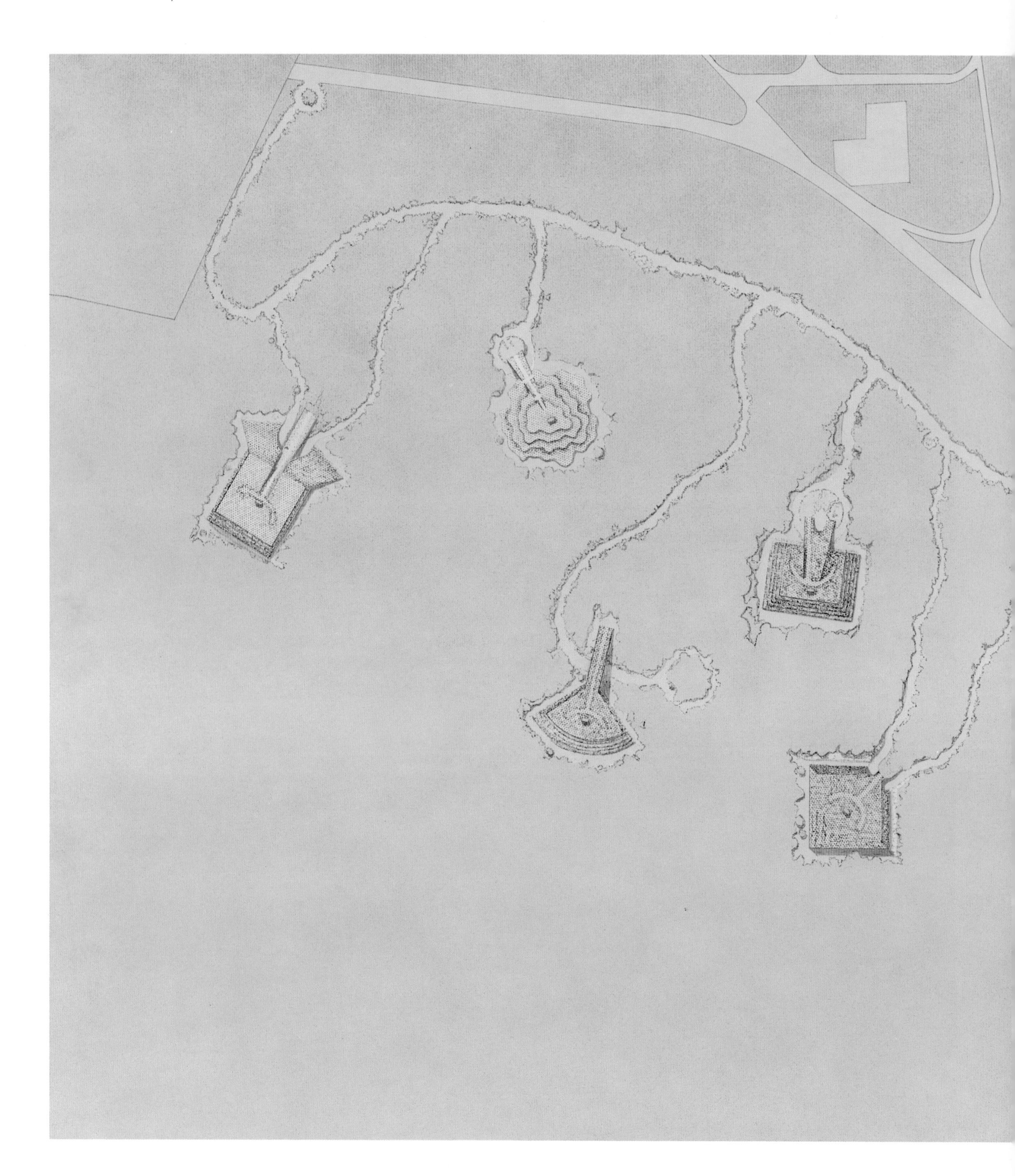

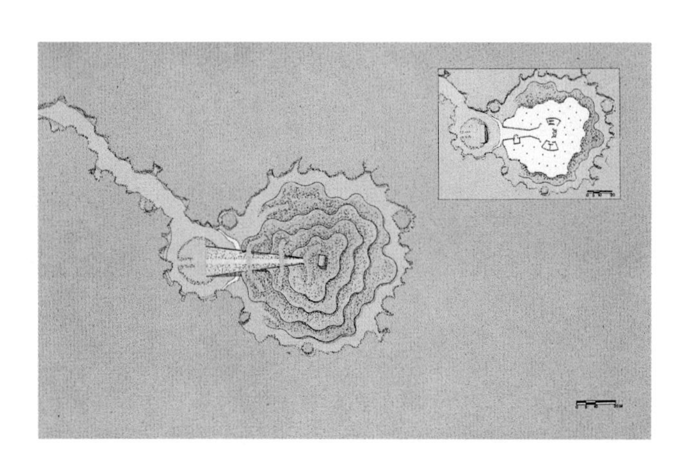

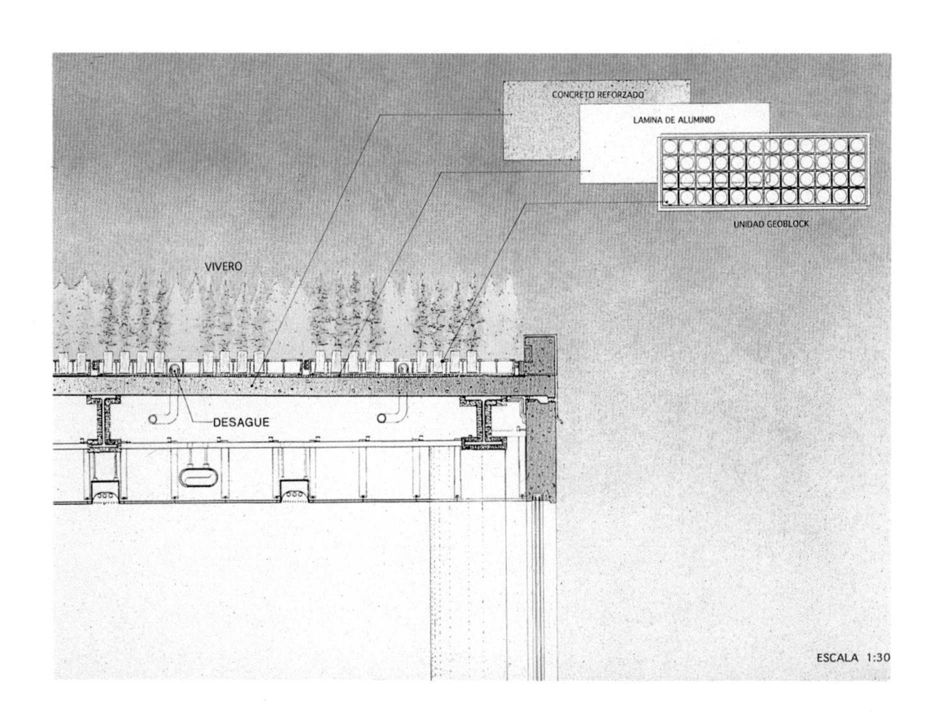
CONCRETO REFORZADO
LAMINA DE ALUMINIO
UNIDAD GEOBLOCK
VIVERO
DESAGUE
ESCALA 1:30

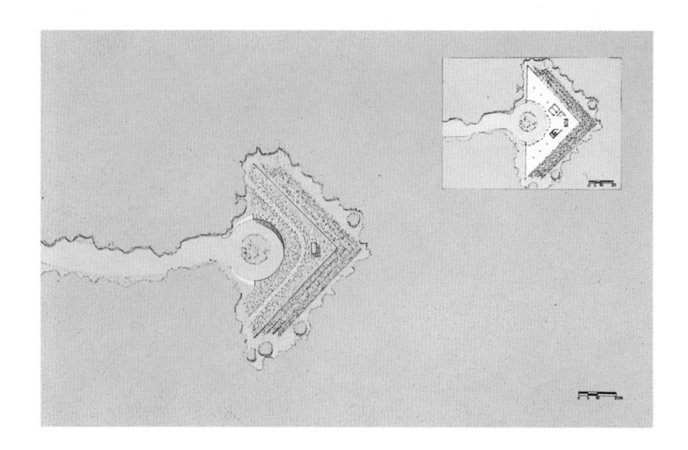

Clean Coal Power Plant

Chubu Prefecture, Japan

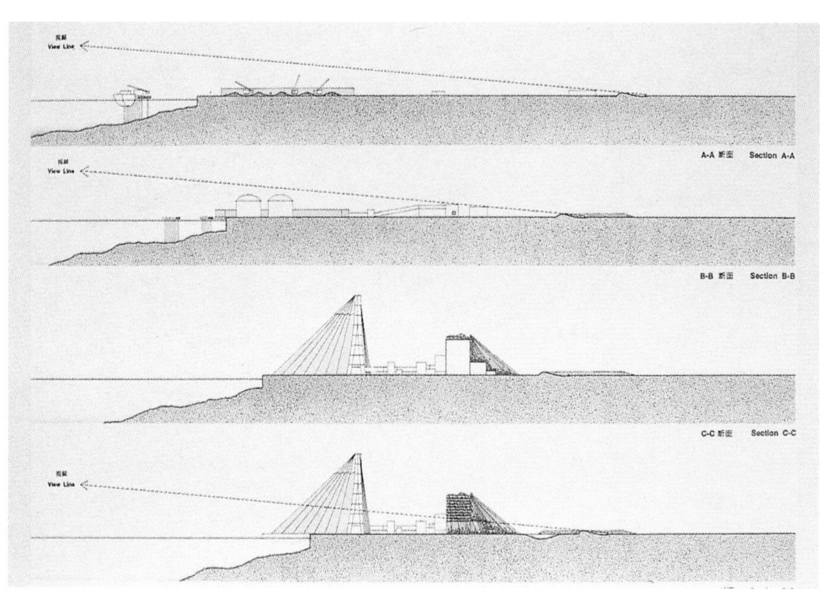
View Line
A-A 断面 Section A-A
View Line
B-B 断面 Section B-B
C-C 断面 Section C-C
View Line

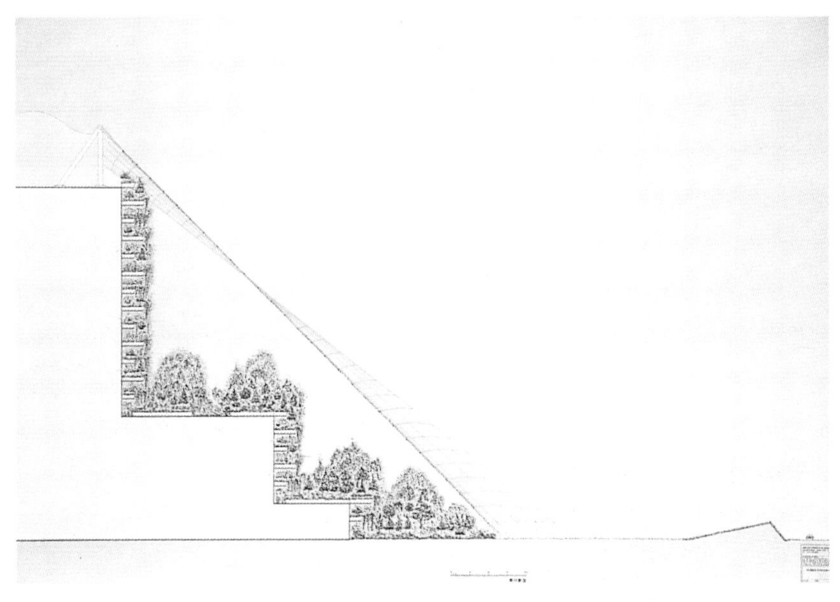

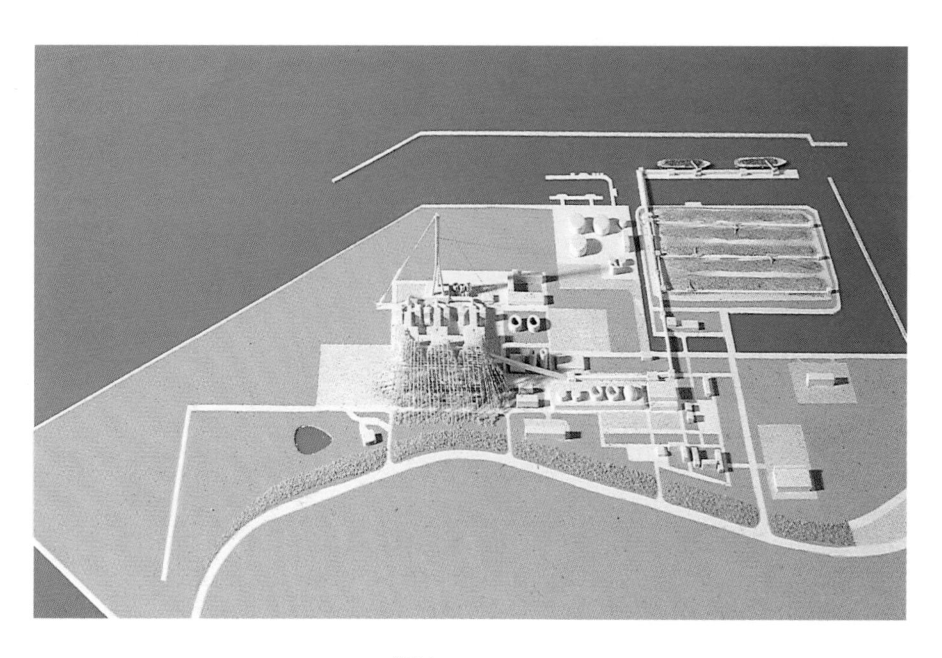

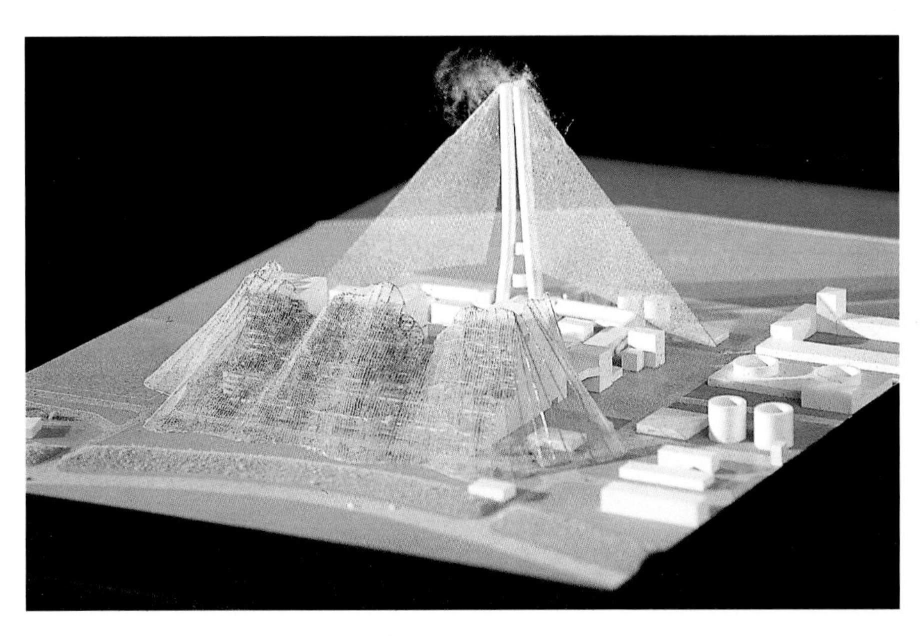

B Ticino Exhibition

Venice, Italy

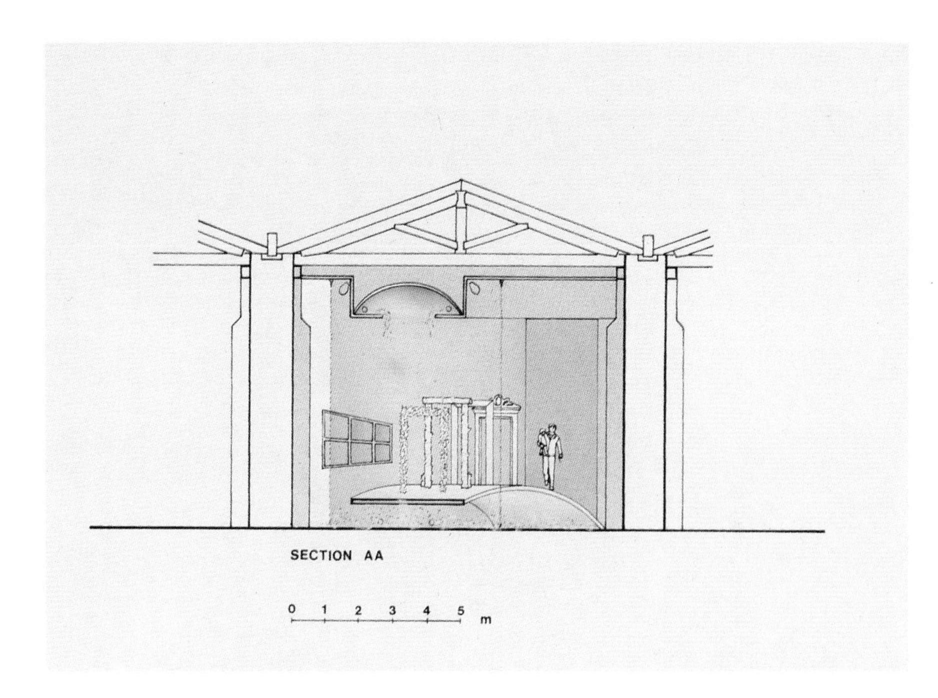
SECTION AA
0 1 2 3 4 5 m

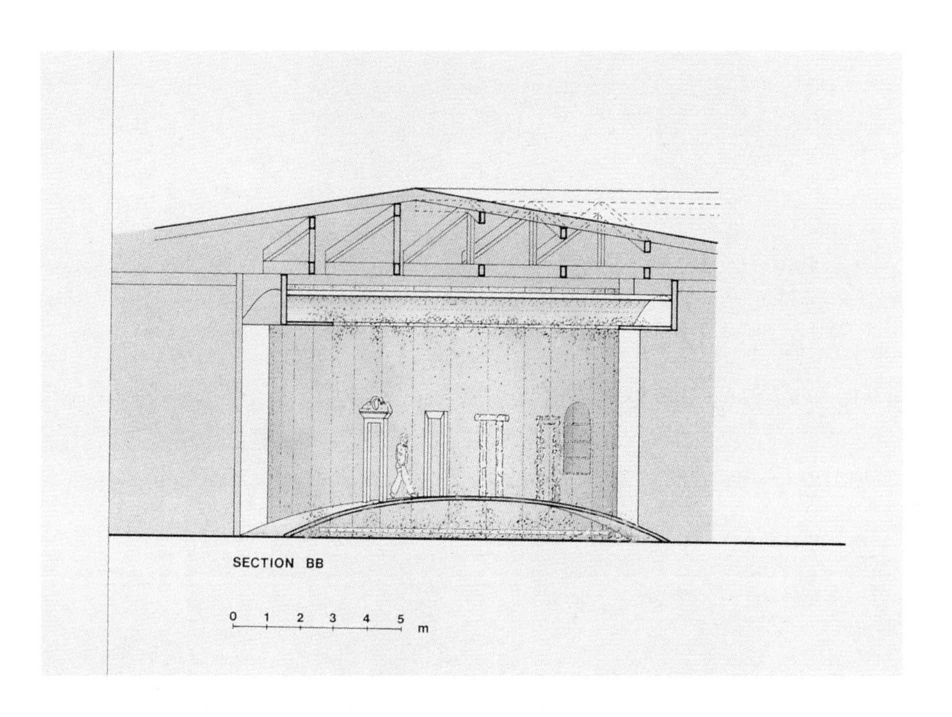
SECTION BB
0 1 2 3 4 5 m

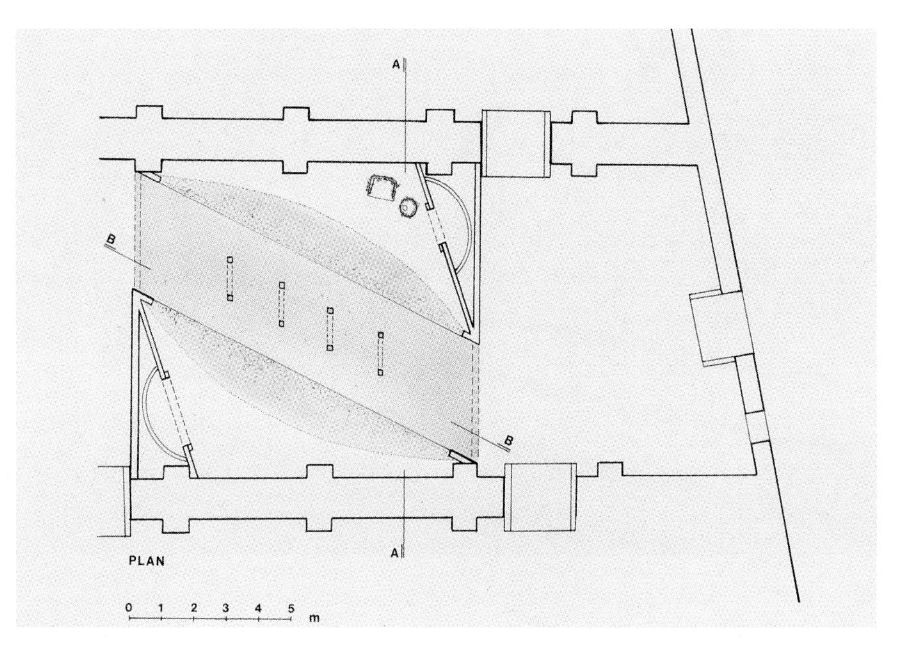
A
B
B
A
PLAN
0 1 2 3 4 5 m

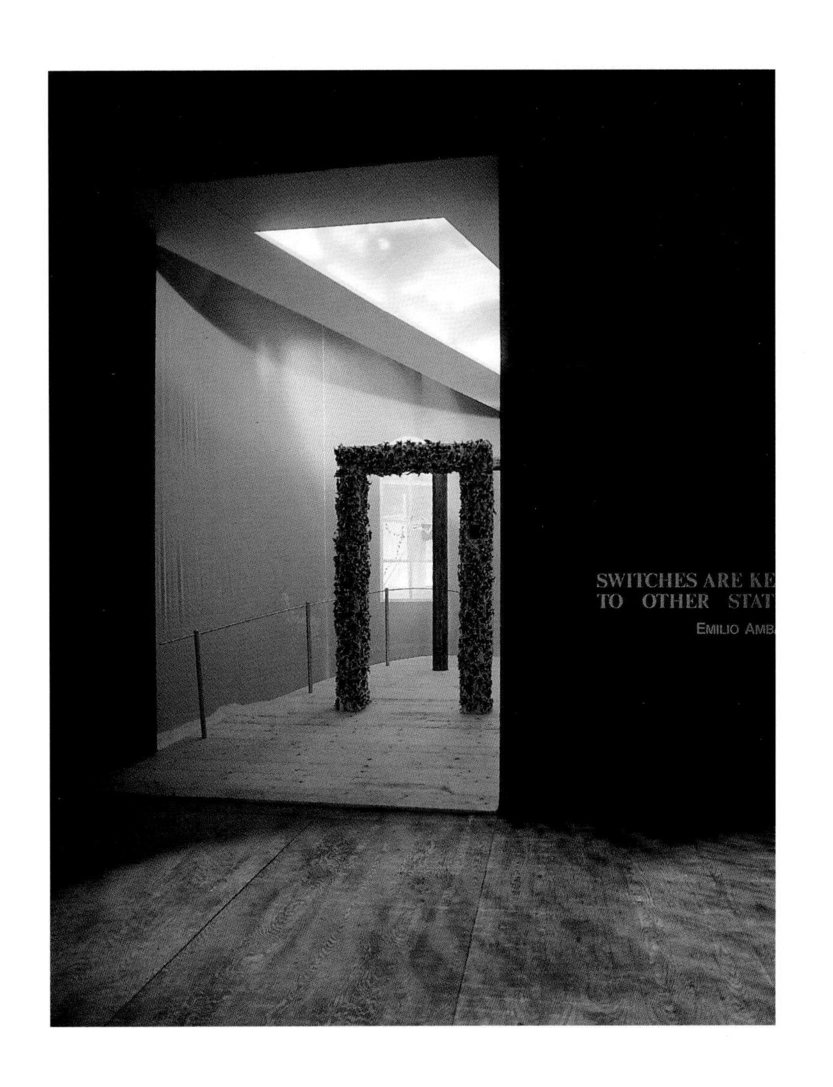
SWITCHES ARE KE
TO OTHER STAT
EMILIO AMBA

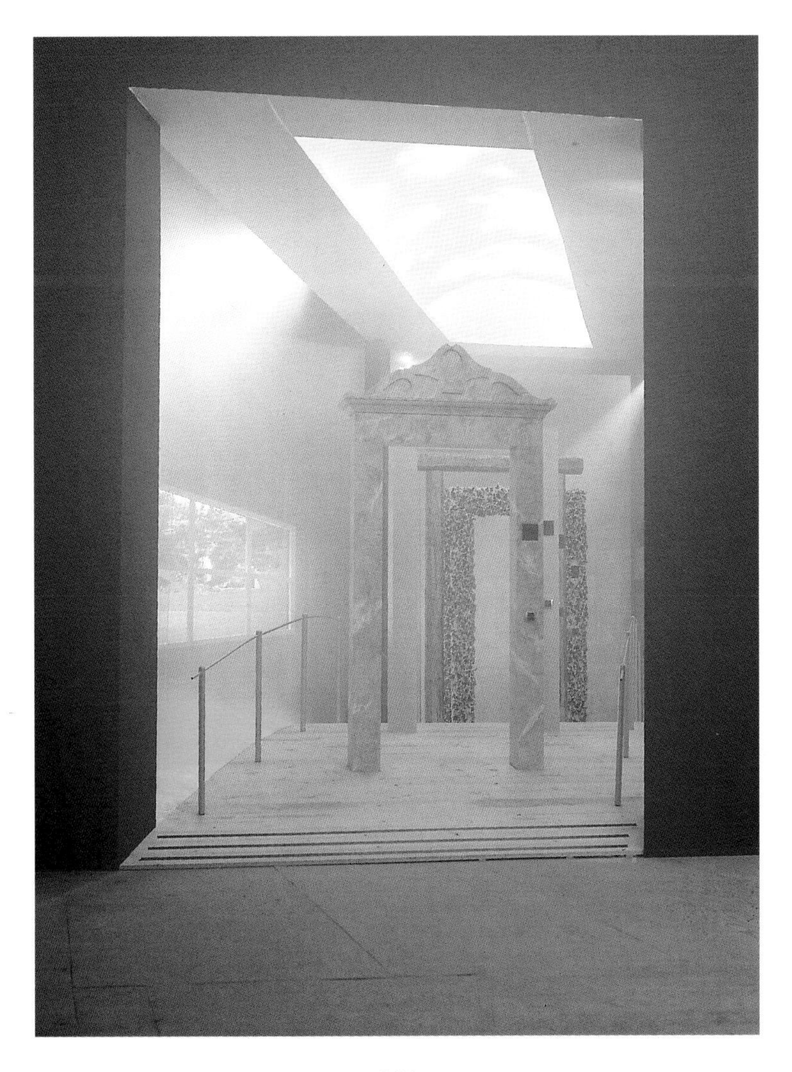

Air Force Memorial

Arlington, Virginia

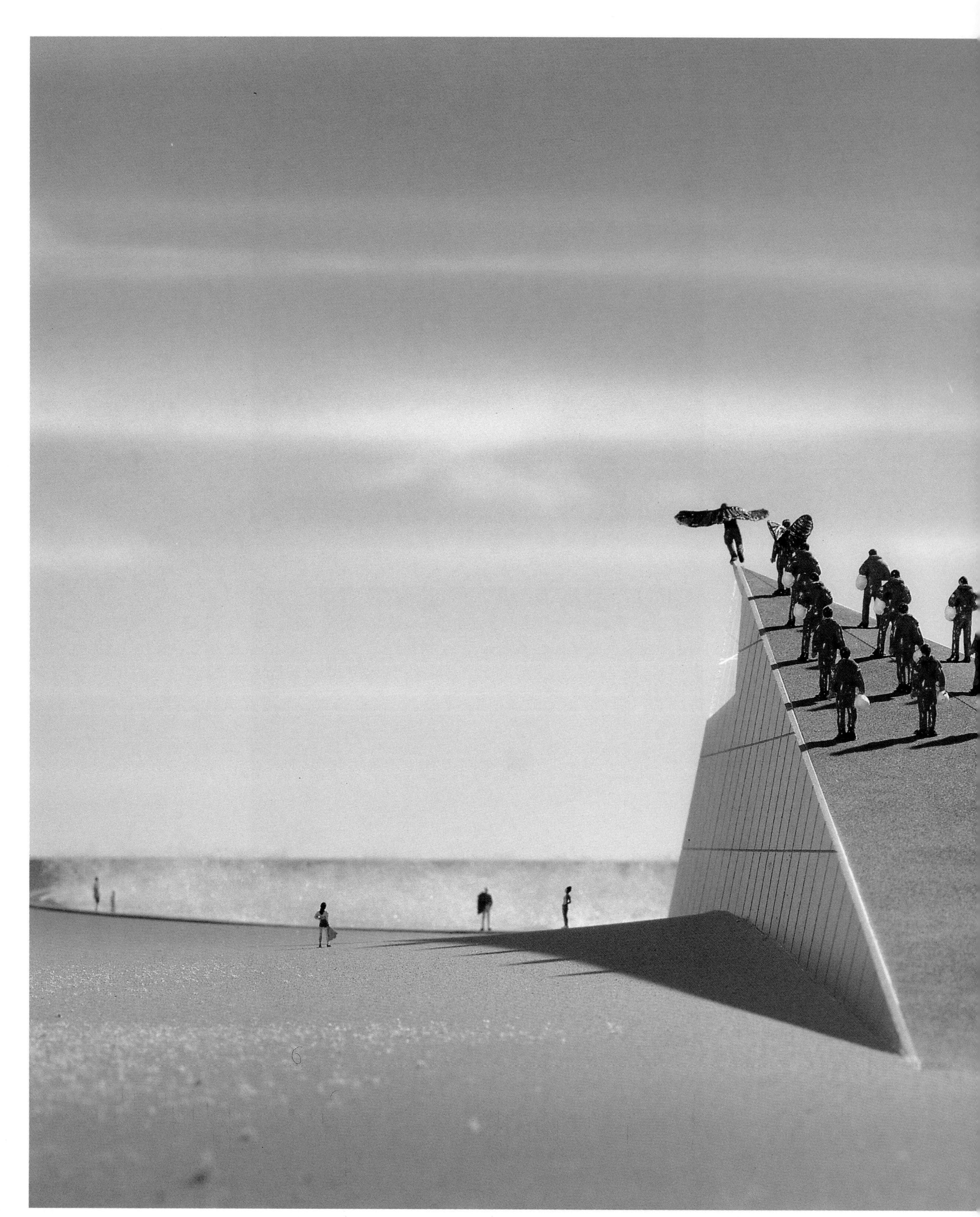

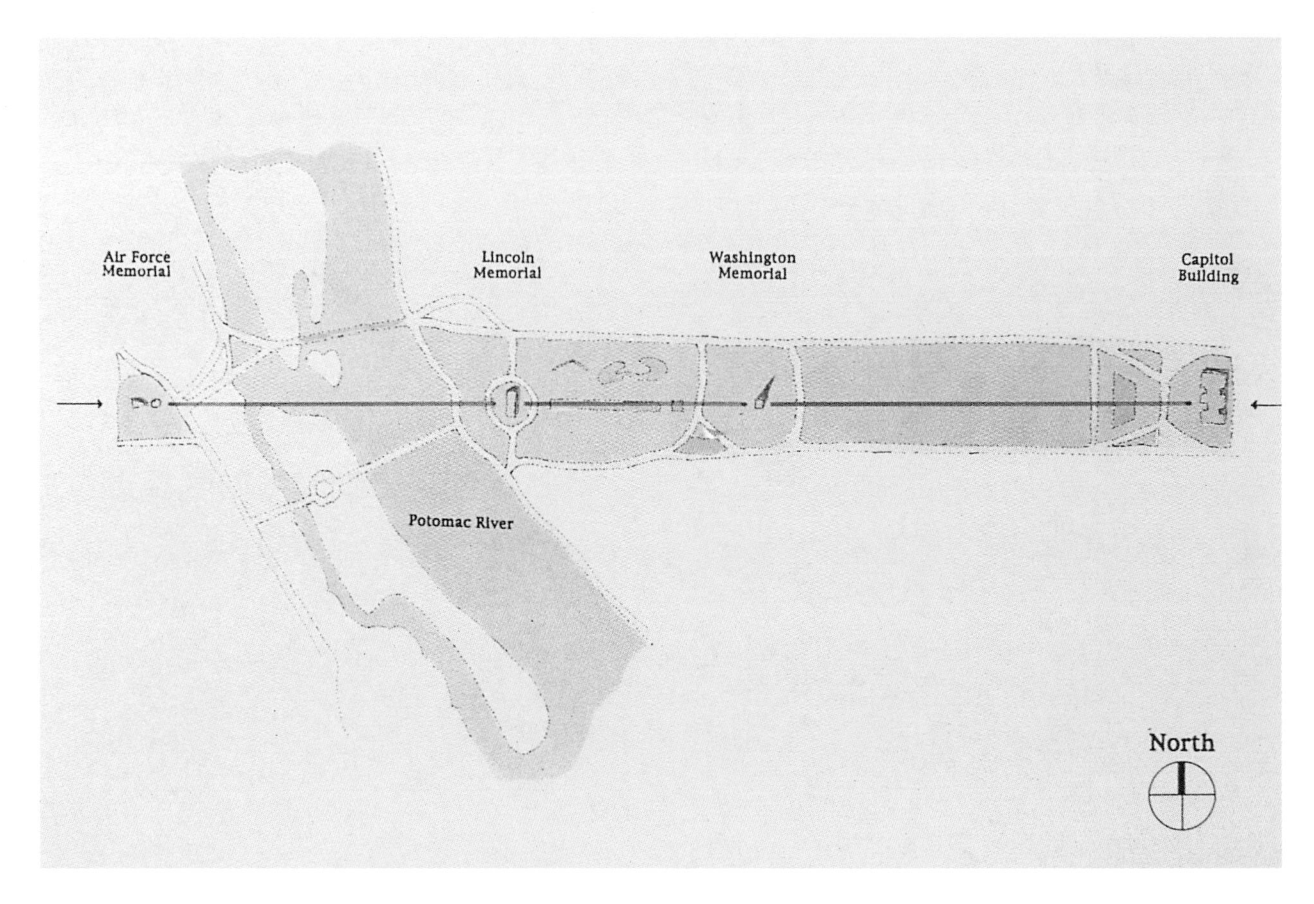
Air Force
Memorial
Lincoln
Memorial
Washington
Memorial
Capitol
Building
Potomac River
North

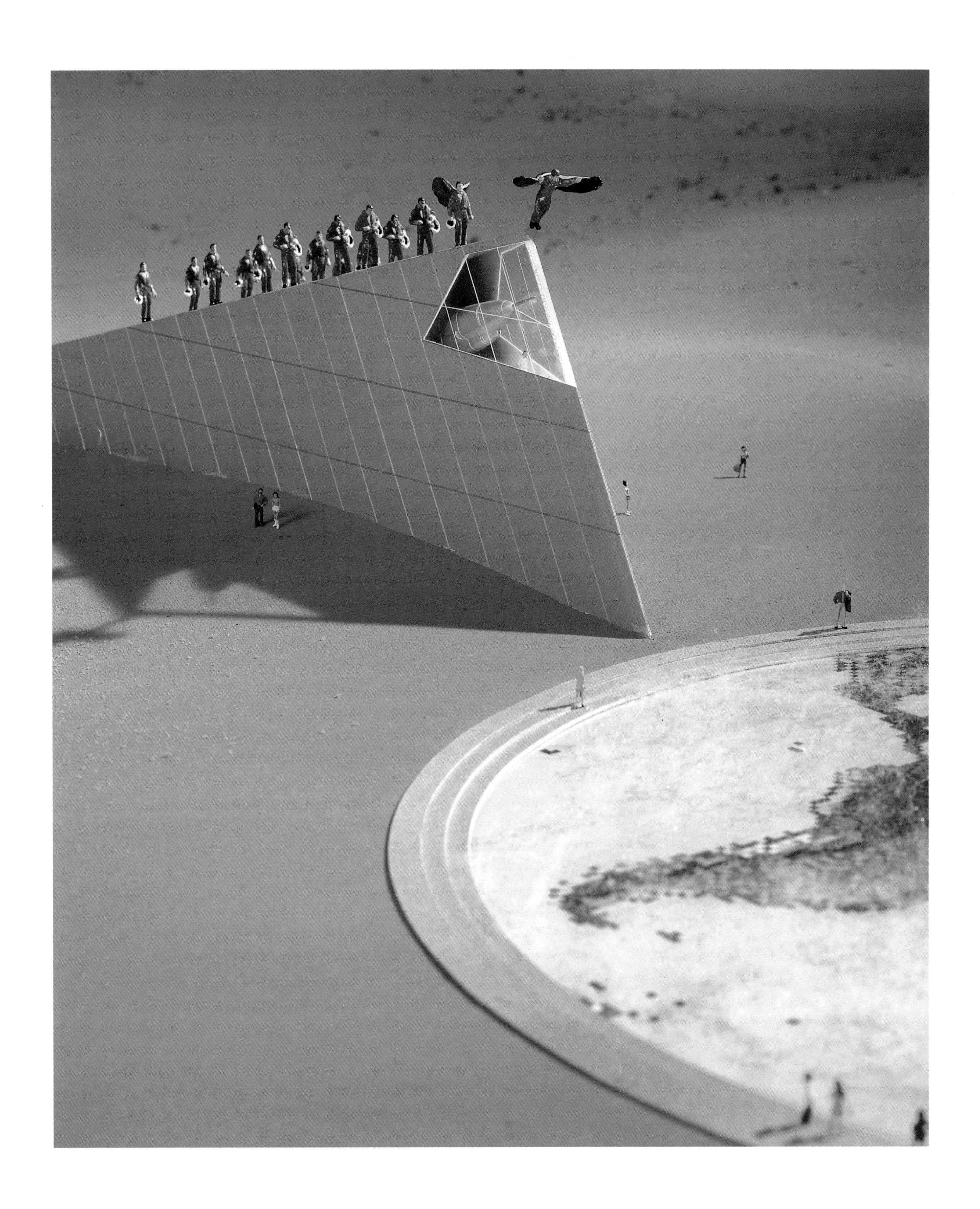

Barbie Doll Museum

Pasadena, California

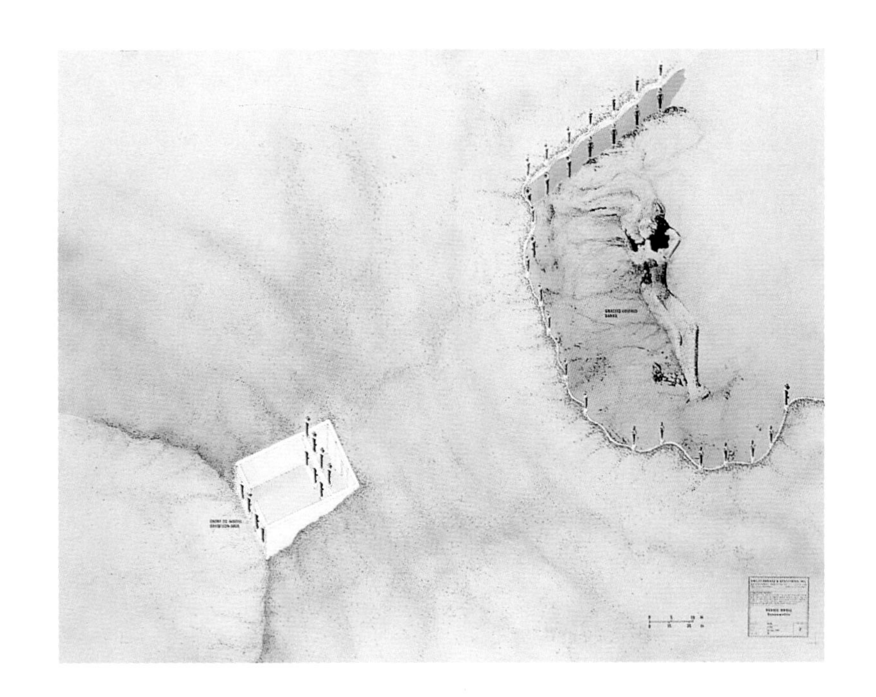

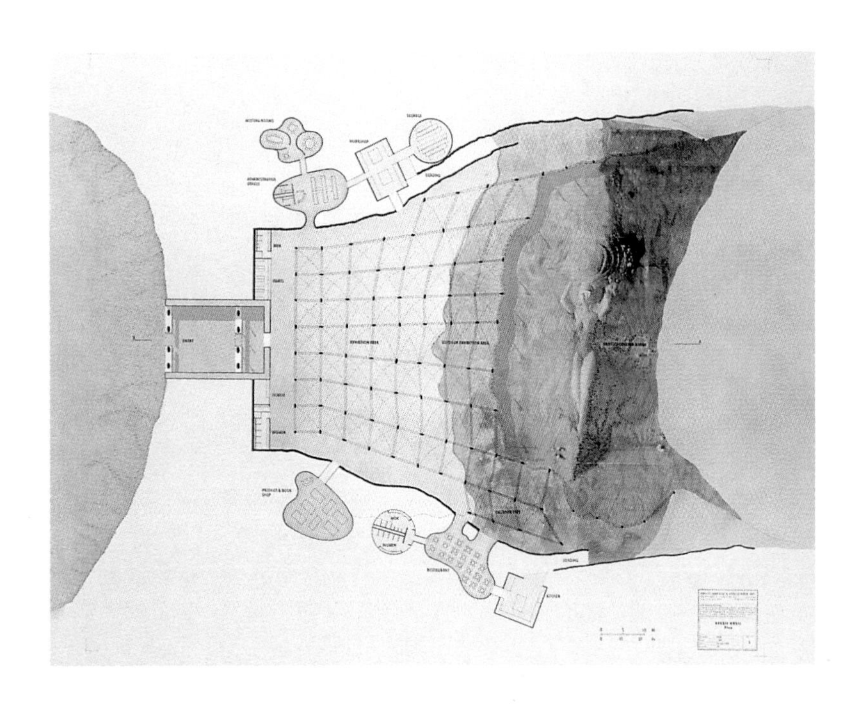

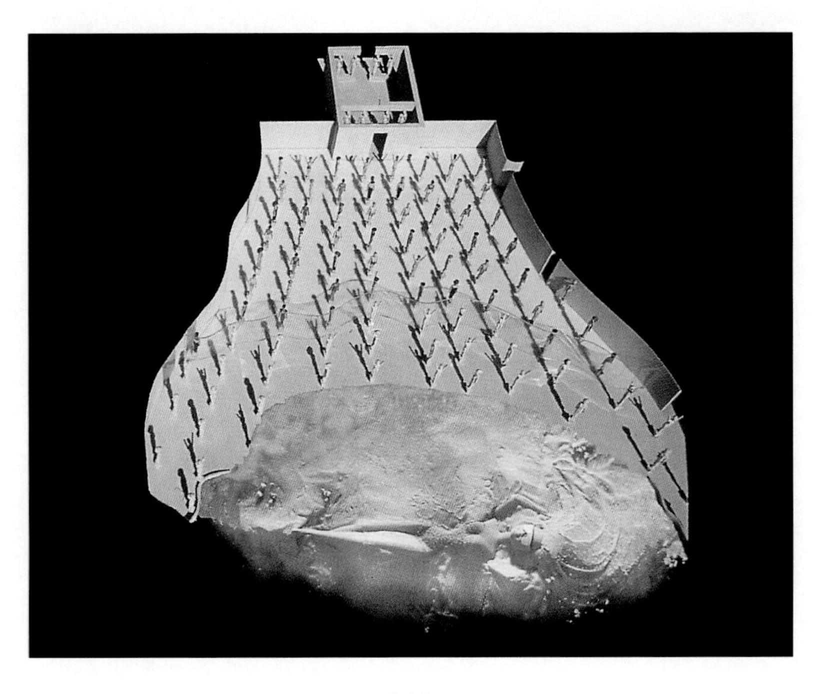

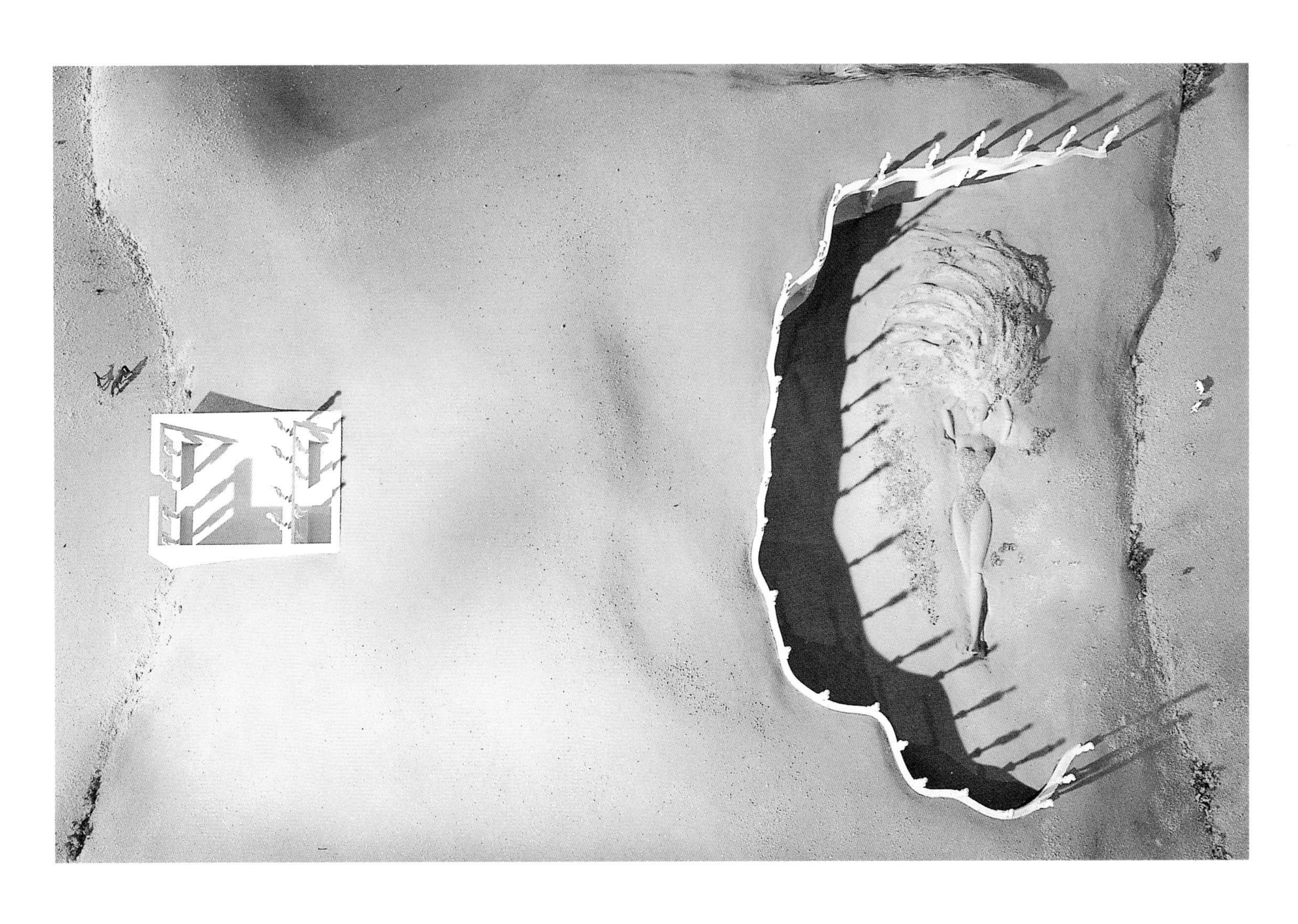

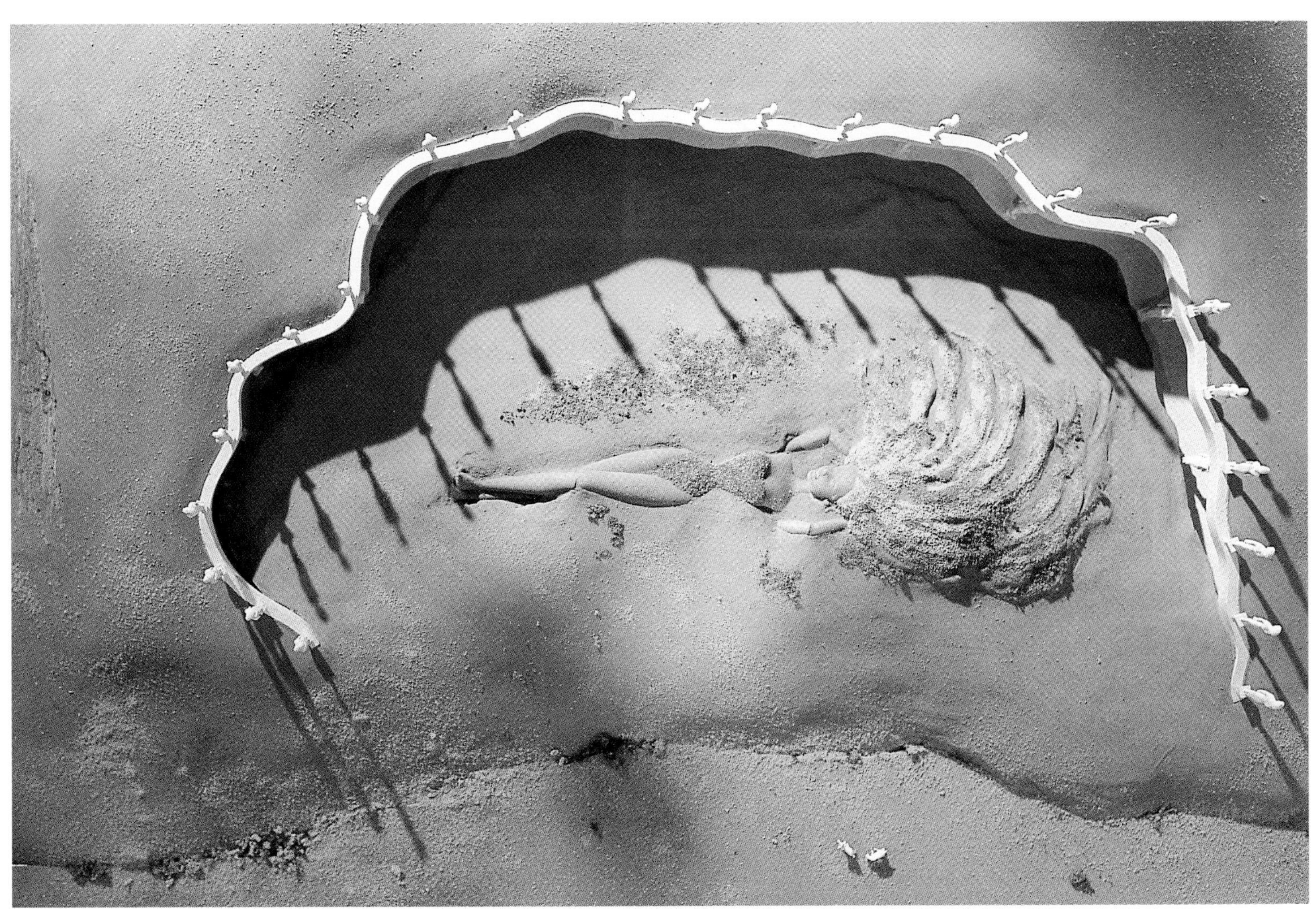

Commercial Office Development

Lisbon, Portugal

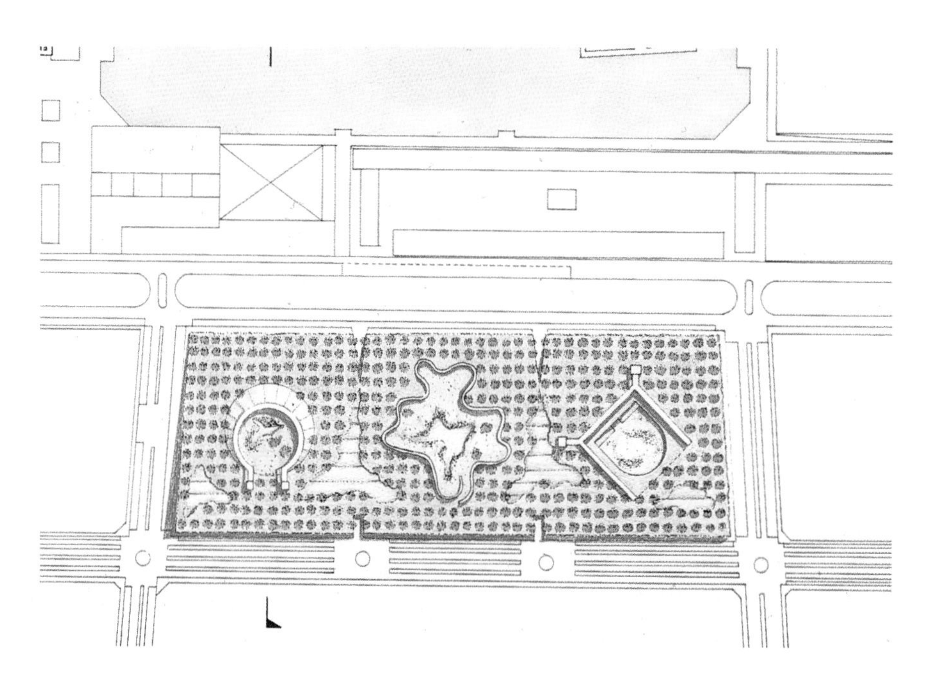

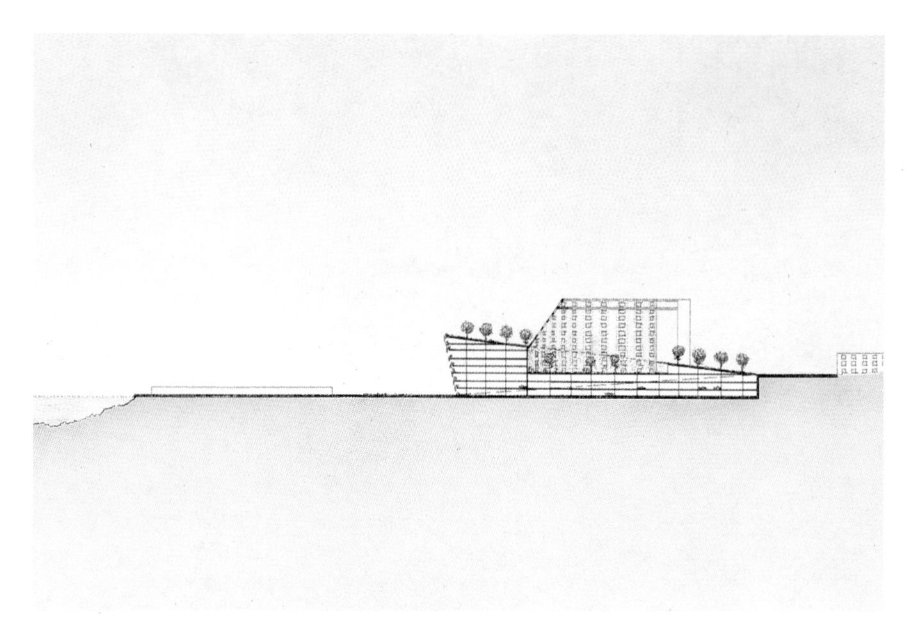

Utrecht, The Netherlands

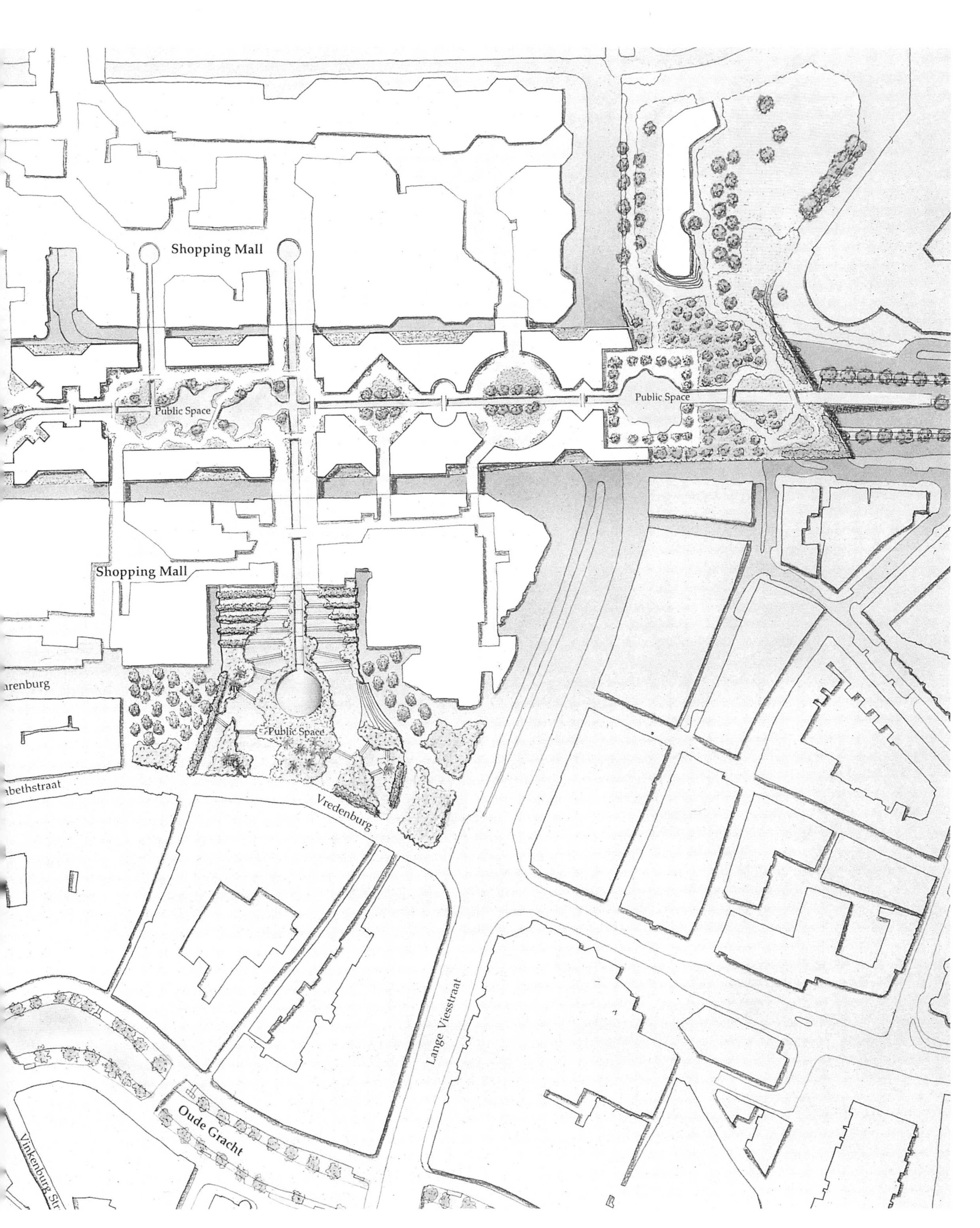
Shopping Mall
Public Space
Public Space
Shopping Mall
arenburg
Public Space
abethstraat
Vredenburg
Lange Viesstraat
Oude Gracht

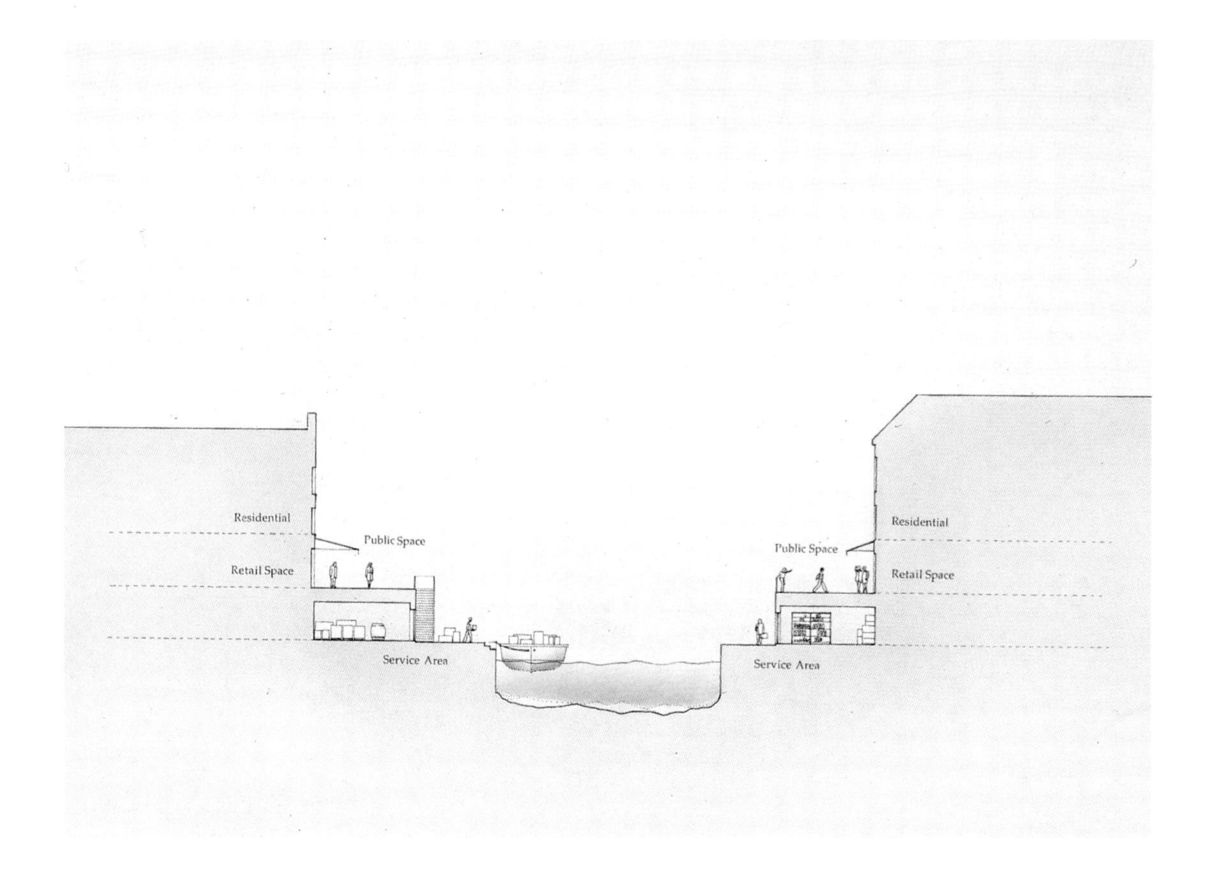
Residential
Public Space
Retail Space
Service Area
Public Space
Residential
Retail Space
Service Area

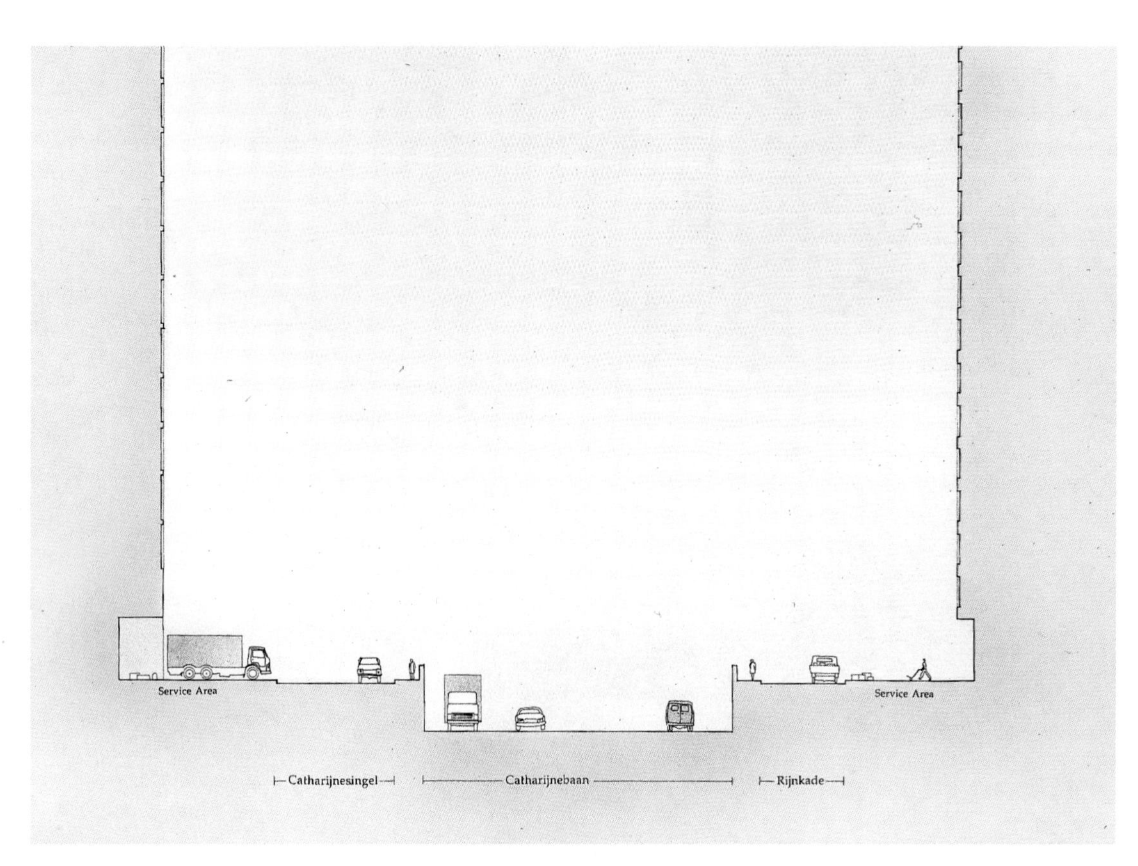
Service Area
Service Area
Catharijnesingel
Catharijnebaan
Rijnkade

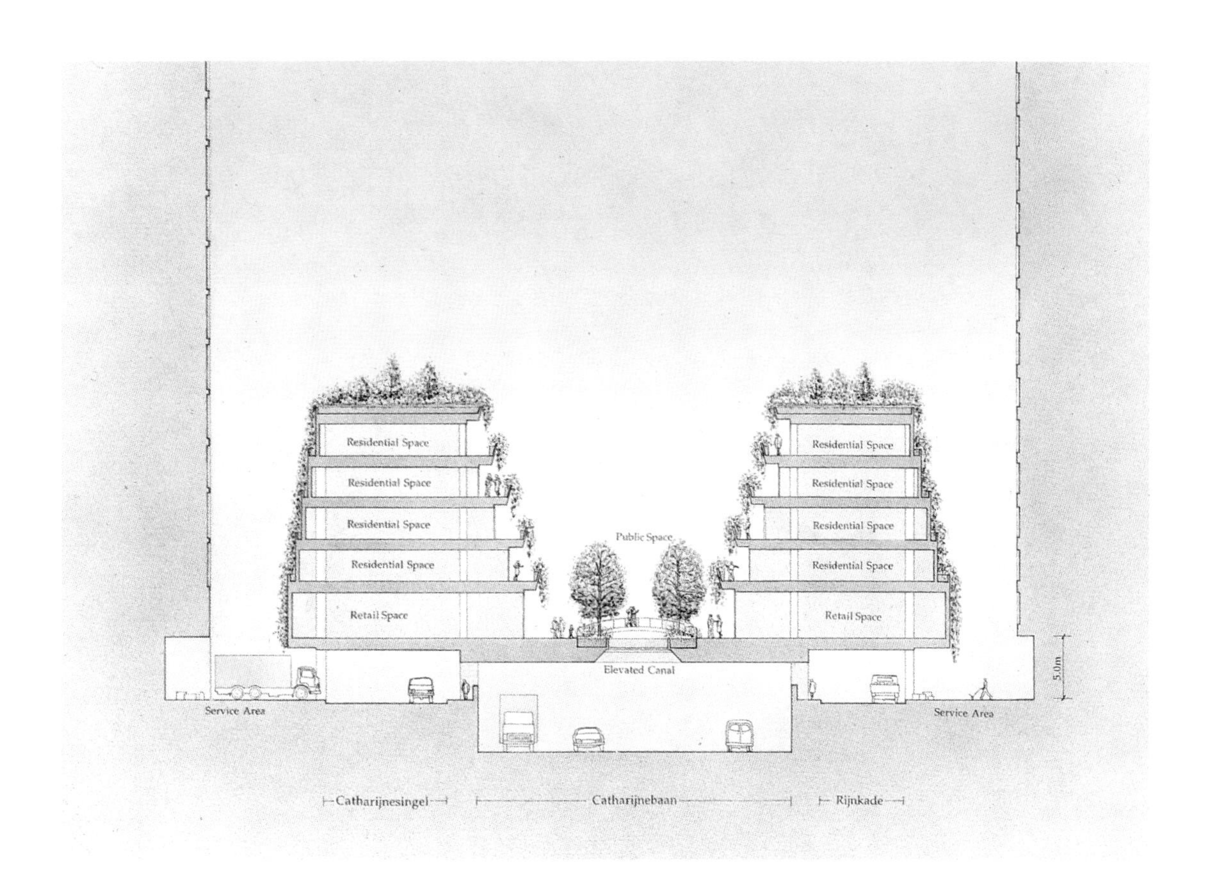
Residential Space
Residential Space
Residential Space
Residential Space
Retail Space
Public Space
Residential Space
Residential Space
Residential Space
Residential Space
Retail Space
Elevated Canal
Service Area
Service Area
5.0m
Catharijnesingel
Catharijnebaan
Rijnkade

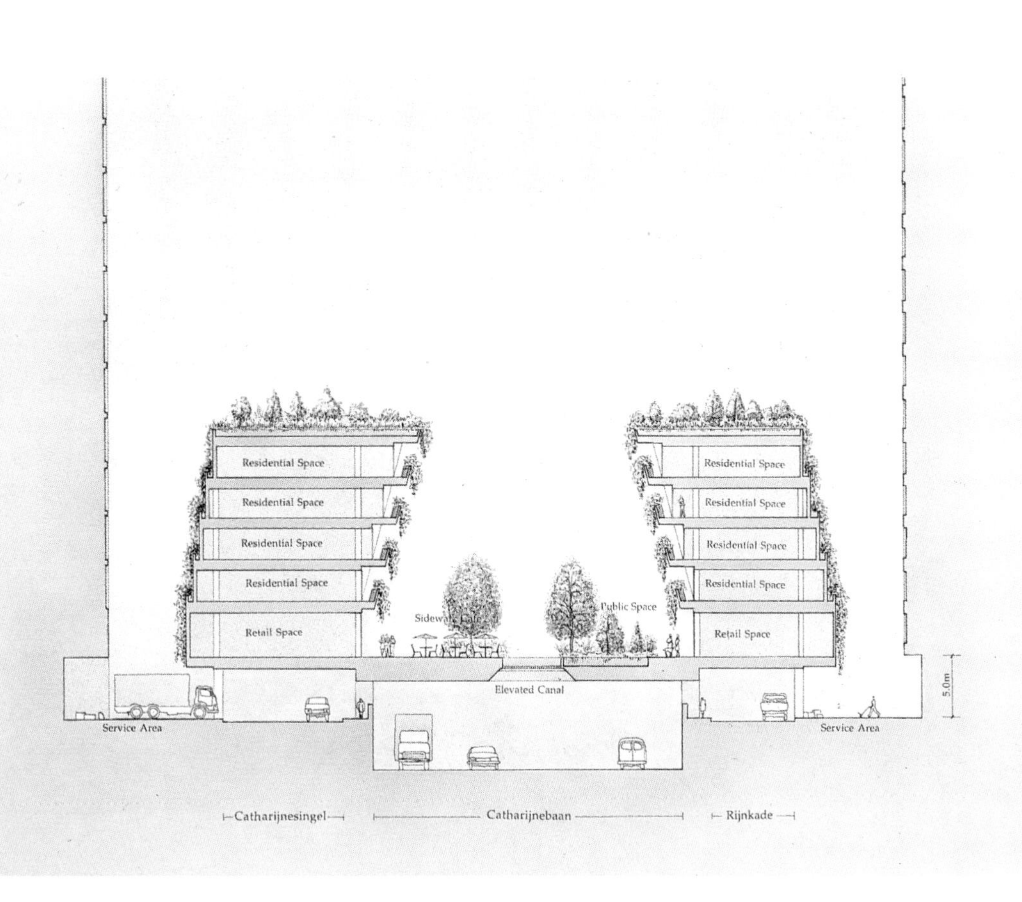
Residential Space
Residential Space
Residential Space
Residential Space
Retail Space
Public Space
Residential Space
Residential Space
Residential Space
Residential Space
Retail Space
Elevated Canal
Service Area
Service Area
5.0m
Catharijnesingel
Catharijnebaan
Rijnkade

Archives of the National Library of Japan

Kaisan Science City, Japan

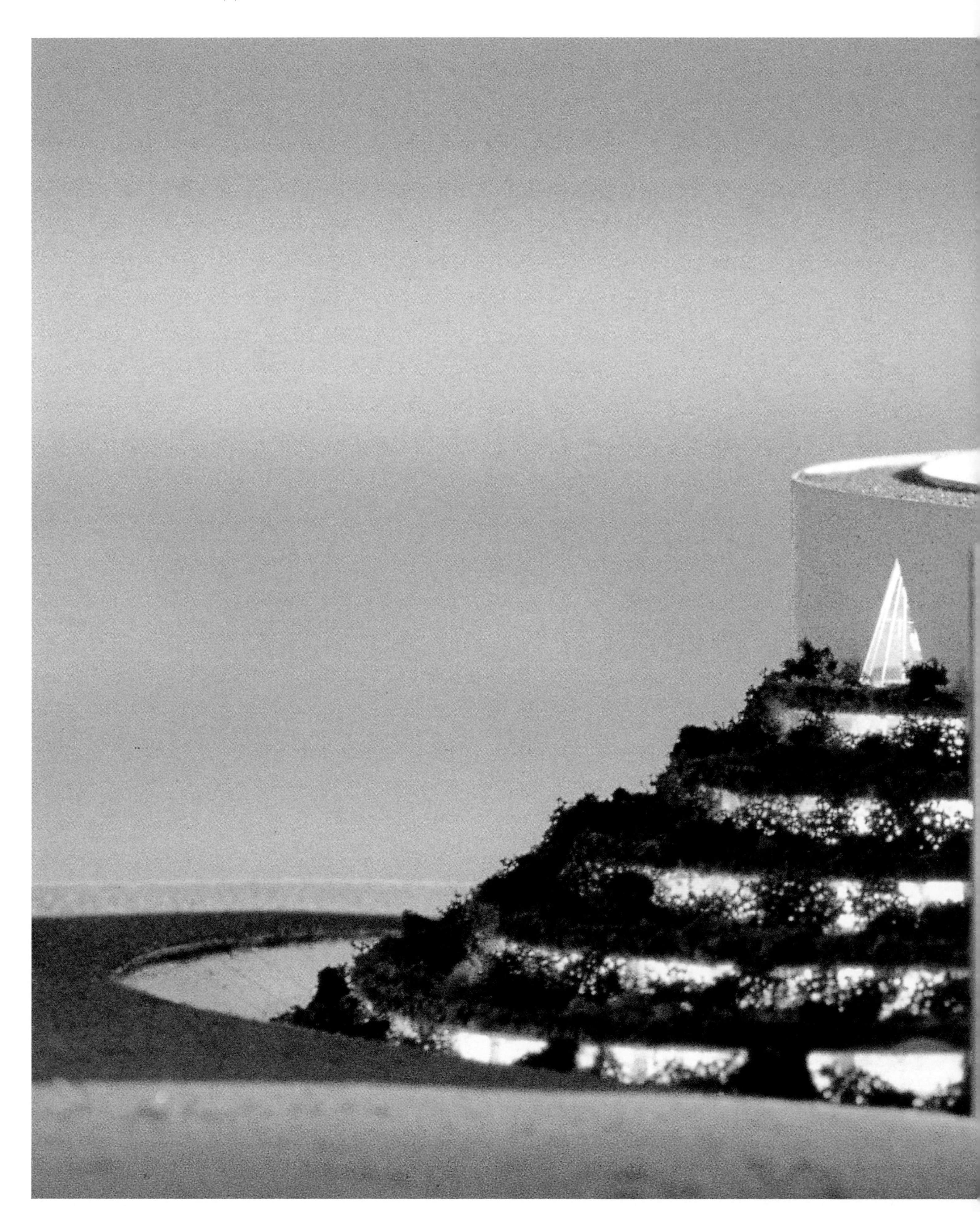

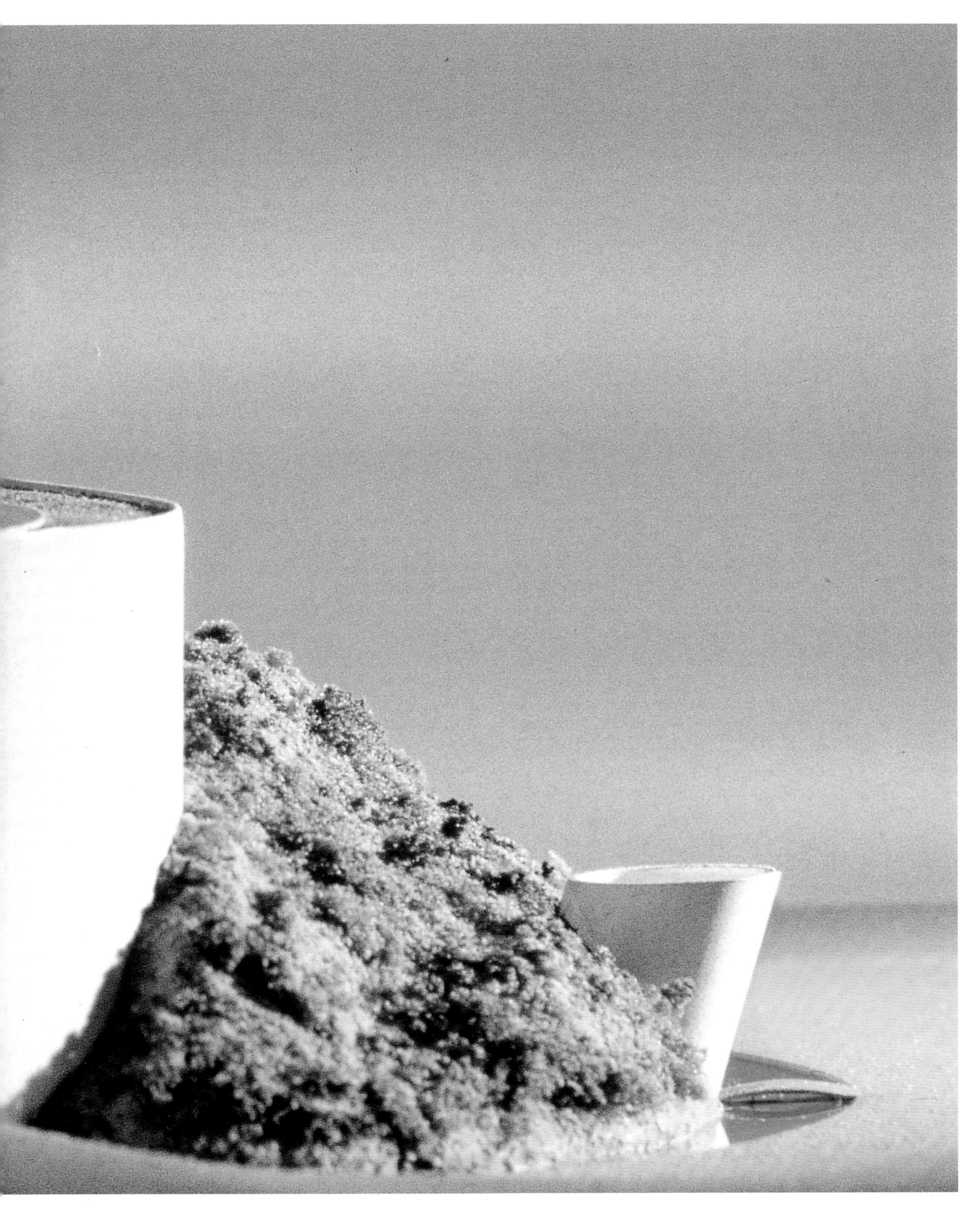

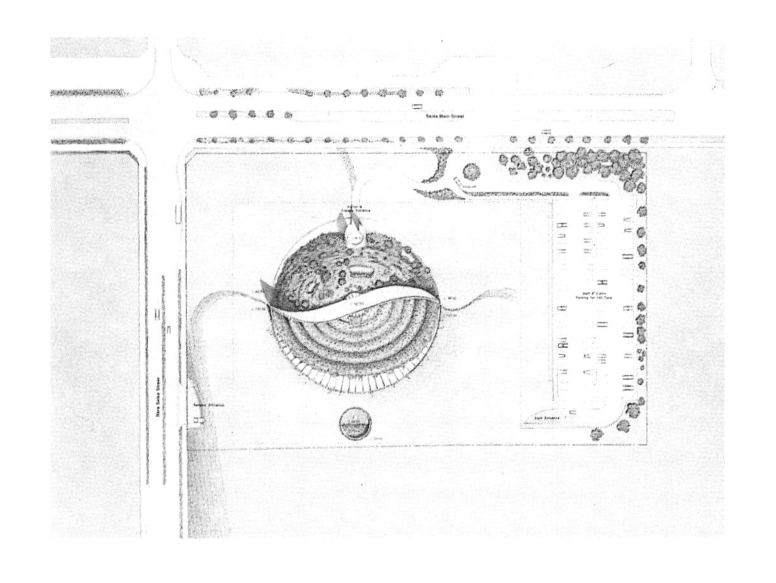

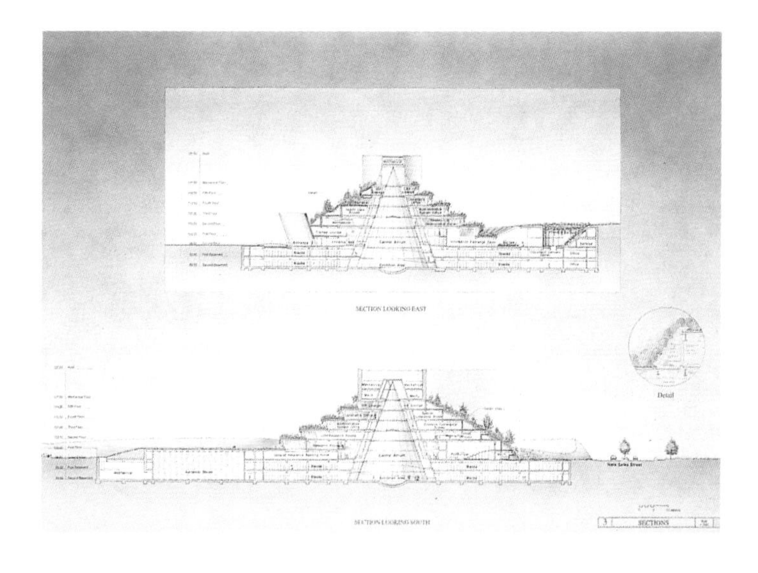
SECTION LOOKING EAST
Detail
SECTION LOOKING SOUTH
3 SECTIONS

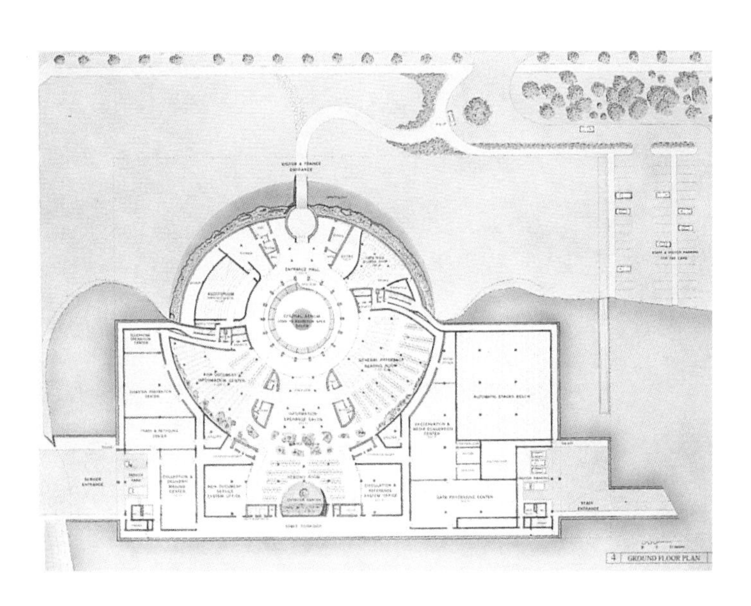
4 GROUND FLOOR PLAN

SOUTH ELEVATION
NORTH ELEVATION
7 ELEVATIONS

The Museum of Modern Art and Cinema

Buenos Aires, Argentina

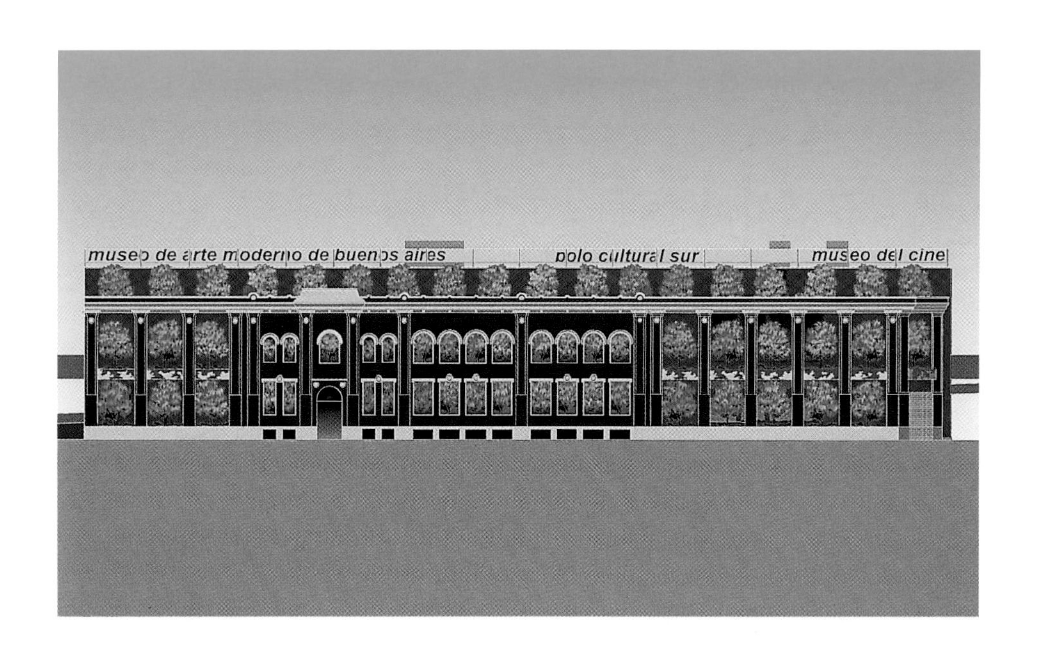
museo de arte moderno de buenos aires
polo cultural sur
museo del cine

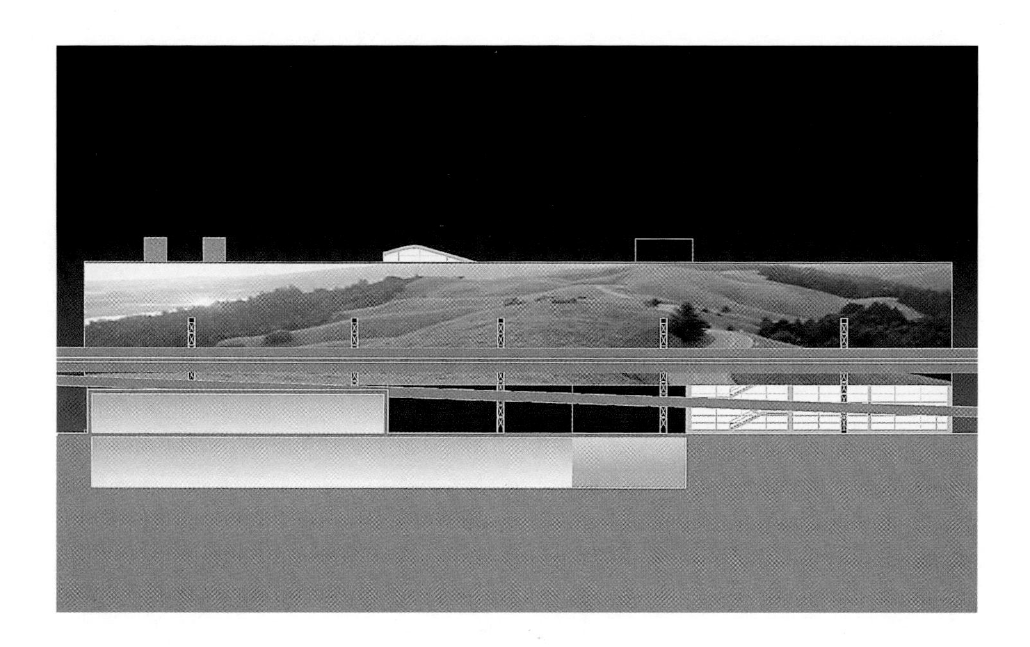

Marina di Bellaria

Bellaria, Italy

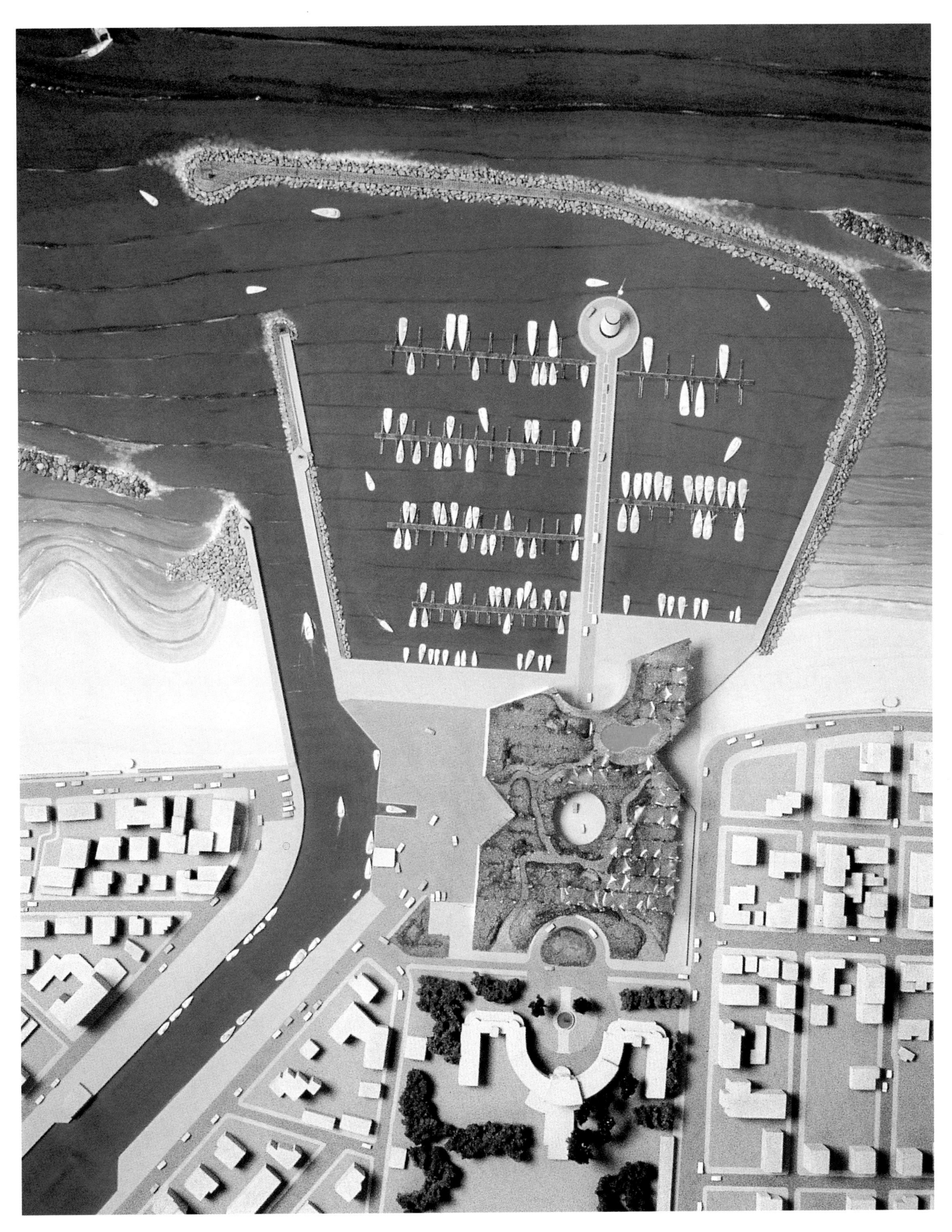

Office Complex

Hilversum, The Netherlands

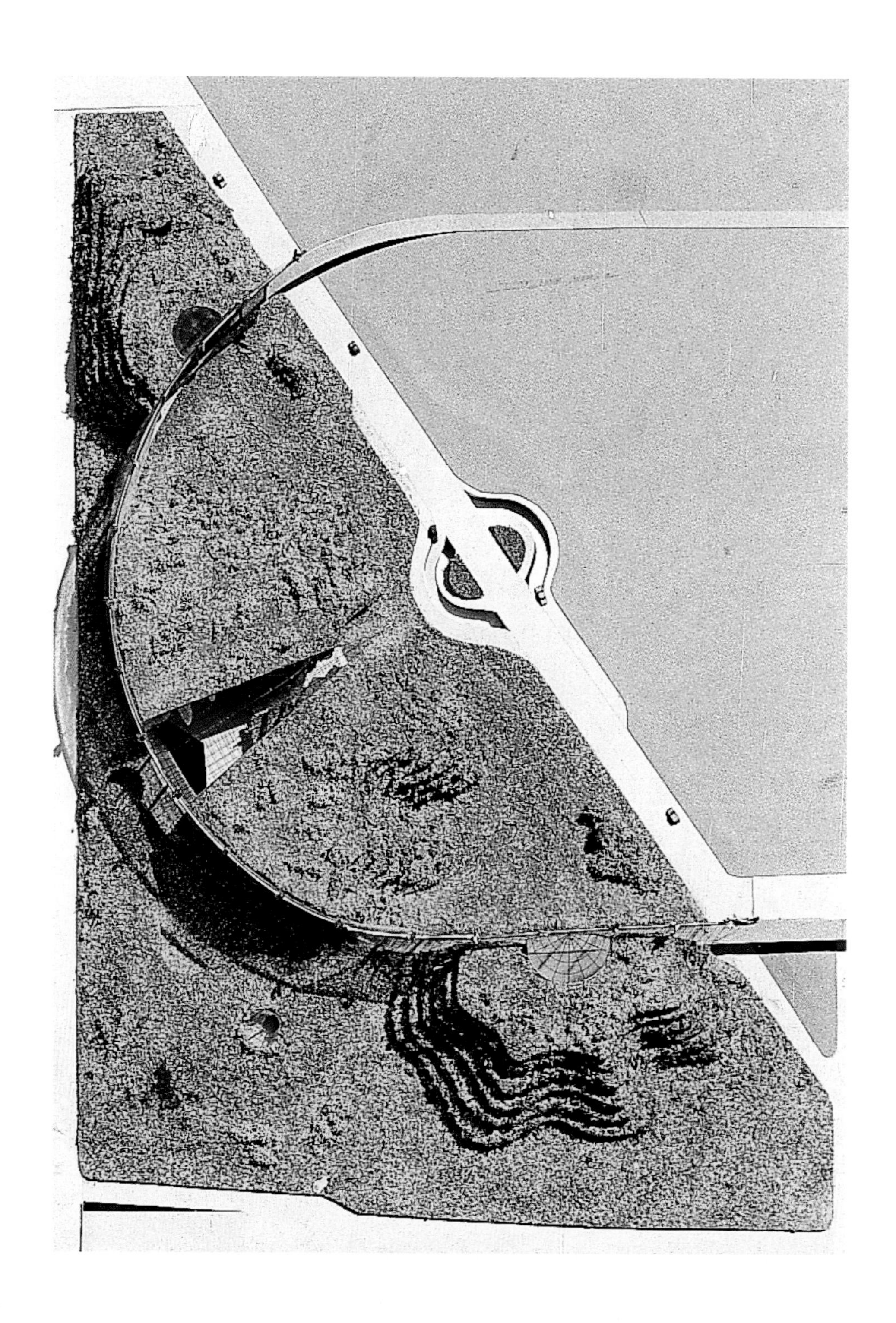

Argentine Pavilion at the Venice Biennale

Venice, Italy

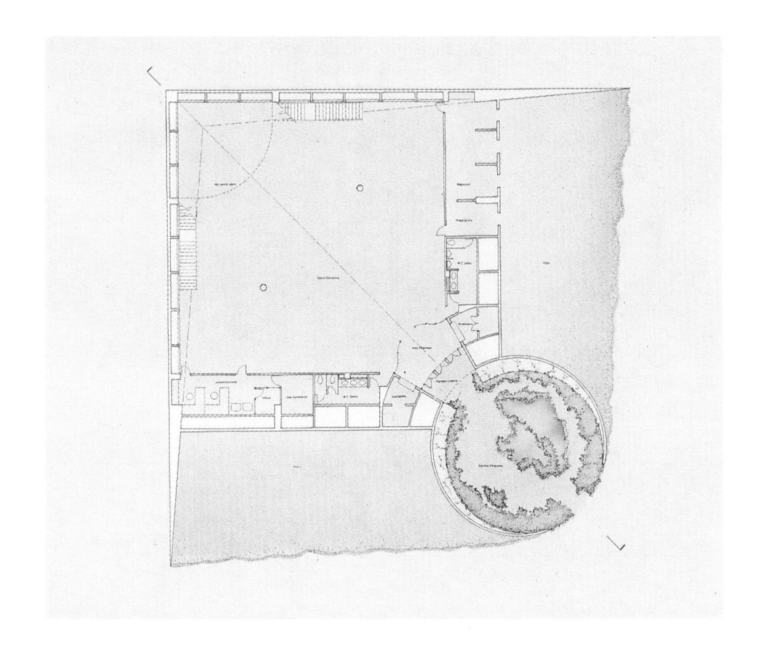

Concert Hall on the Waterfront

Copenhagen, Denmark

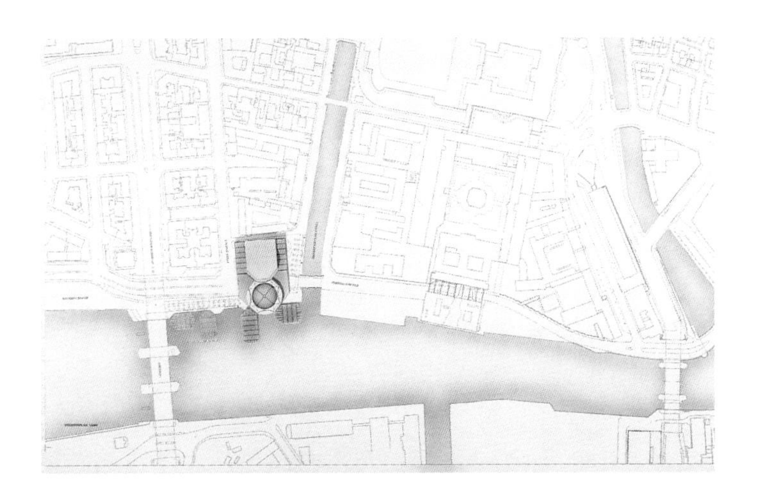

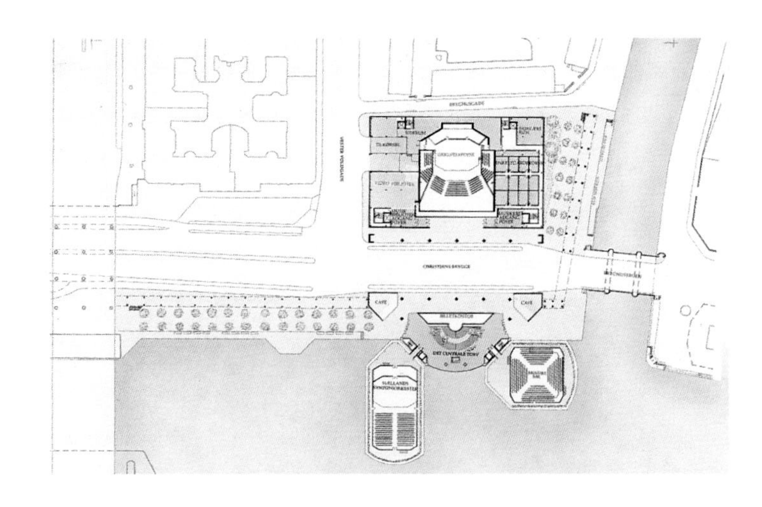
CHRISTIANS BRYGGE
CAFE
CAFE
DET CENTRALE TORV

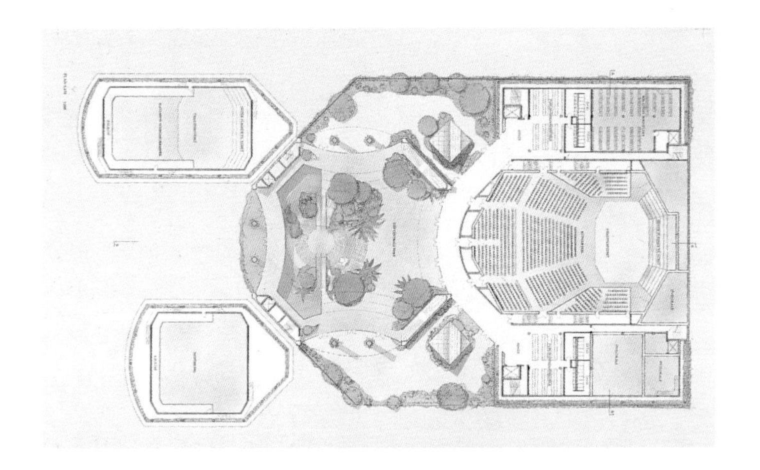

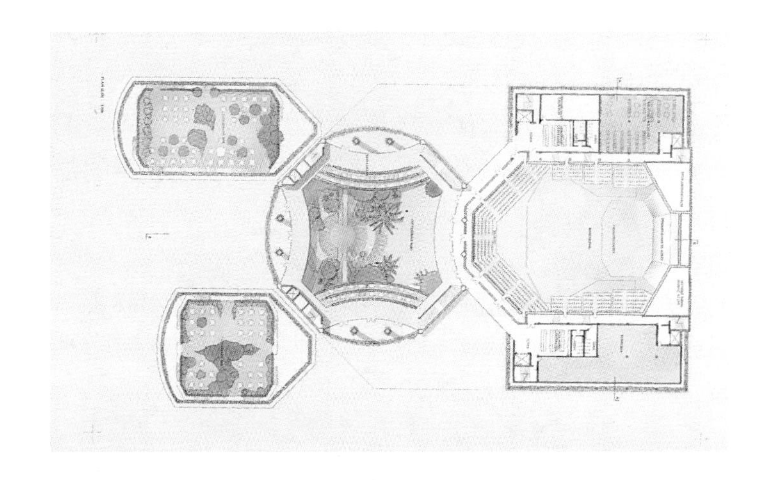

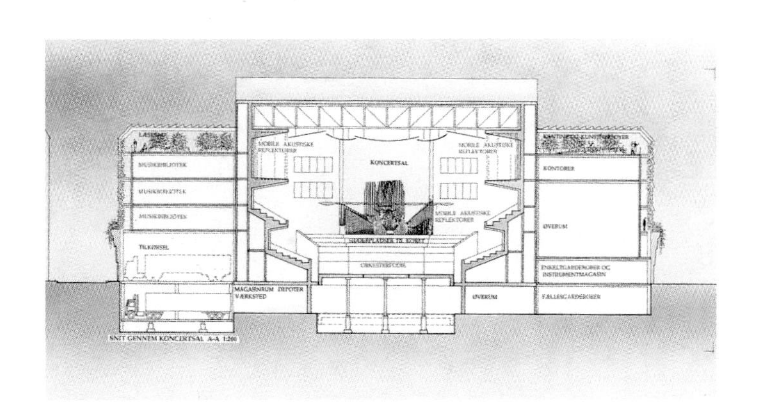
MOBILE AKUSTISKE
REFLEKTORER
KONCERTSAL
MOBILE AKUSTISKE
REFLEKTORER
MUSIKBIBLIOTEK
KONTORER
ØVERUM
TILKØRSEL
ORKESTERFODE
ENKELTGARDEROBER OG
INSTRUMENTMAGASIN
MAGASINRUM DEPOTER
VÆRKSTED
ØVERUM
FÆLLESGARDEROBER
SNIT GENNEM KONCERTSAL A-A 1:200

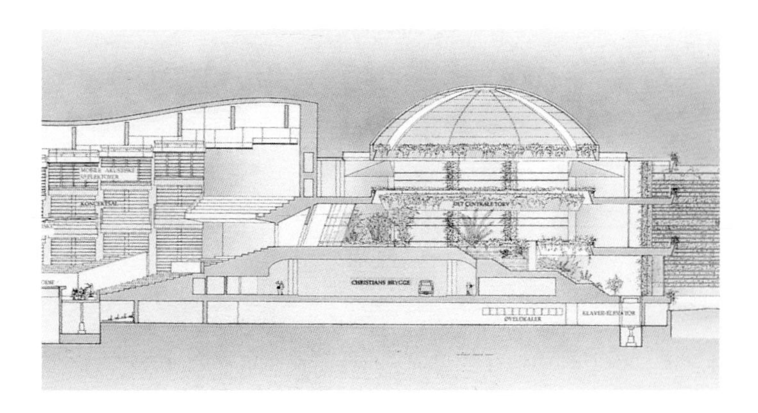
KONCERTSAL
DET CENTRALE TORV
CHRISTIANS BRYGGE
ØVELOKALER

Thermal Gardens

Sirmione, Italy

Covered pool with 16° C water
Covered pool with 34° C water
Uncovered pool
Entrance
Section C-C

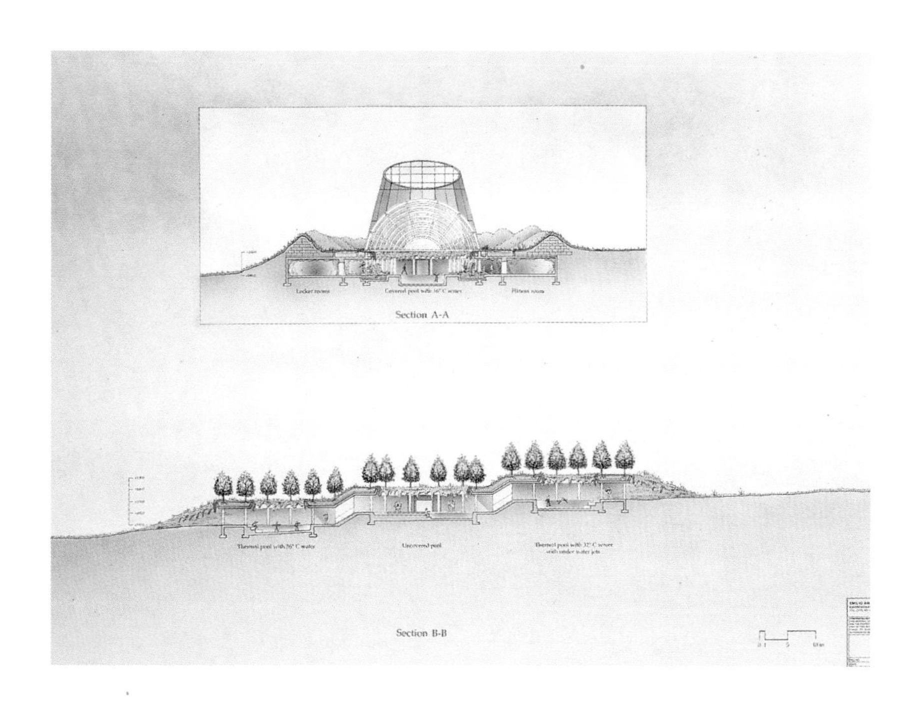
Locker rooms
Covered pool with 36° C water
Fitness room
Section A-A
Thermal pool with 36° C water
Uncovered pool
Thermal pool with 32° C water
with under water jets
Section B-B

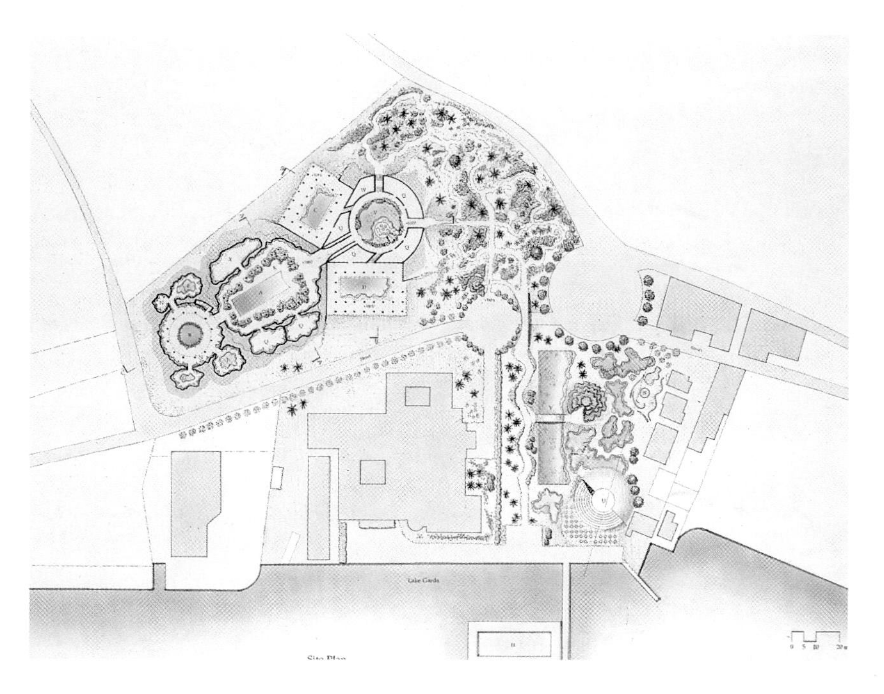

Complex of Research Laboratories and Offices

Environment Park, Torino, Italy

Redesign of ENI Headquarter's Office

Rome (EUR), Italy

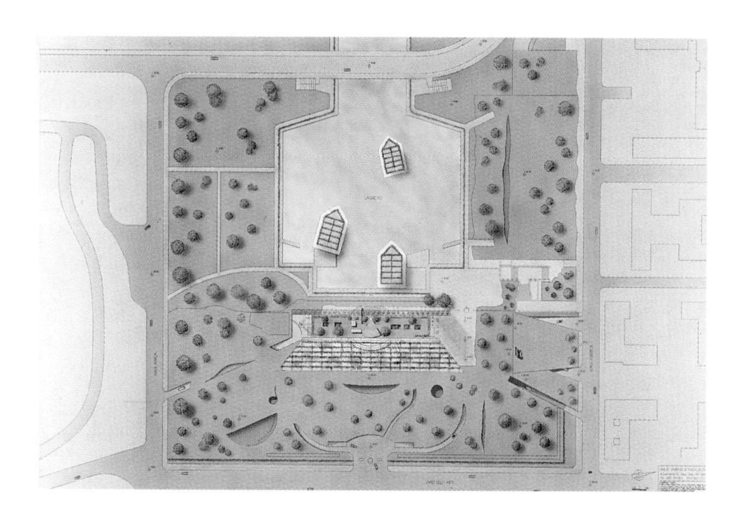

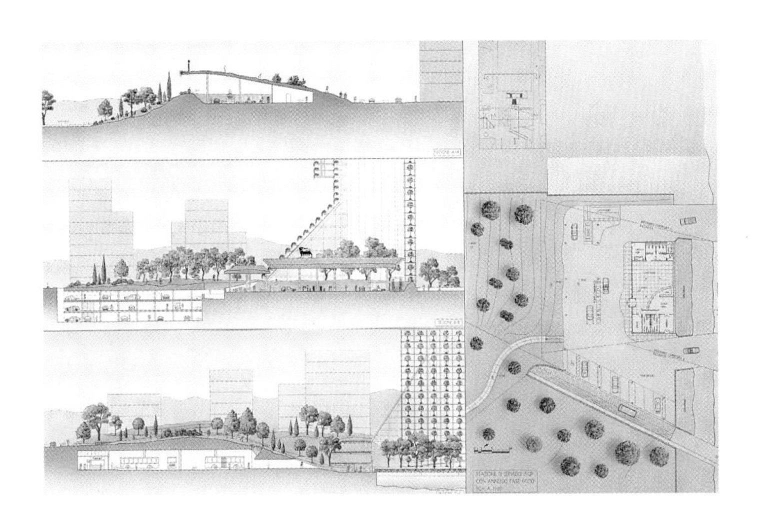

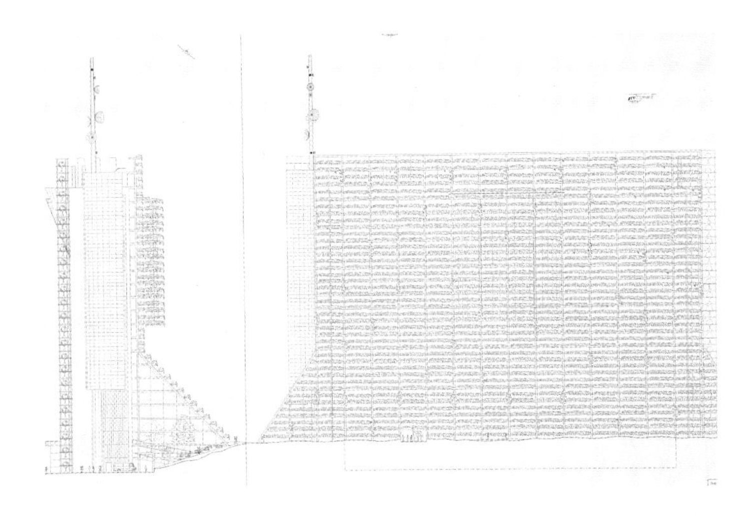

Agip

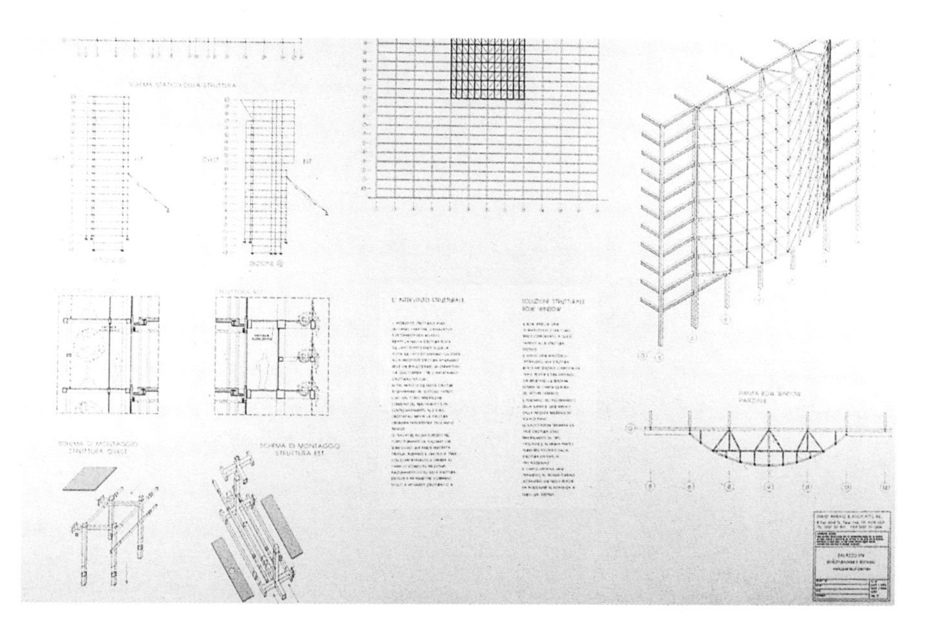

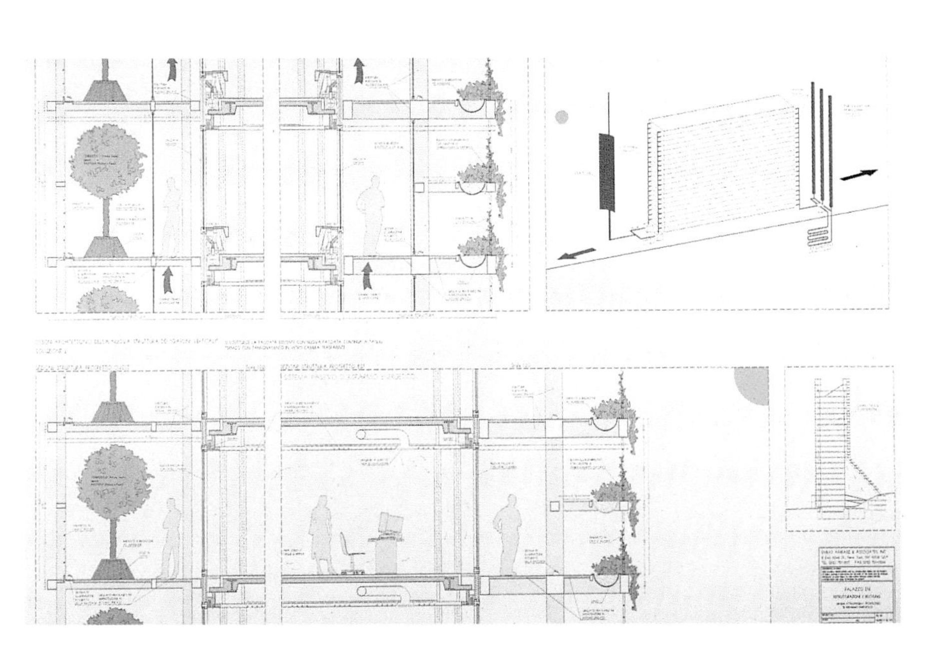

Glory Art Museum

Hsinchu, Taiwan, R.O.C.

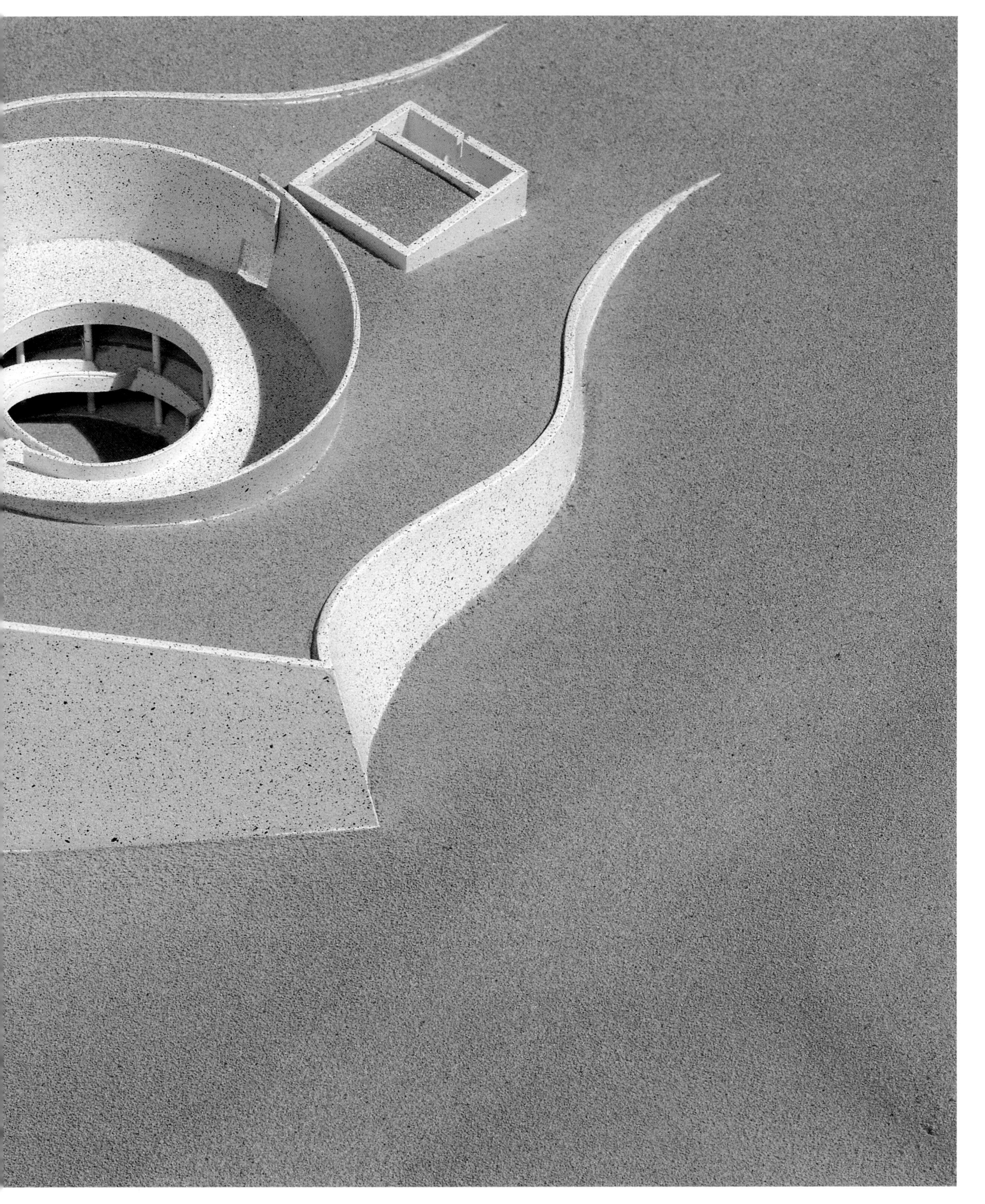

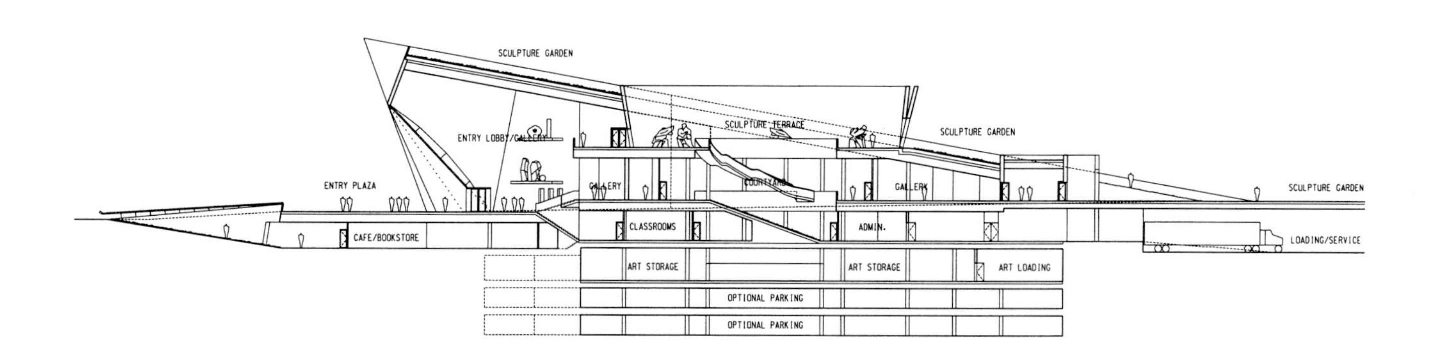
SCULPTURE GARDEN
ENTRY LOBBY/GALLERY
SCULPTURE TERRACE
SCULPTURE GARDEN
ENTRY PLAZA
GALLERY
COURTYARD
GALLERY
SCULPTURE GARDEN
CAFE/BOOKSTORE
CLASSROOMS
ADMIN.
LOADING/SERVICE
ART STORAGE
ART STORAGE
ART LOADING
OPTIONAL PARKING
OPTIONAL PARKING

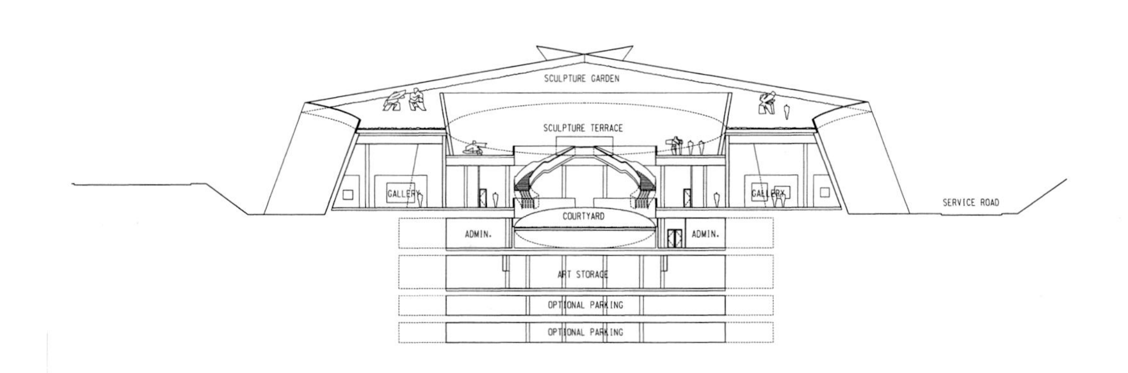
SCULPTURE GARDEN
SCULPTURE TERRACE
GALLERY
GALLERY
SERVICE ROAD
COURTYARD
ADMIN.
ADMIN.
ART STORAGE
OPTIONAL PARKING
OPTIONAL PARKING

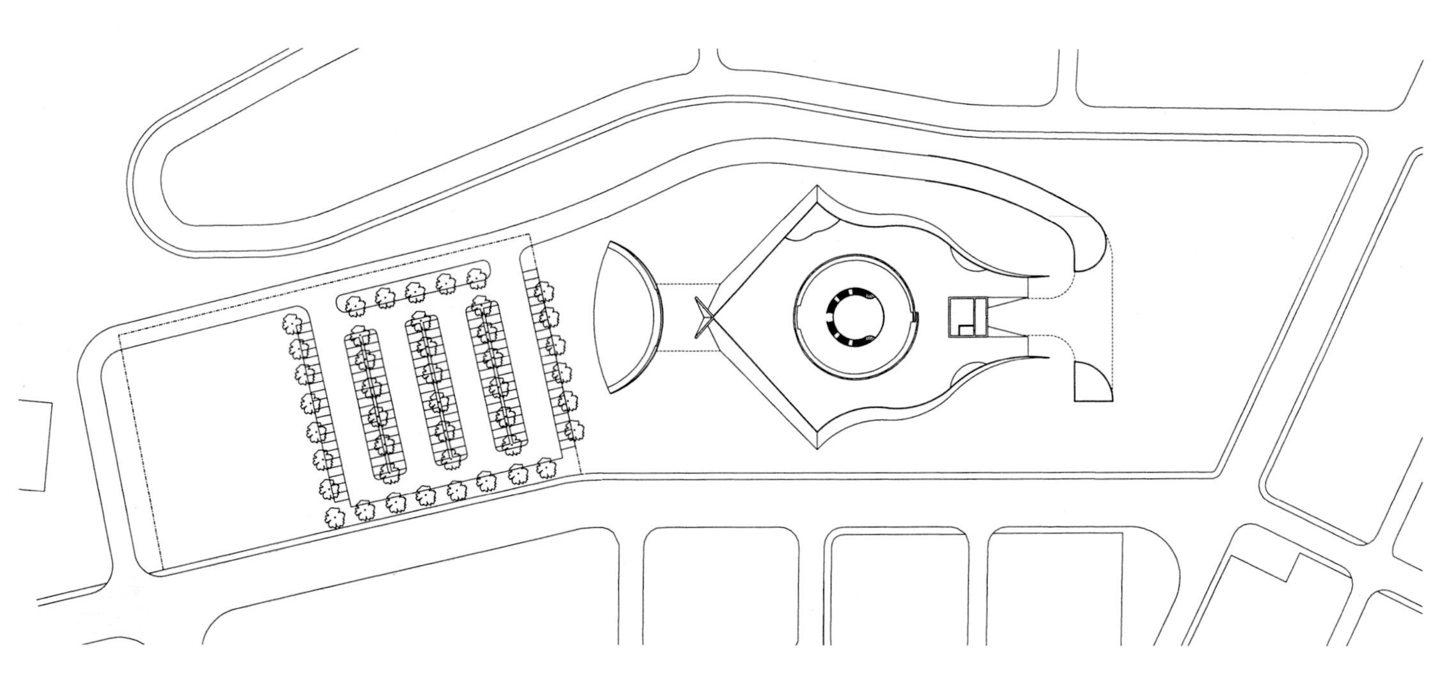

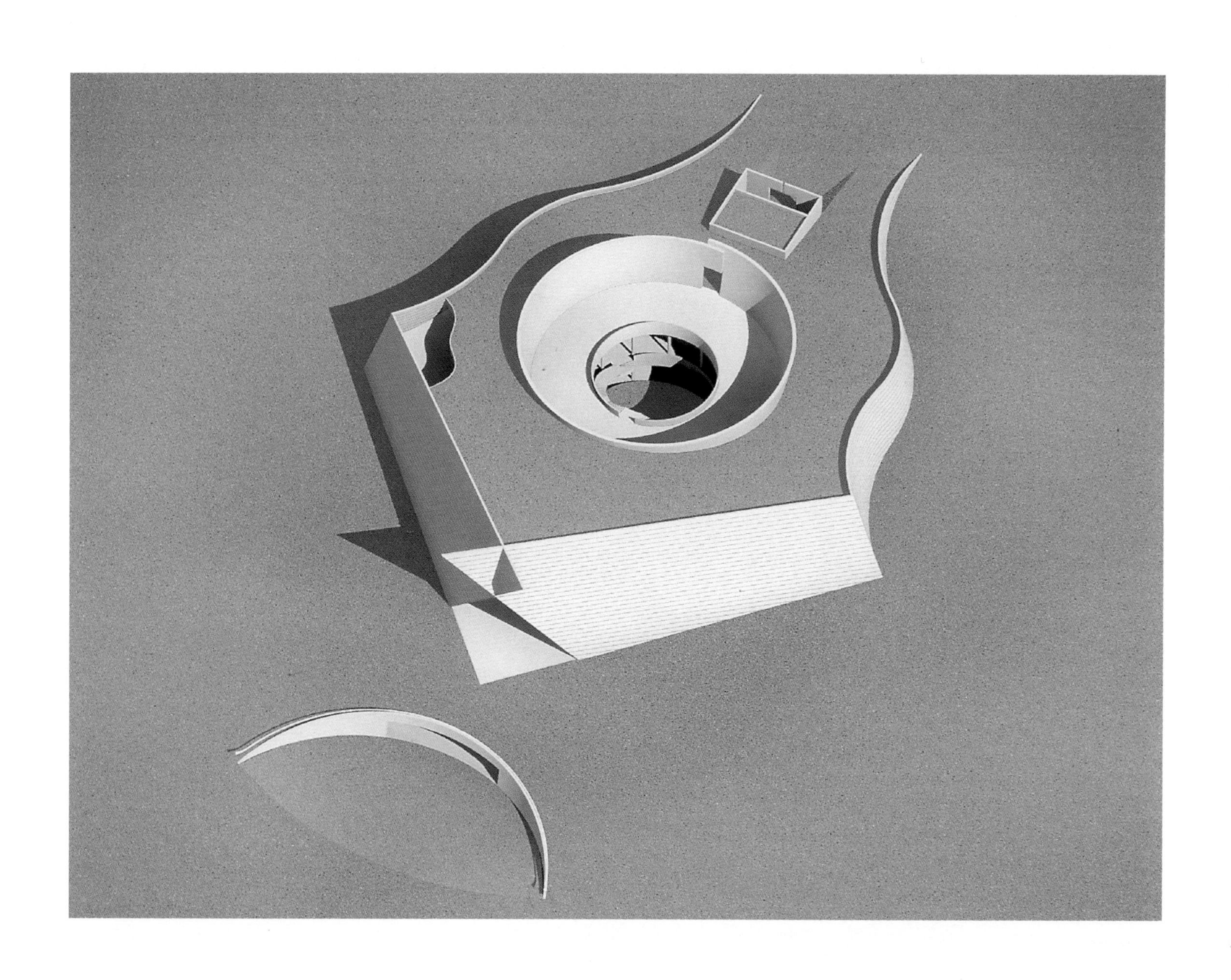

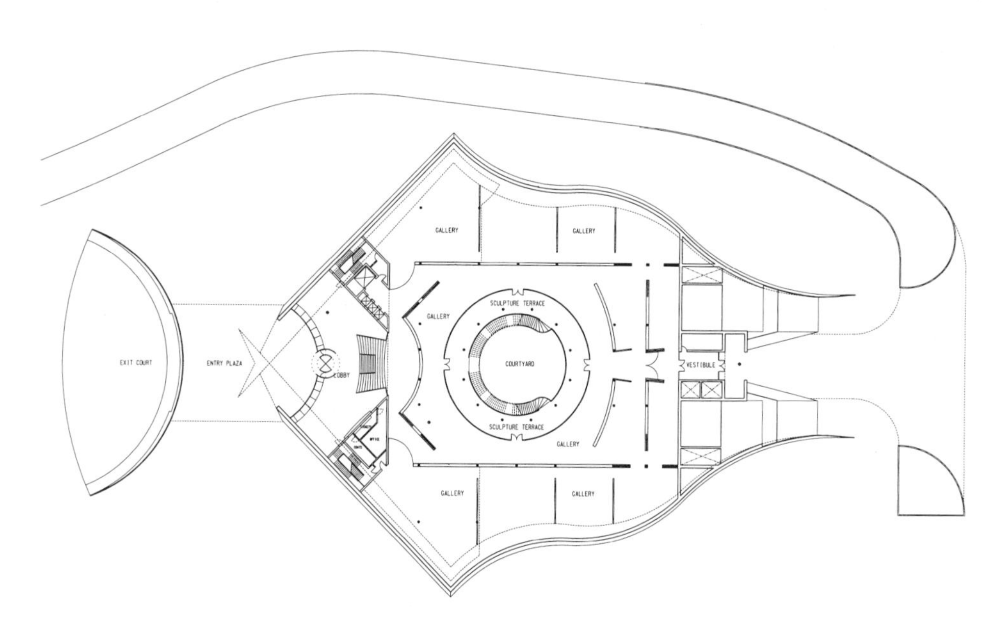
GALLERY
GALLERY
SCULPTURE TERRACE
GALLERY
EXIT COURT
ENTRY PLAZA
LOBBY
COURTYARD
VESTIBULE
SCULPTURE TERRACE
GALLERY
GALLERY
GALLERY

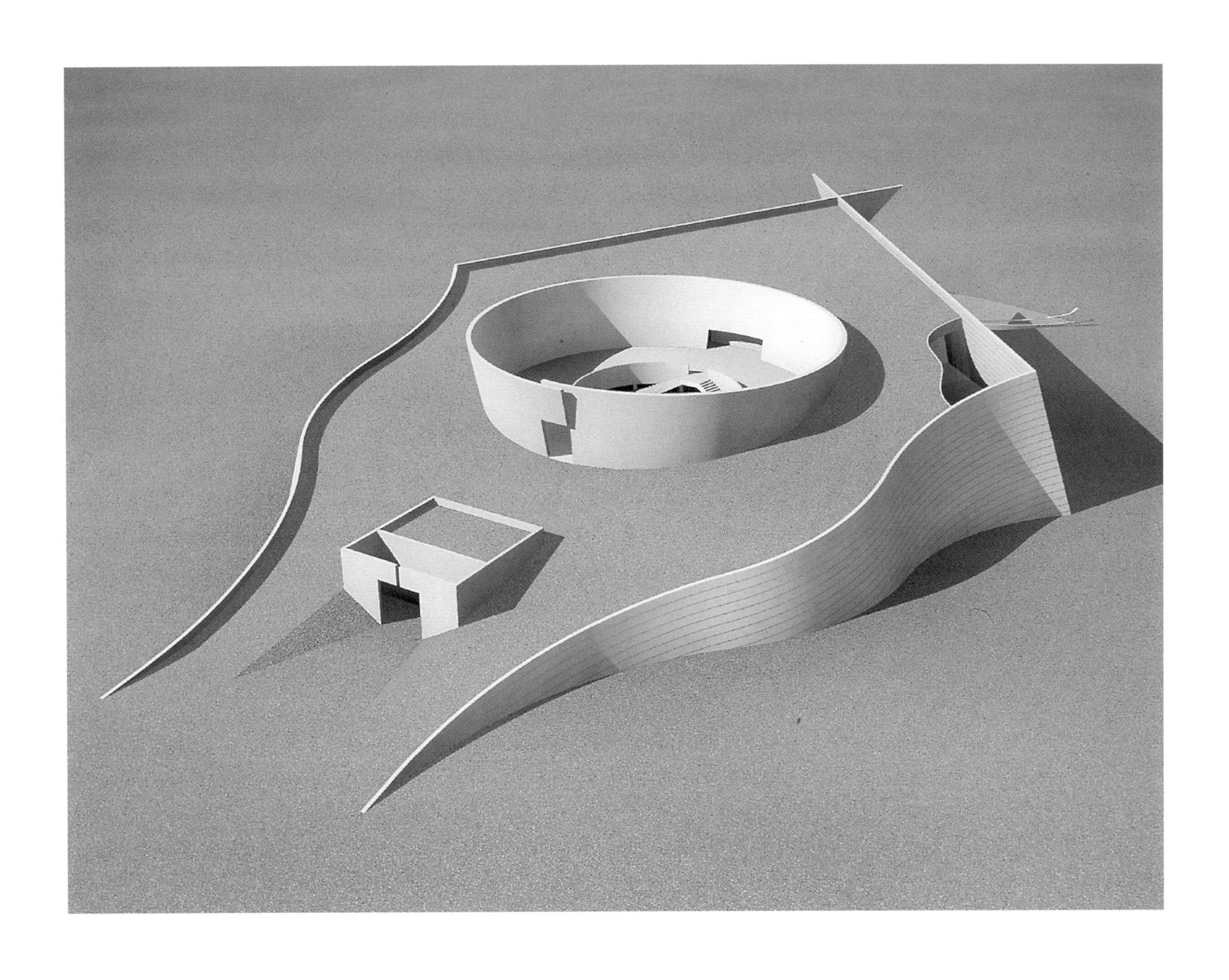

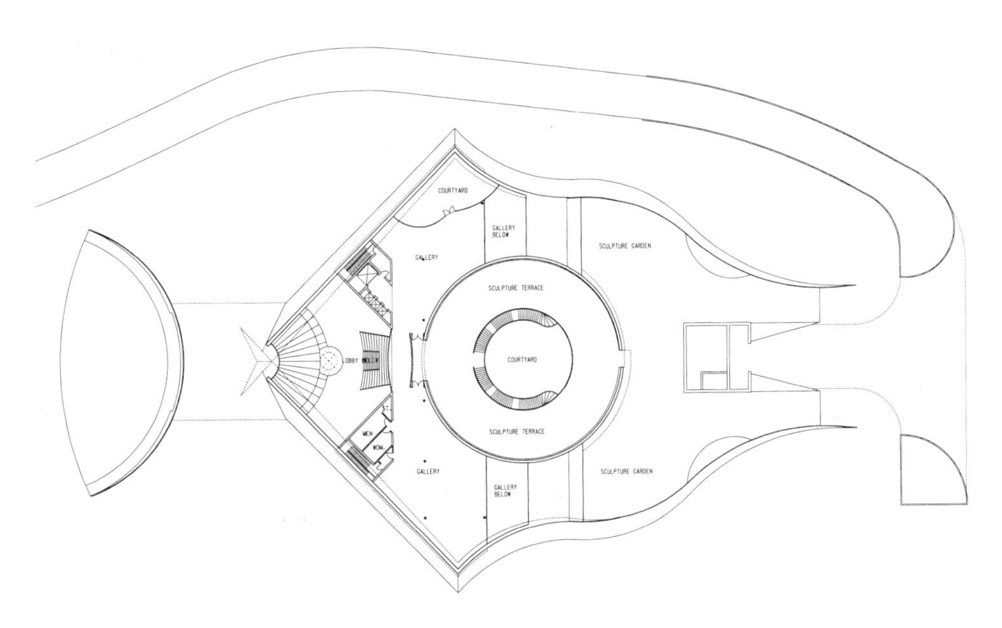
COURTYARD
GALLERY BELOW
GALLERY
SCULPTURE GARDEN
SCULPTURE TERRACE
COURTYARD
ST.
MEN
WOM.
SCULPTURE TERRACE
GALLERY
GALLERY BELOW
SCULPTURE GARDEN

Shopping Center

Amersfoort, The Netherlands

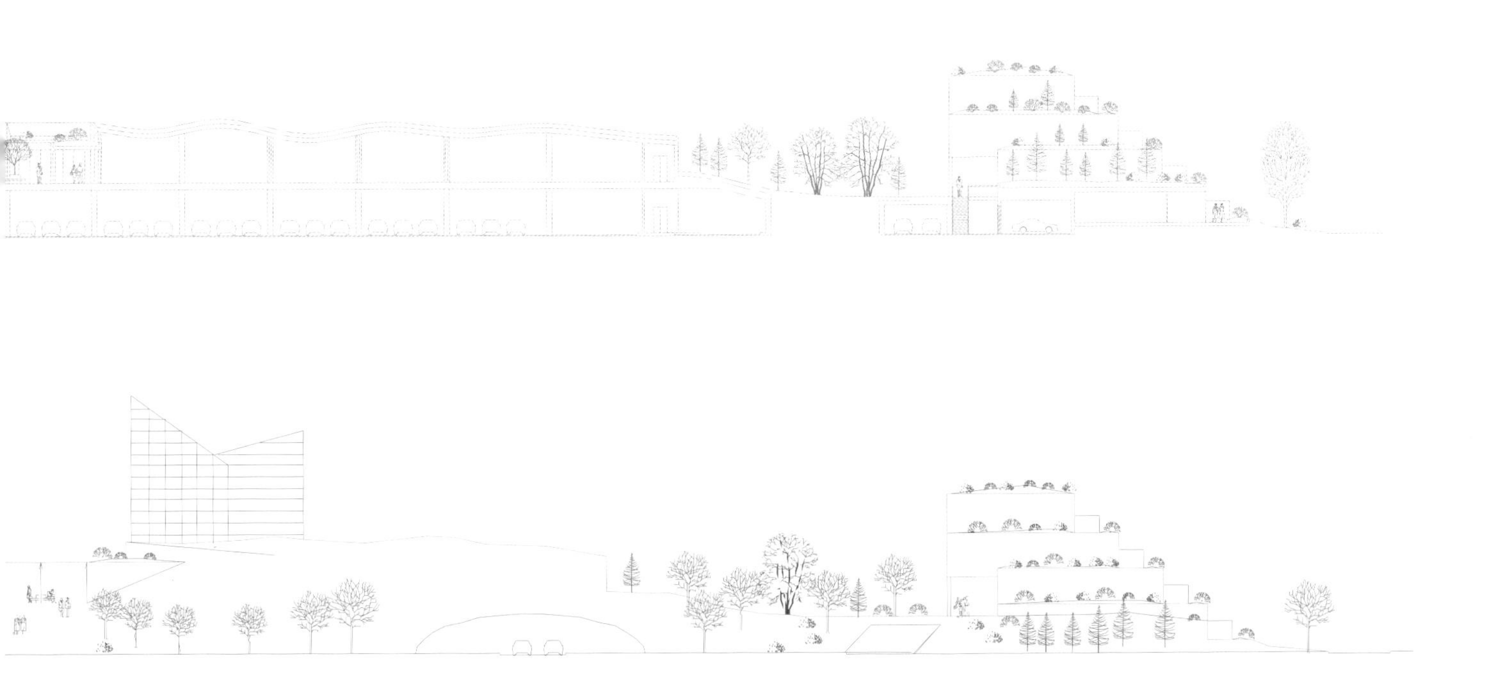

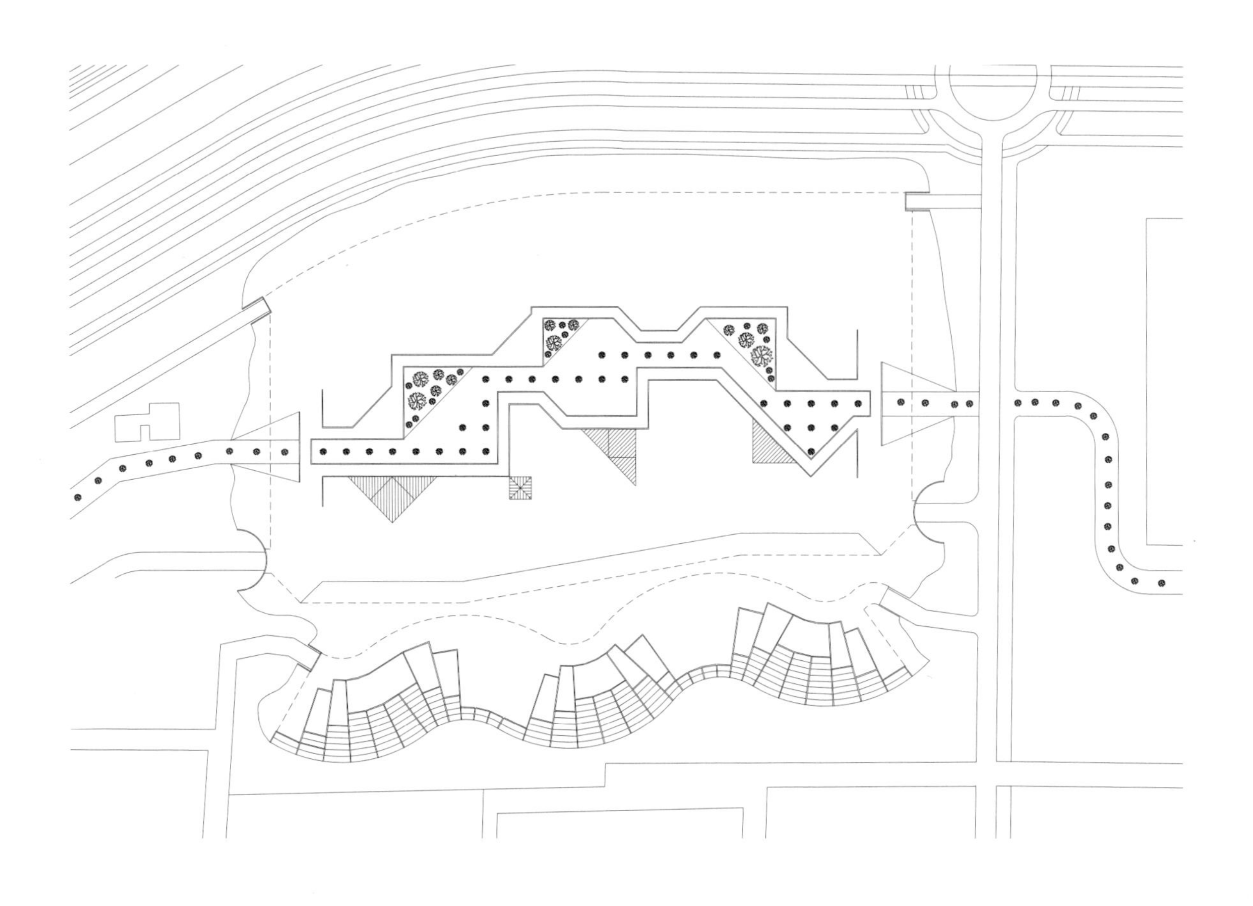

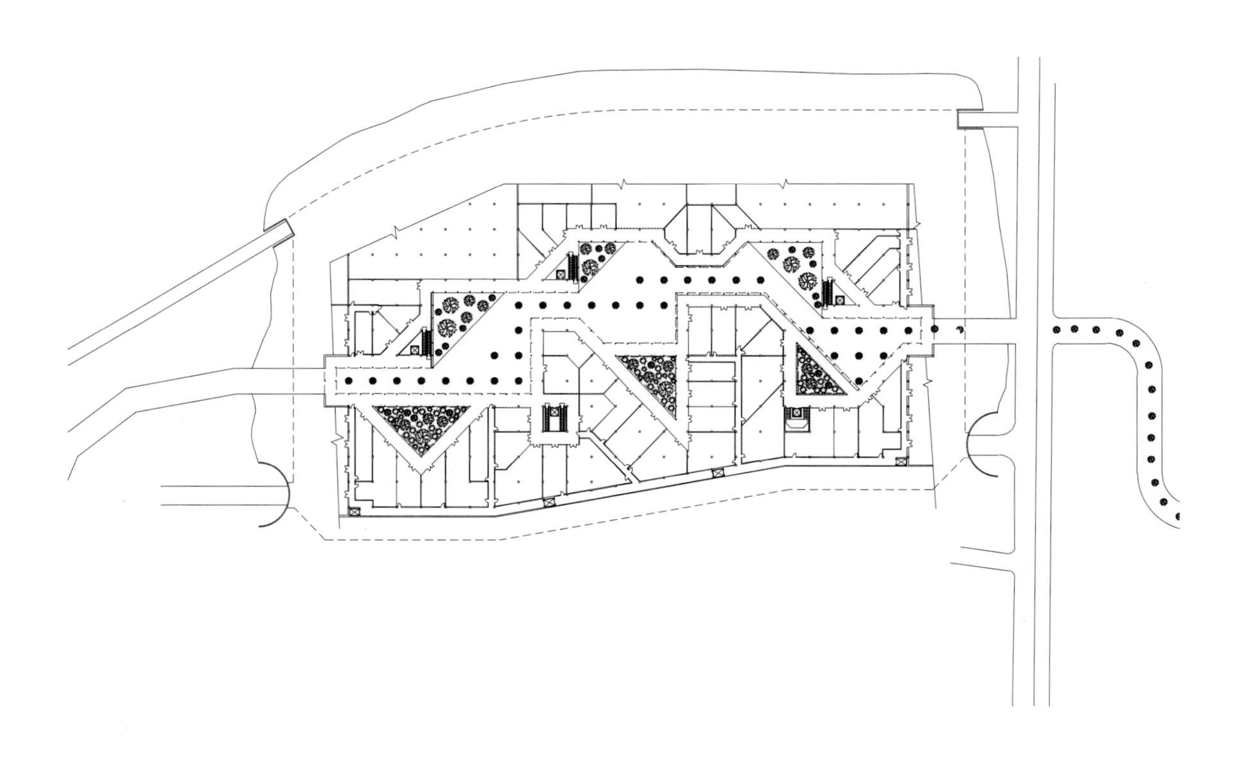

Master Plan

Barletta, Italy

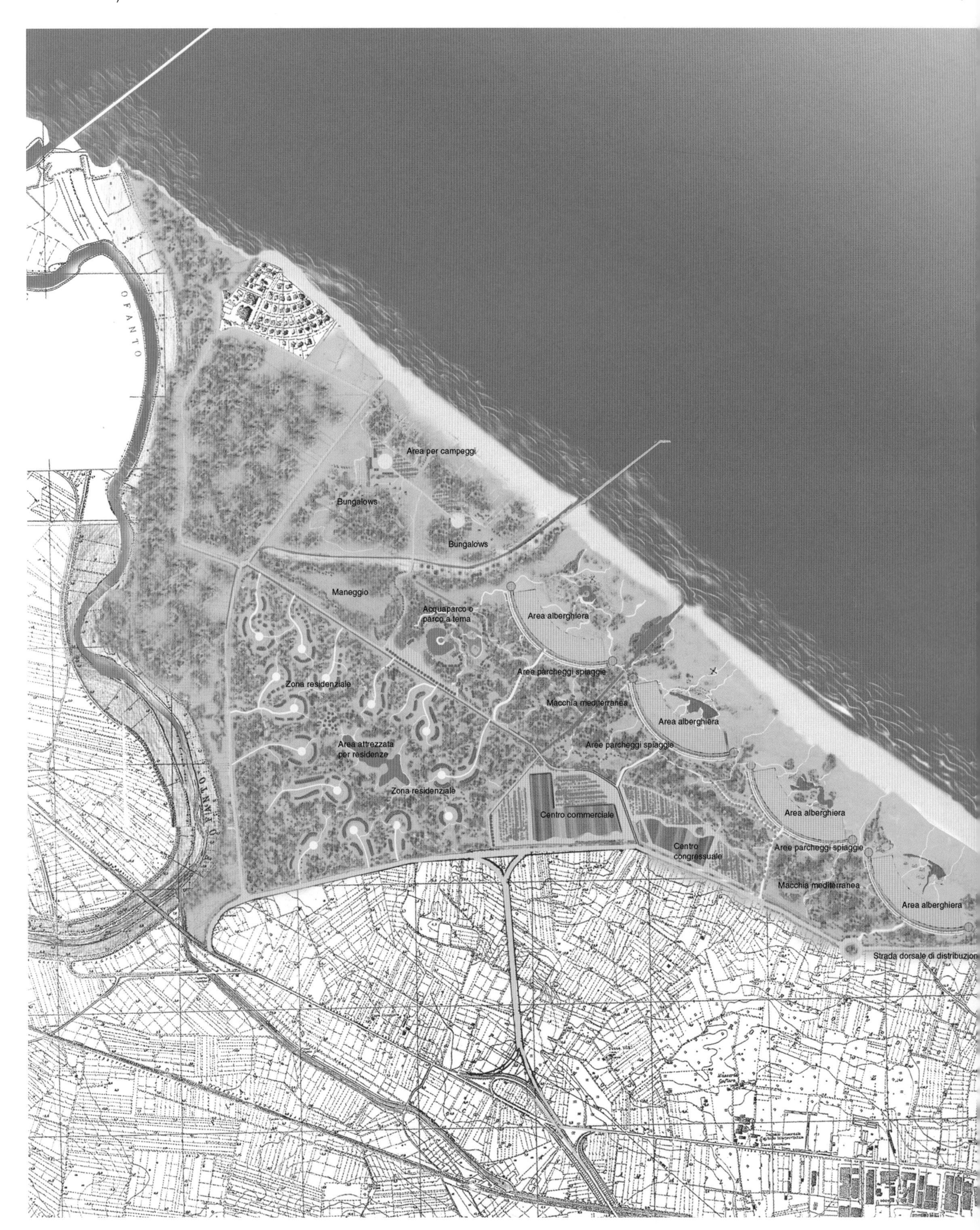

Parco urbano
Parcheggio coperto di ponente
Asse viario coperto cementeria-porto
Verde pubblico
Parcheggio coperto di levante
Area per grandi manifestazioni
Parco urbano attrezzato

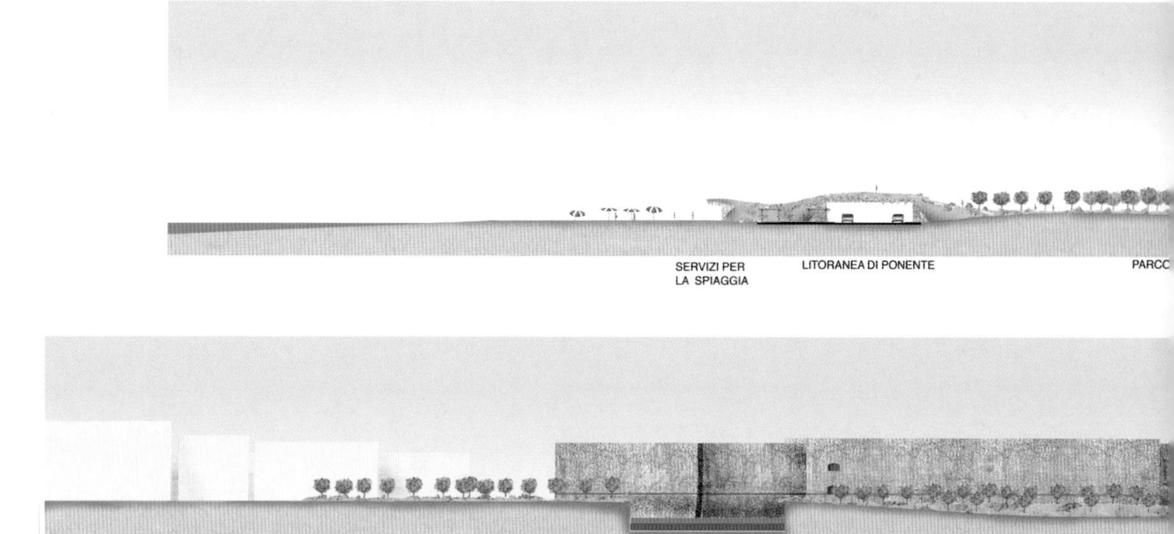
SERVIZI PER
LA SPIAGGIA
LITORANEA DI PONENTE
PIAZZA CASTELLO

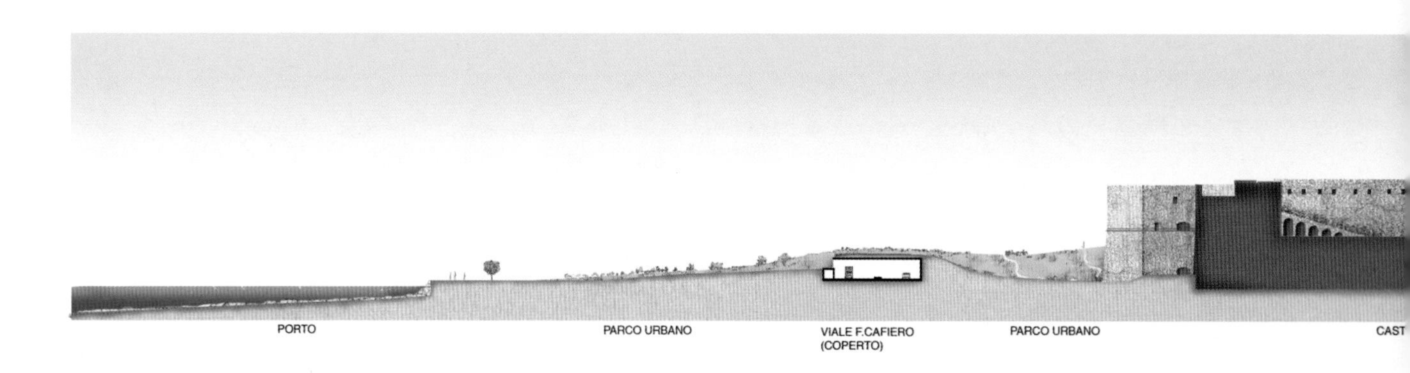
PORTO
PARCO URBANO
VIALE F.CAFIERO
(COPERTO)
PARCO URBANO

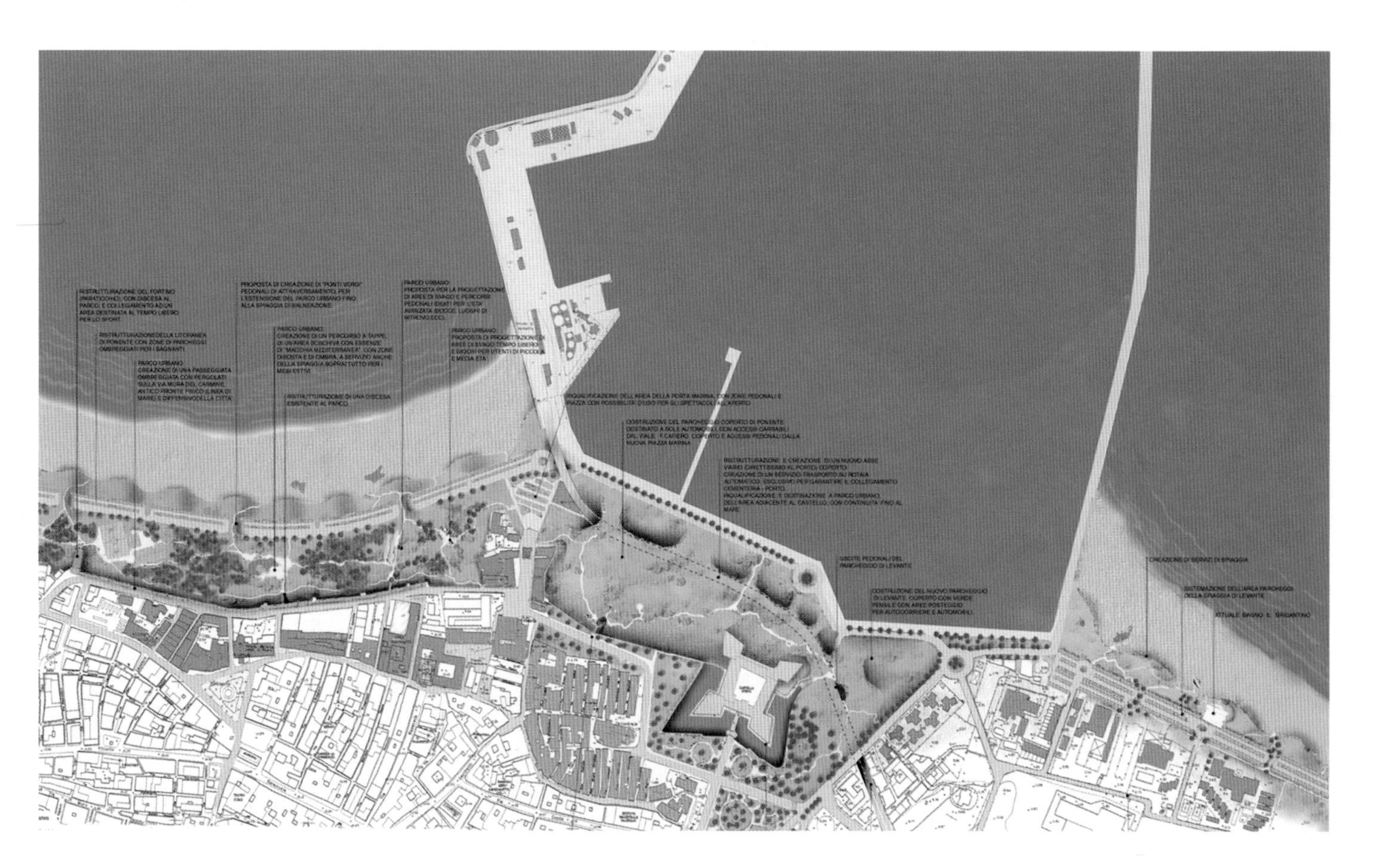

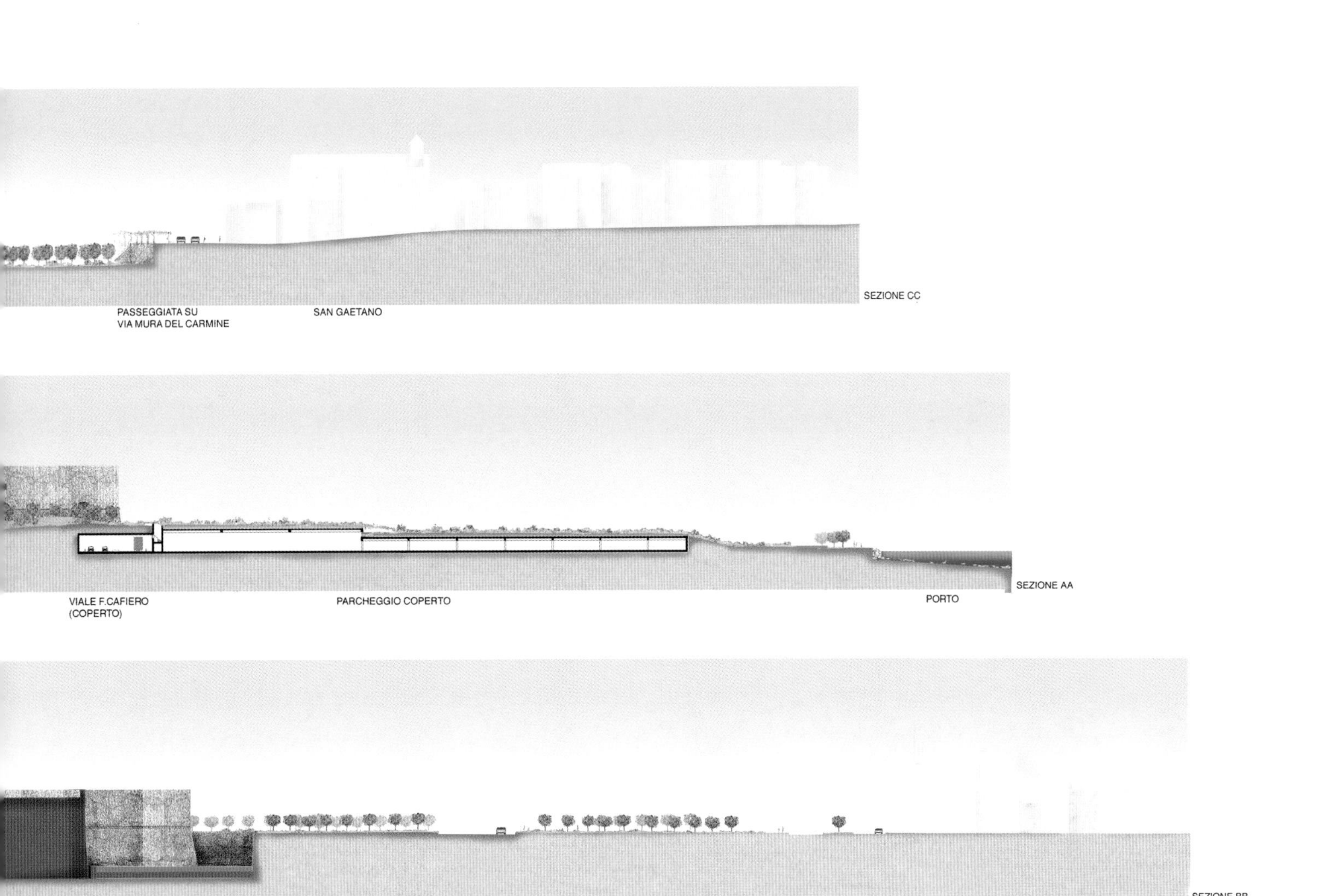
PASSEGGIATA SU
VIA MURA DEL CARMINE
SAN GAETANO
SEZIONE CC
VIALE F.CAFIERO
(COPERTO)
PARCHEGGIO COPERTO
PORTO
SEZIONE AA
PIAZZA CASTELLO
SEZIONE BB

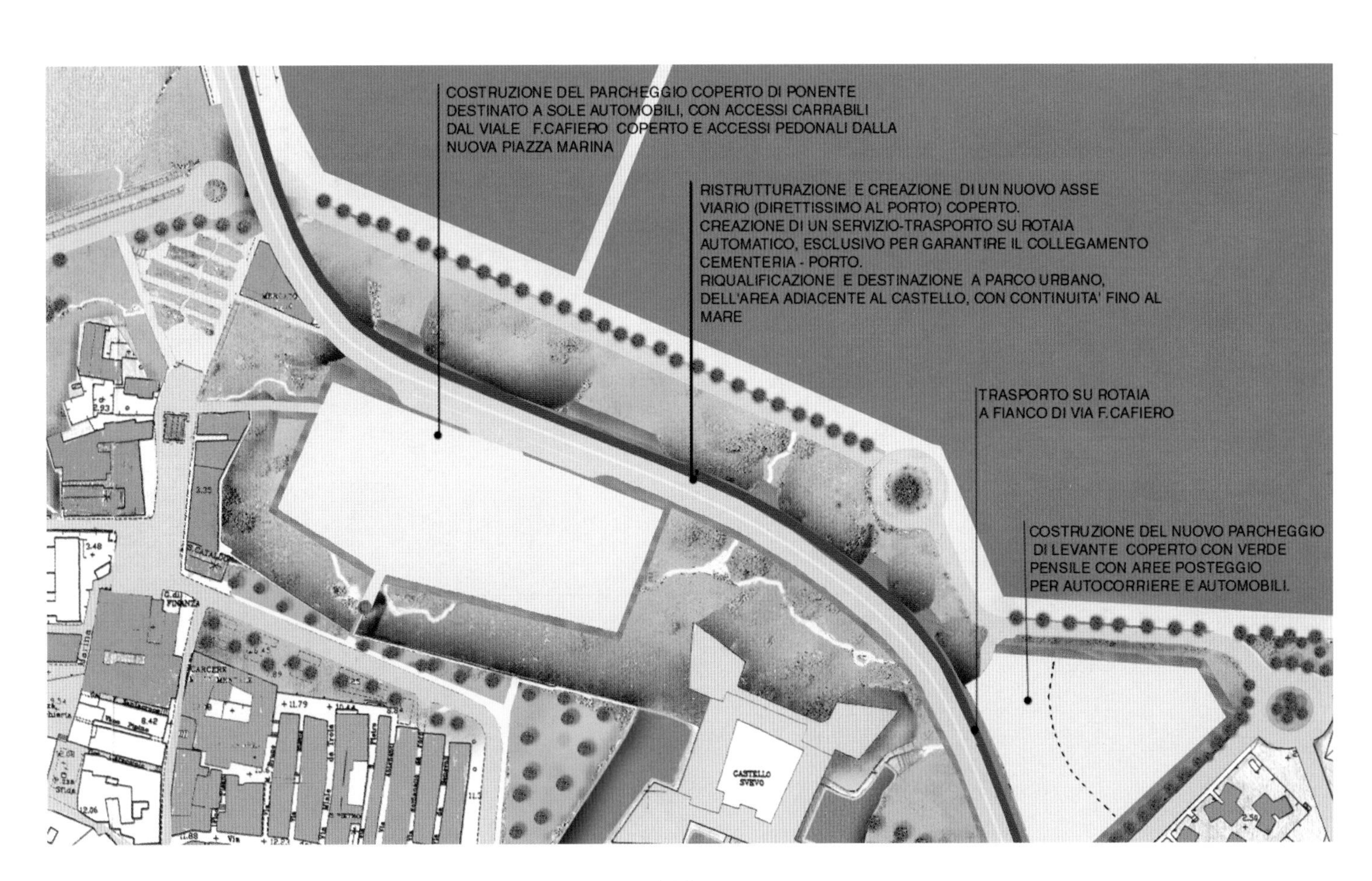
COSTRUZIONE DEL PARCHEGGIO COPERTO DI PONENTE
DESTINATO A SOLE AUTOMOBILI, CON ACCESSI CARRABILI
DAL VIALE F.CAFIERO COPERTO E ACCESSI PEDONALI DALLA
NUOVA PIAZZA MARINA
RISTRUTTURAZIONE E CREAZIONE DI UN NUOVO ASSE
VIARIO (DIRETTISSIMO AL PORTO) COPERTO.
CREAZIONE DI UN SERVIZIO-TRASPORTO SU ROTAIA
AUTOMATICO, ESCLUSIVO PER GARANTIRE IL COLLEGAMENTO
CEMENTERIA - PORTO.
RIQUALIFICAZIONE E DESTINAZIONE A PARCO URBANO,
DELL'AREA ADIACENTE AL CASTELLO, CON CONTINUITA' FINO AL
MARE
TRASPORTO SU ROTAIA
A FIANCO DI VIA F.CAFIERO
COSTRUZIONE DEL NUOVO PARCHEGGIO
DI LEVANTE COPERTO CON VERDE
PENSILE CON AREE POSTEGGIO
PER AUTOCORRIERE E AUTOMOBILI.
CASTELLO
SVEVO

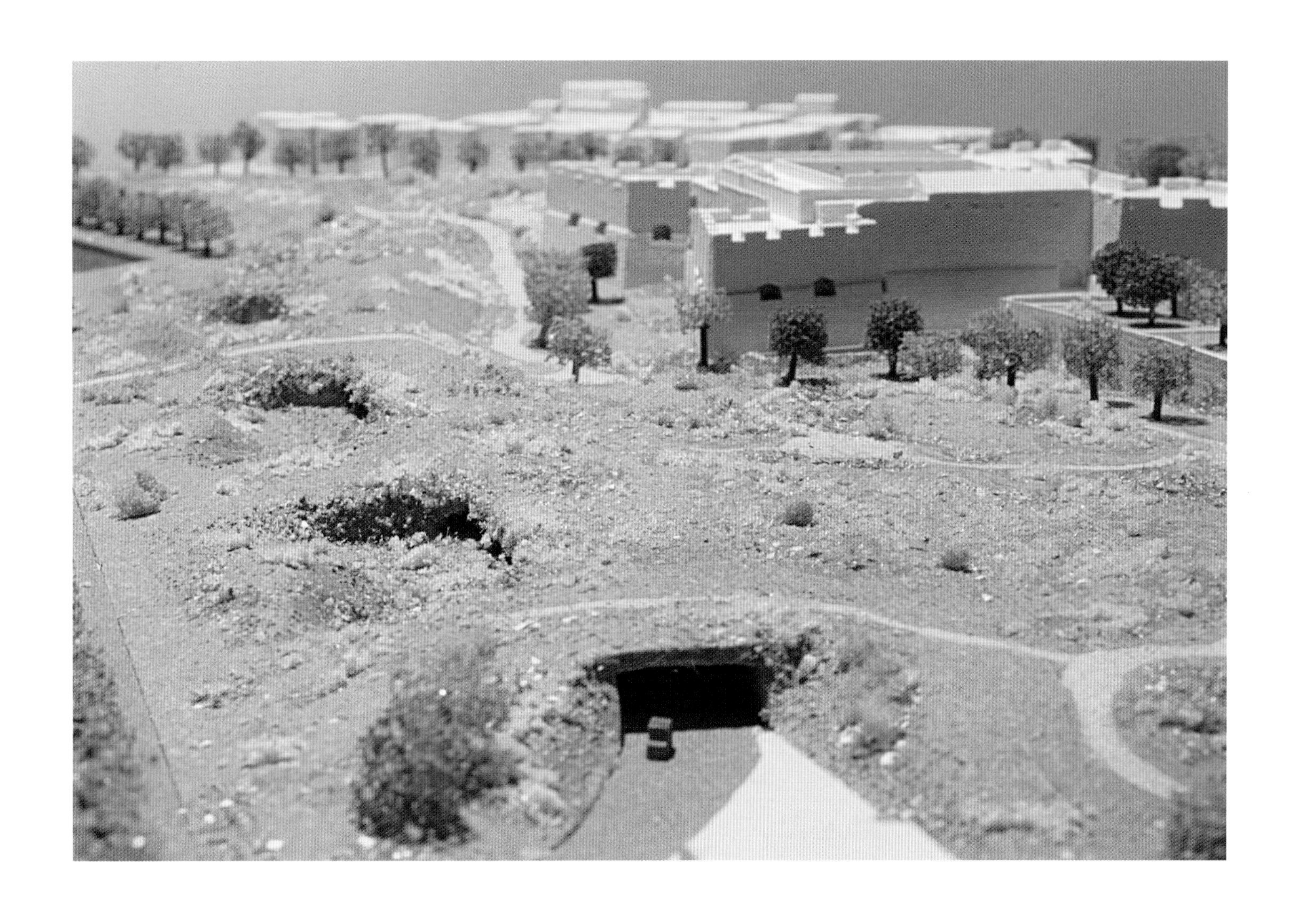

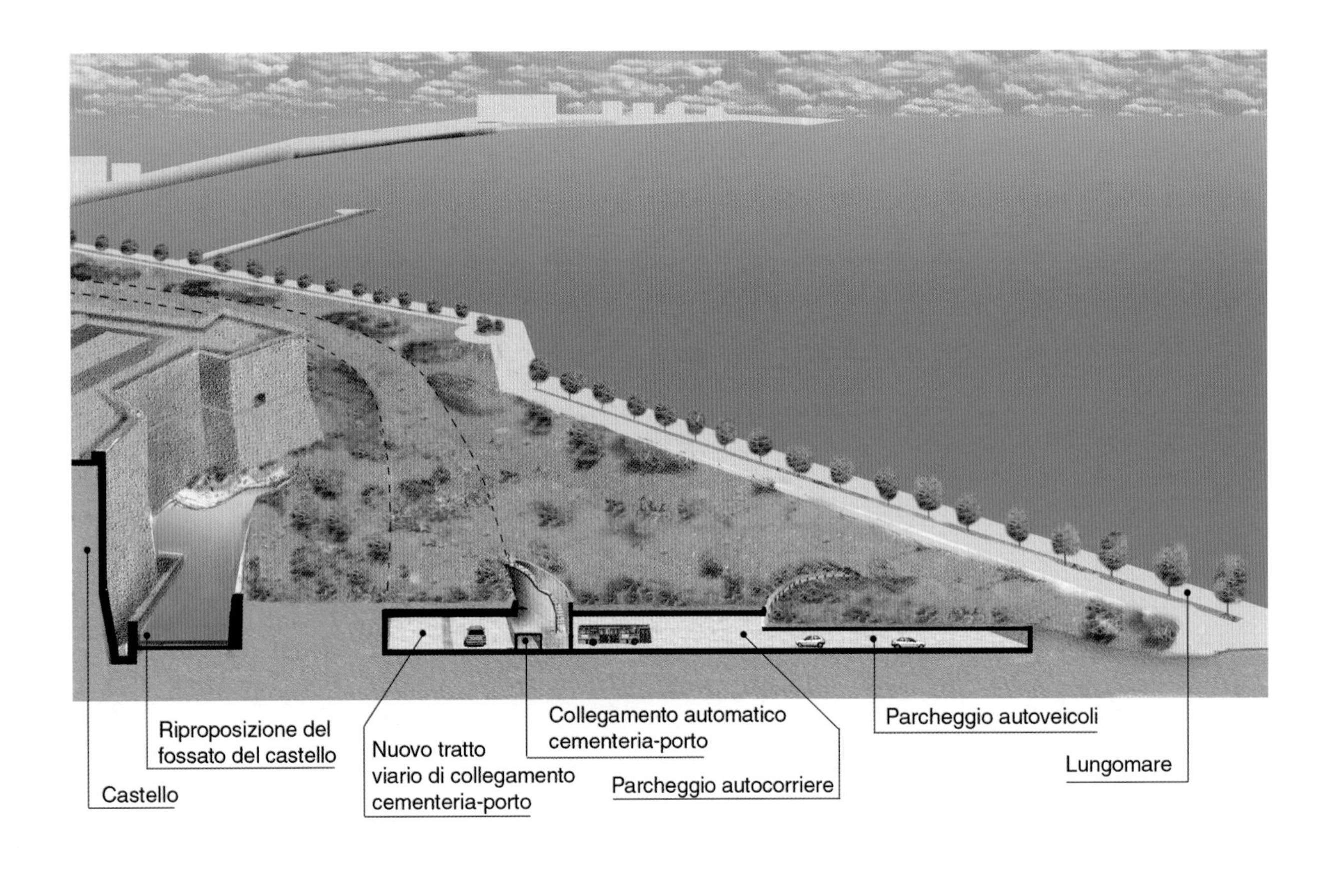

Cavalcavia pedonali attrezzati a verde colleganti il parco urbano e la spiaggia di ponente
Parcheggi ombreggiati lungo la litoranea di ponente

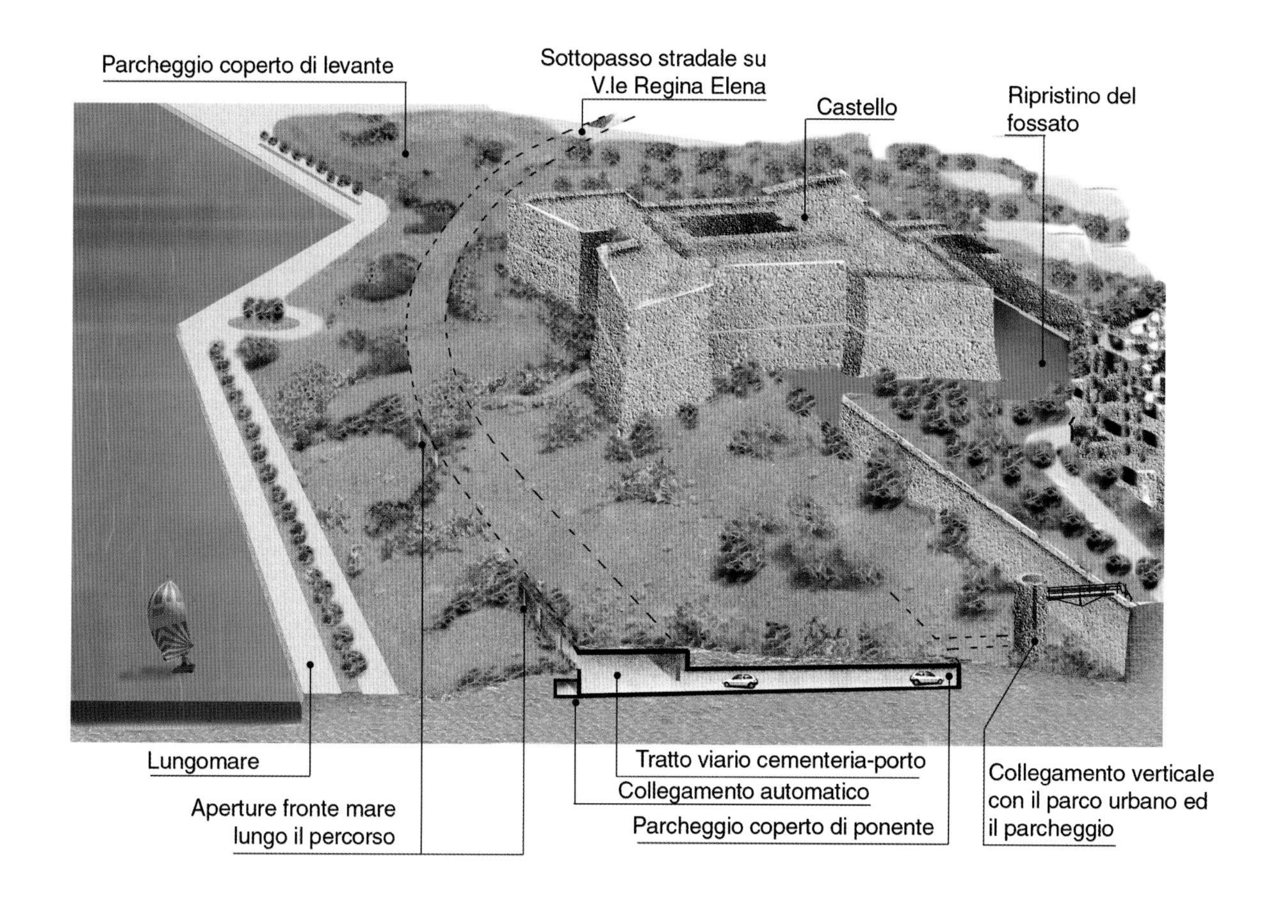

Servizi di spiaggia

Public Park and Residences

Old Port, Monte Carlo

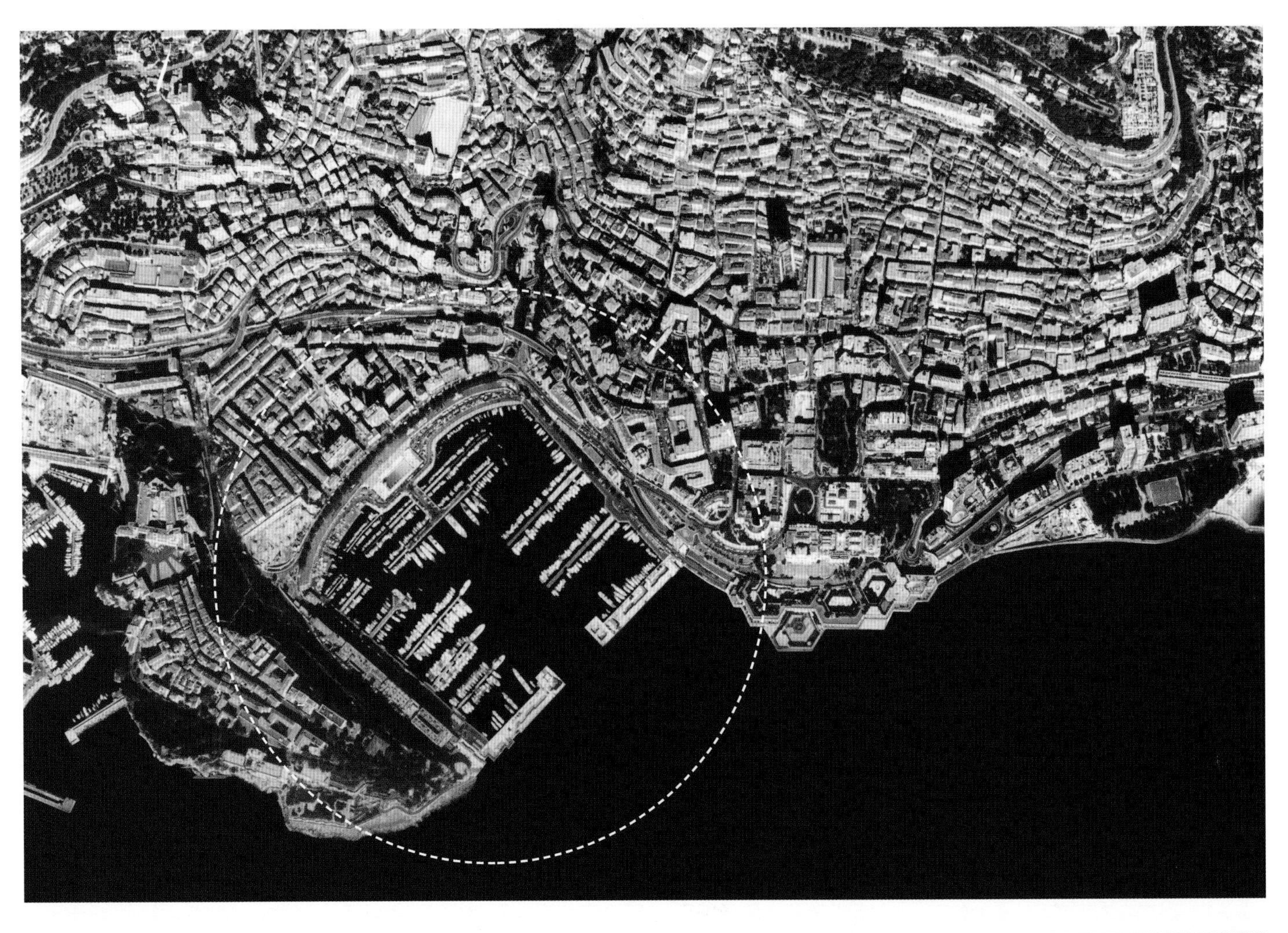

Winnisook Lodge

Catskill Mountains, New York

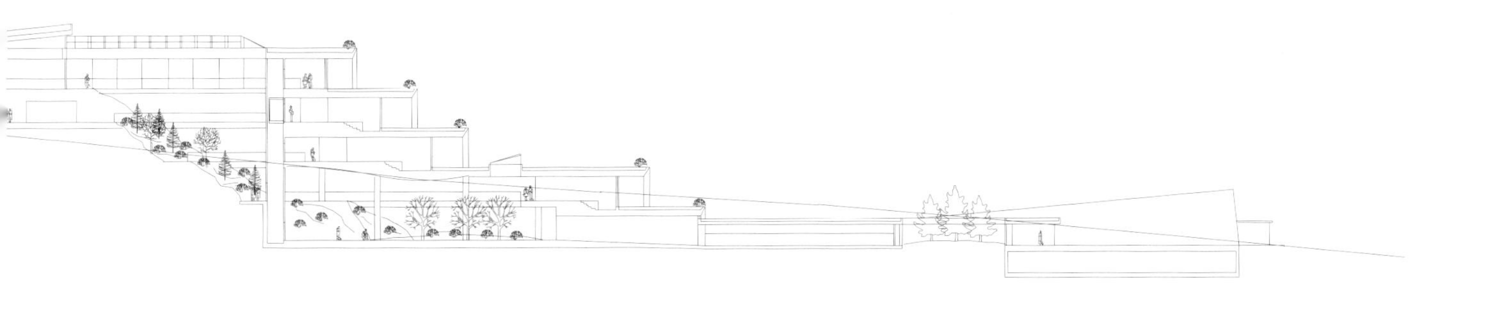

Monument Tower Offices

Phoenix, Arizona

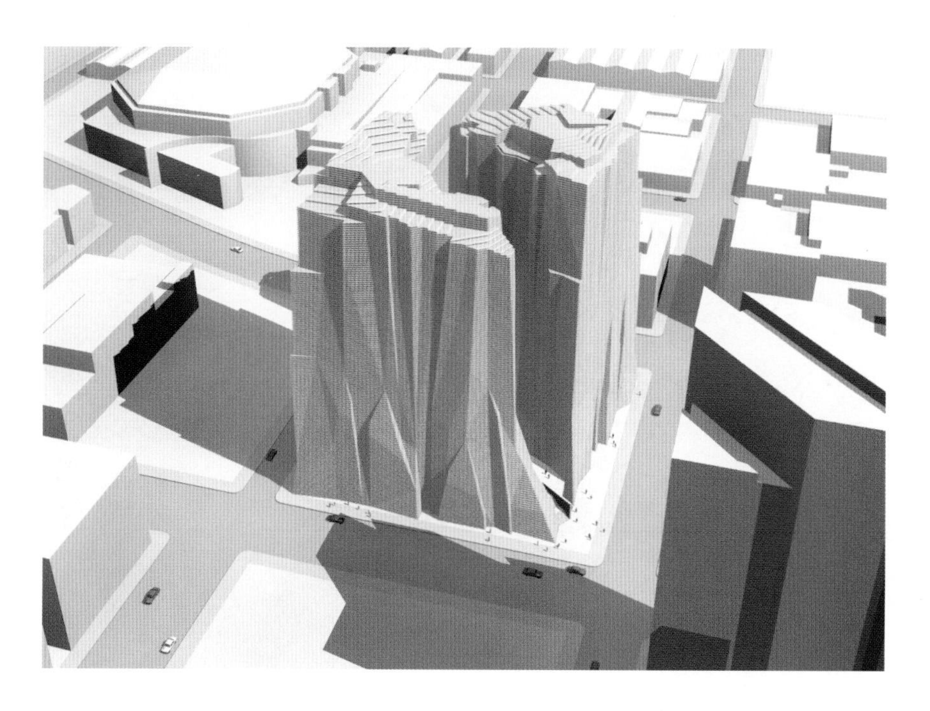

Industrial & Graphic Design Projects

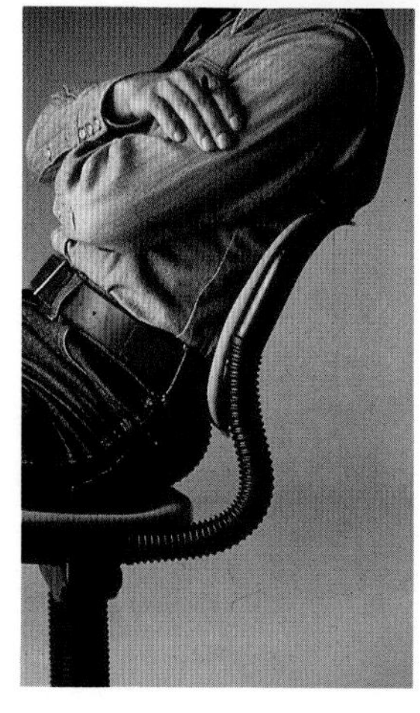

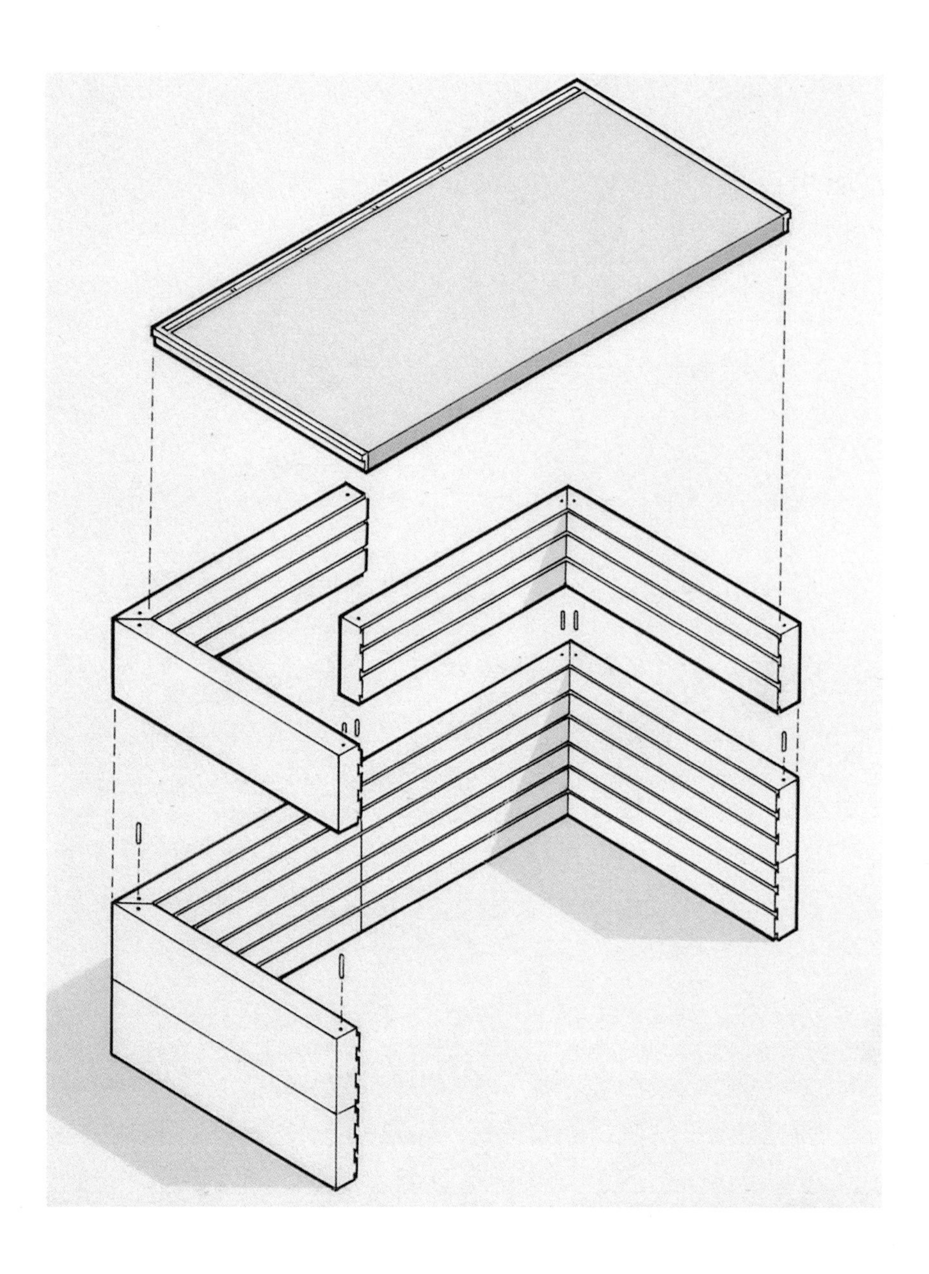

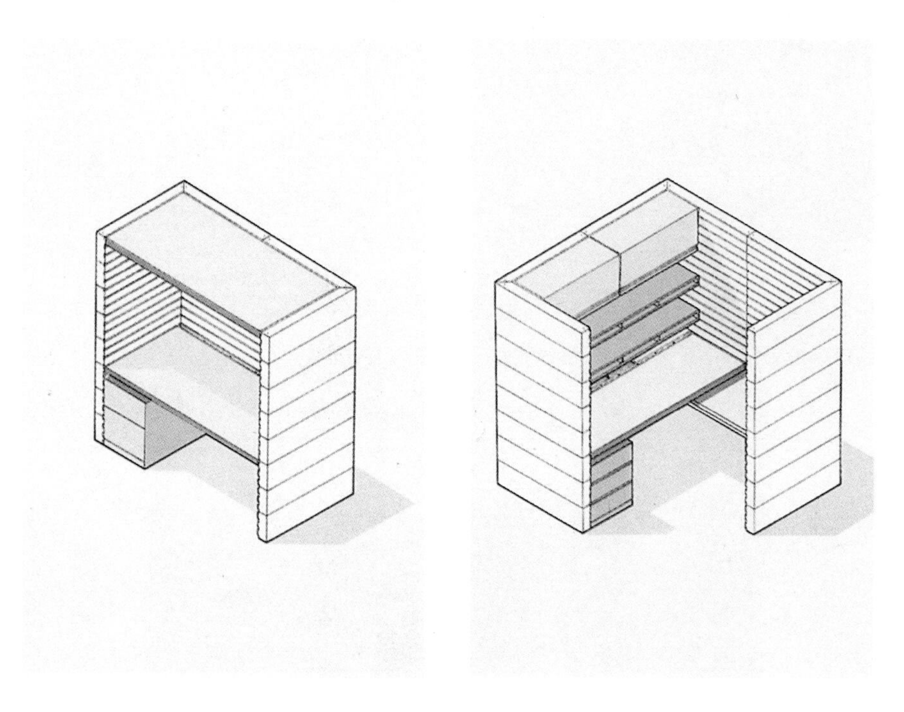

ERCO
75
60
45
30

ERCO

ERCO

ERCO

ERCO

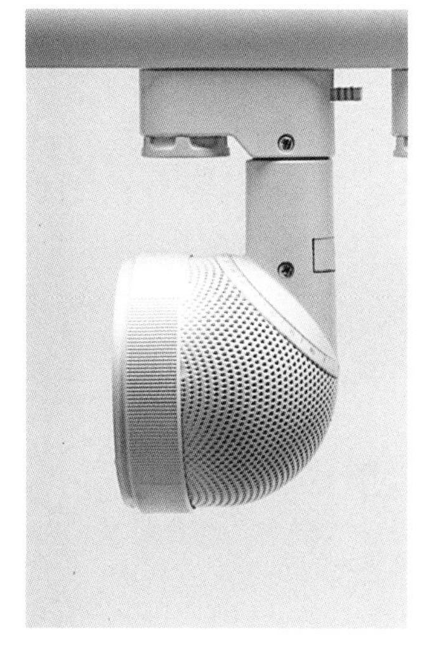

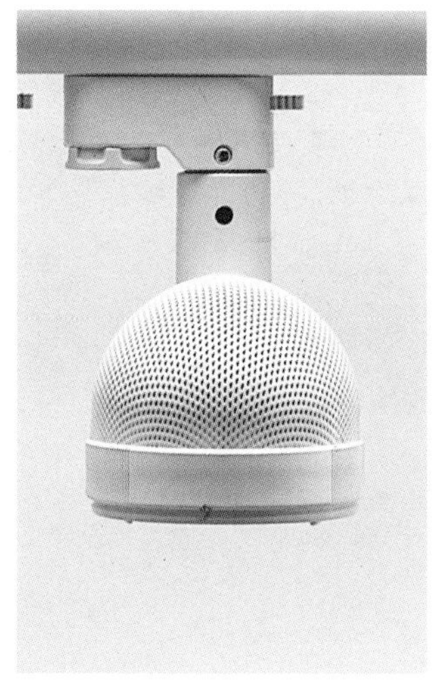

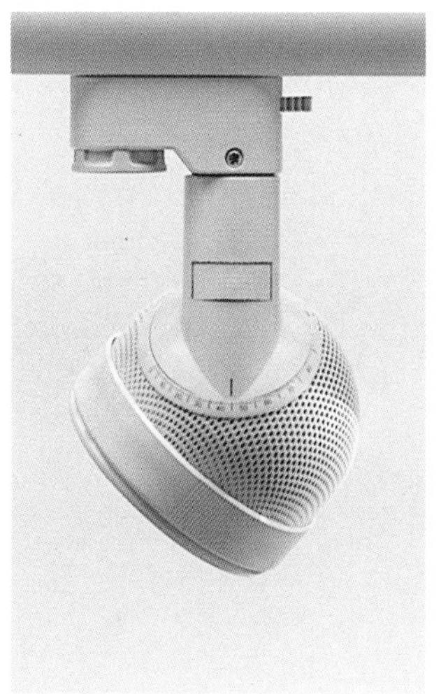

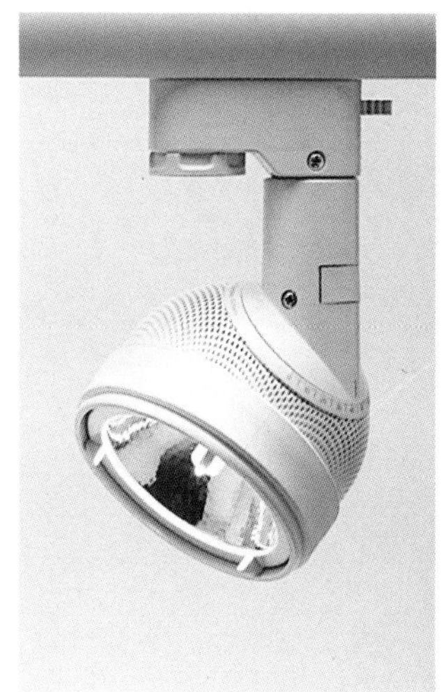

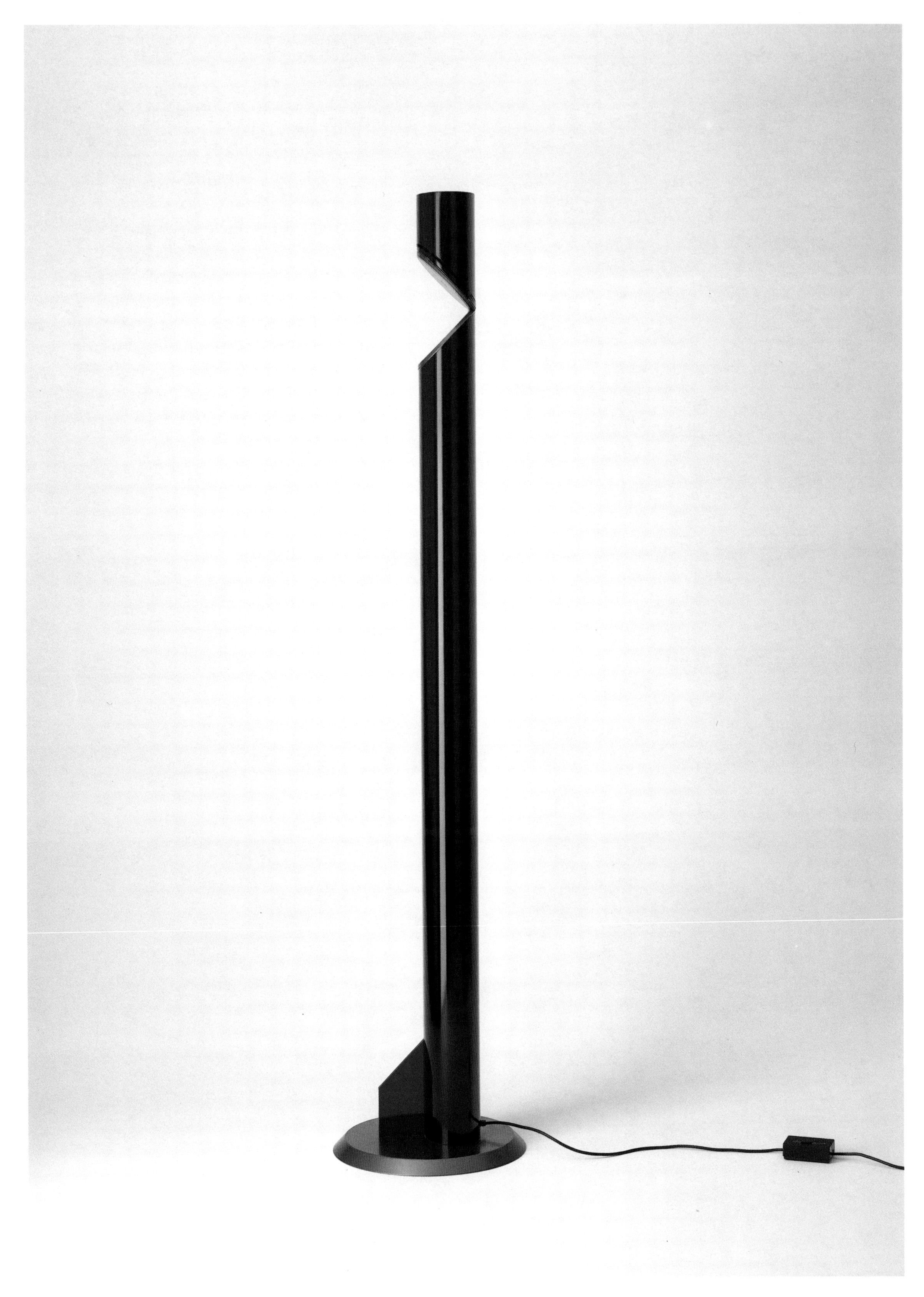

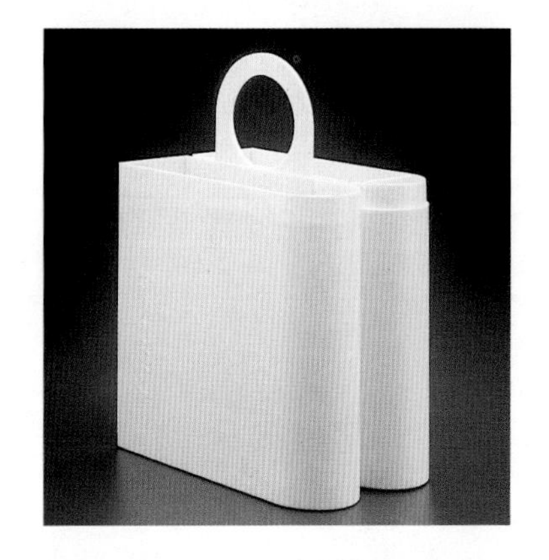

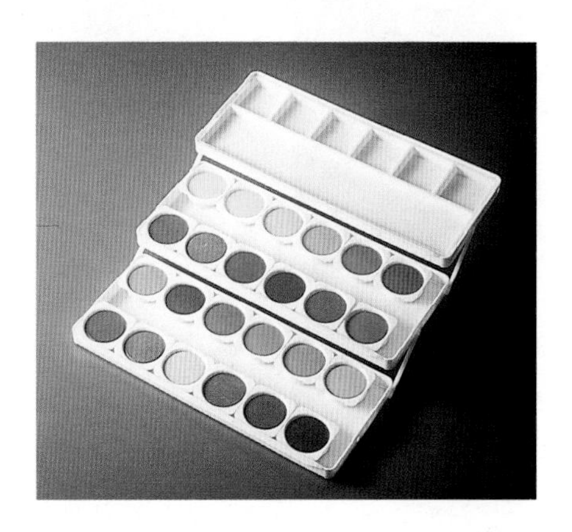

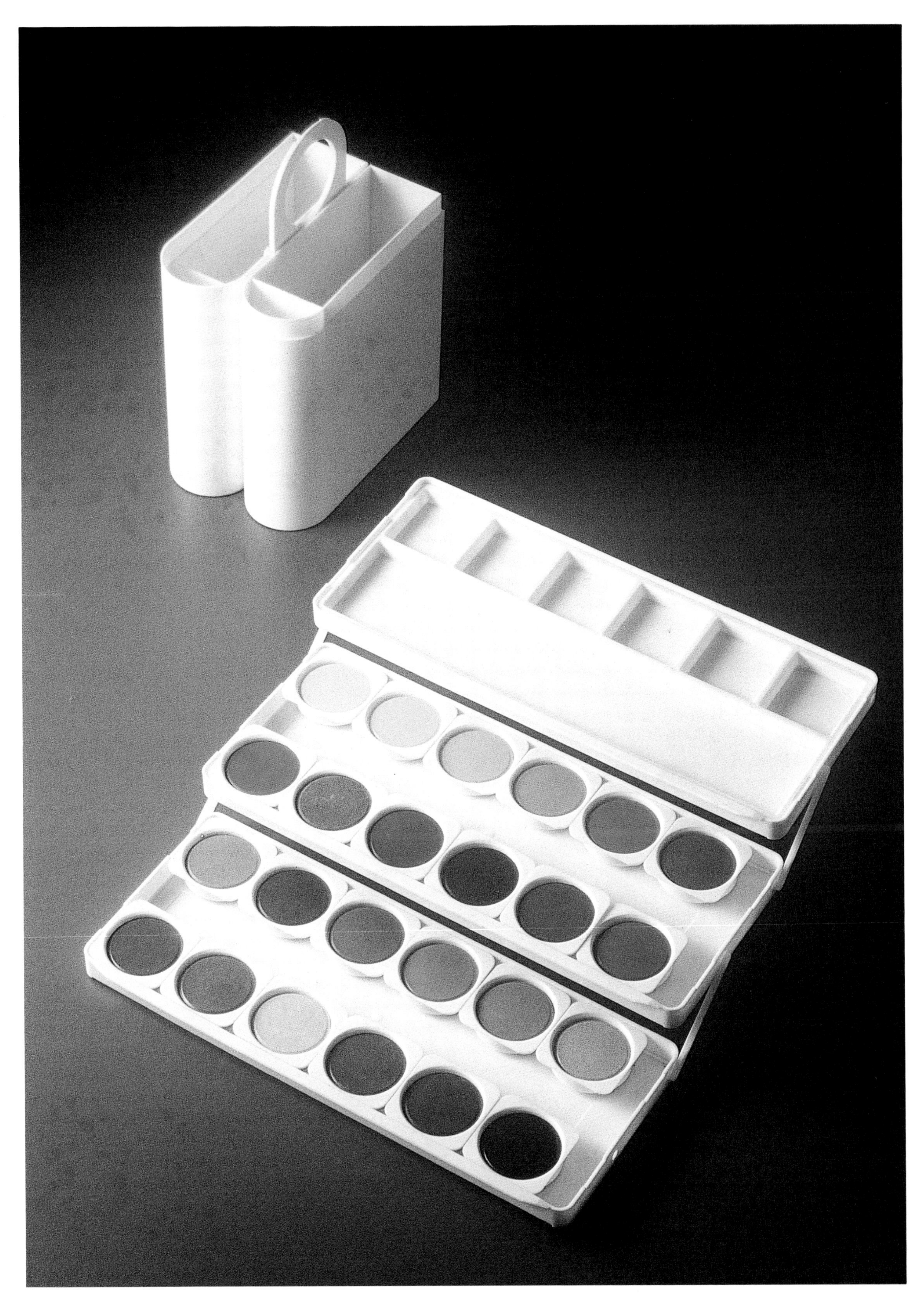

9
78
d l m m g v s
1 2
3 4 5 6 7 8 9
10 11 12 13 14 15 16
17 18 19 20 21 22 23
24 25 26 27 28 29 30

12/74
11/74
1/74
s m t w t f s
1 2 3 4 5
6 7 8 9 10 11 12
13 14 15 16 17 18 19
20 21 22 23 24 25 26
27 28 29 30 31
Calendar/ Memory - Bag:
To keep those days one wants to remember, and dispose of thos
to forget; to collect newspaper clippings, snapshots, leaves, locks,

Desk Set

15
Saturday
August
1992
Sunday
August
1992
16
Monday, August 17

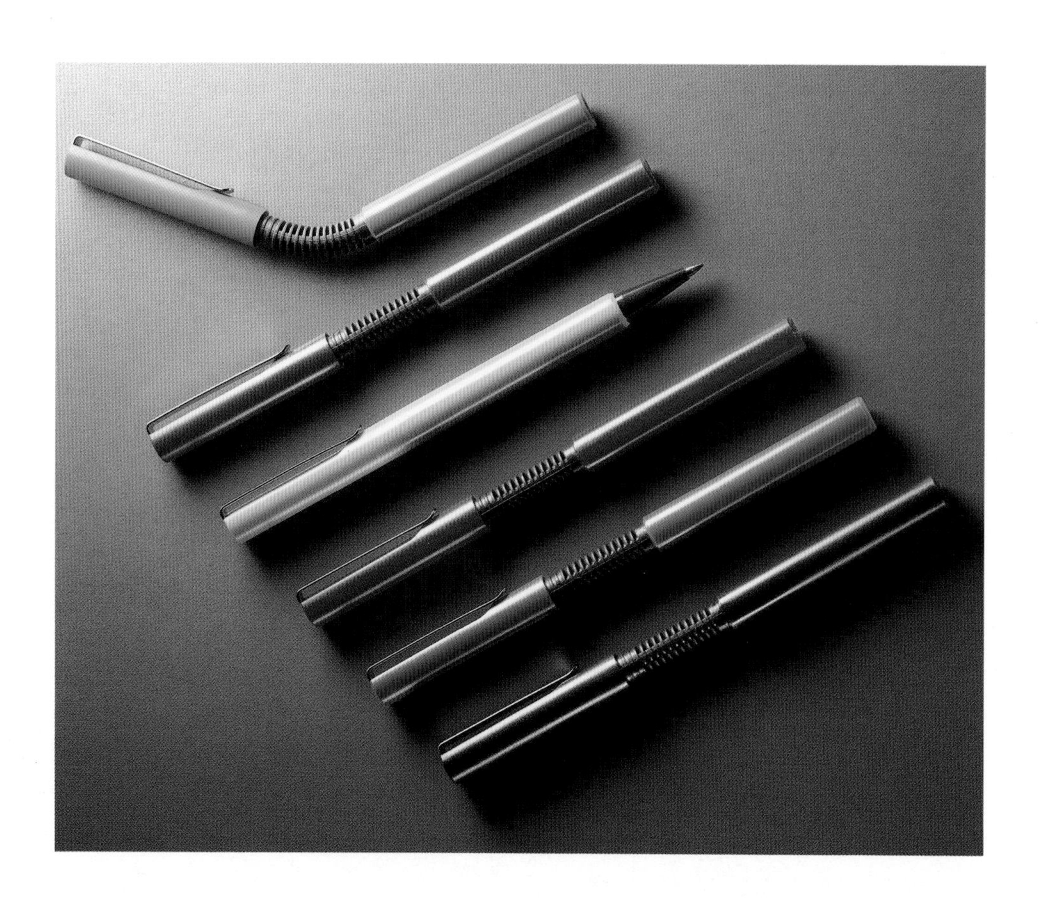

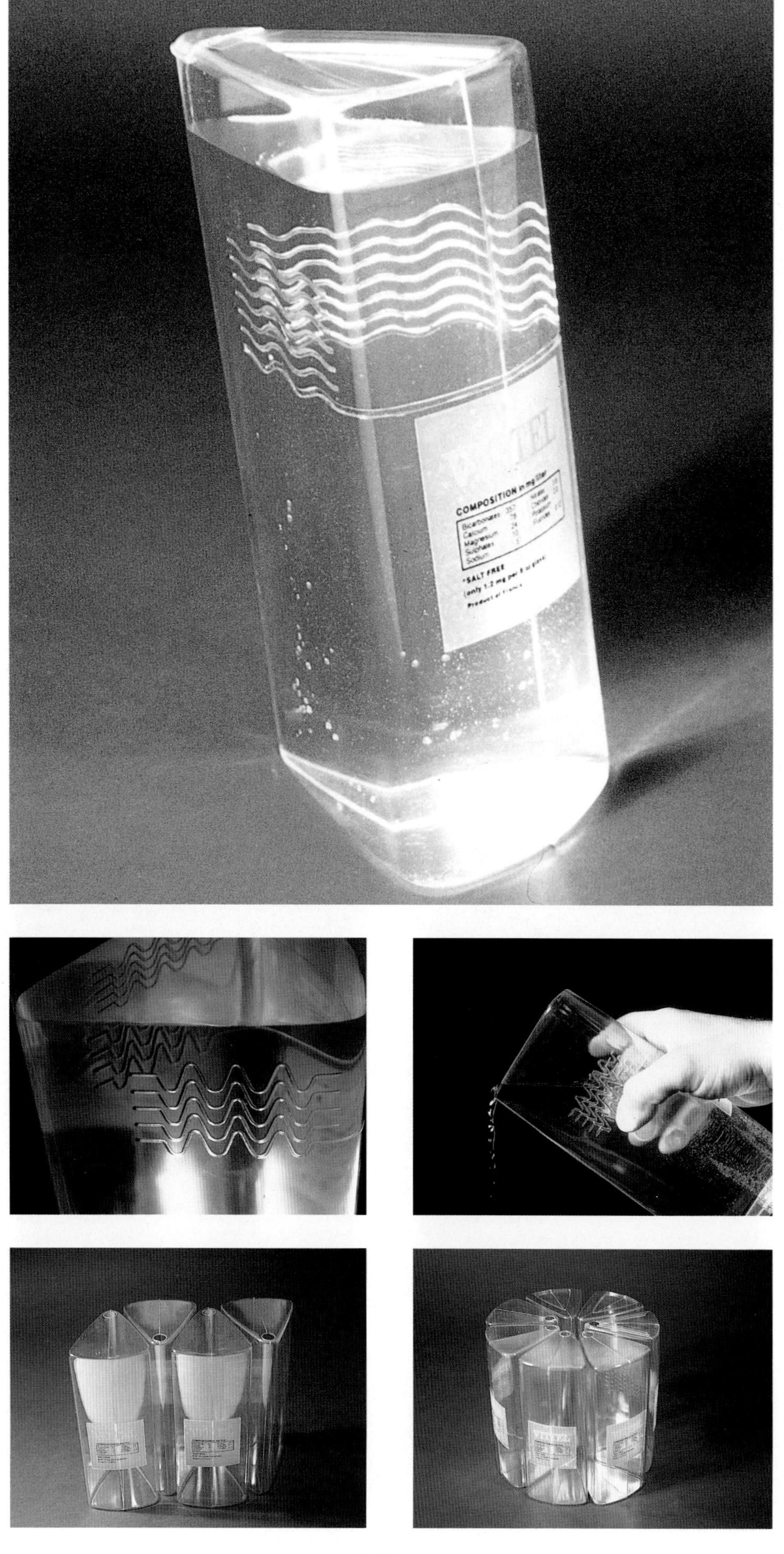
COMPOSITION
Bicarbonates
Calcium
Magnesium
Sulphates
Sodium
*SALT FREE

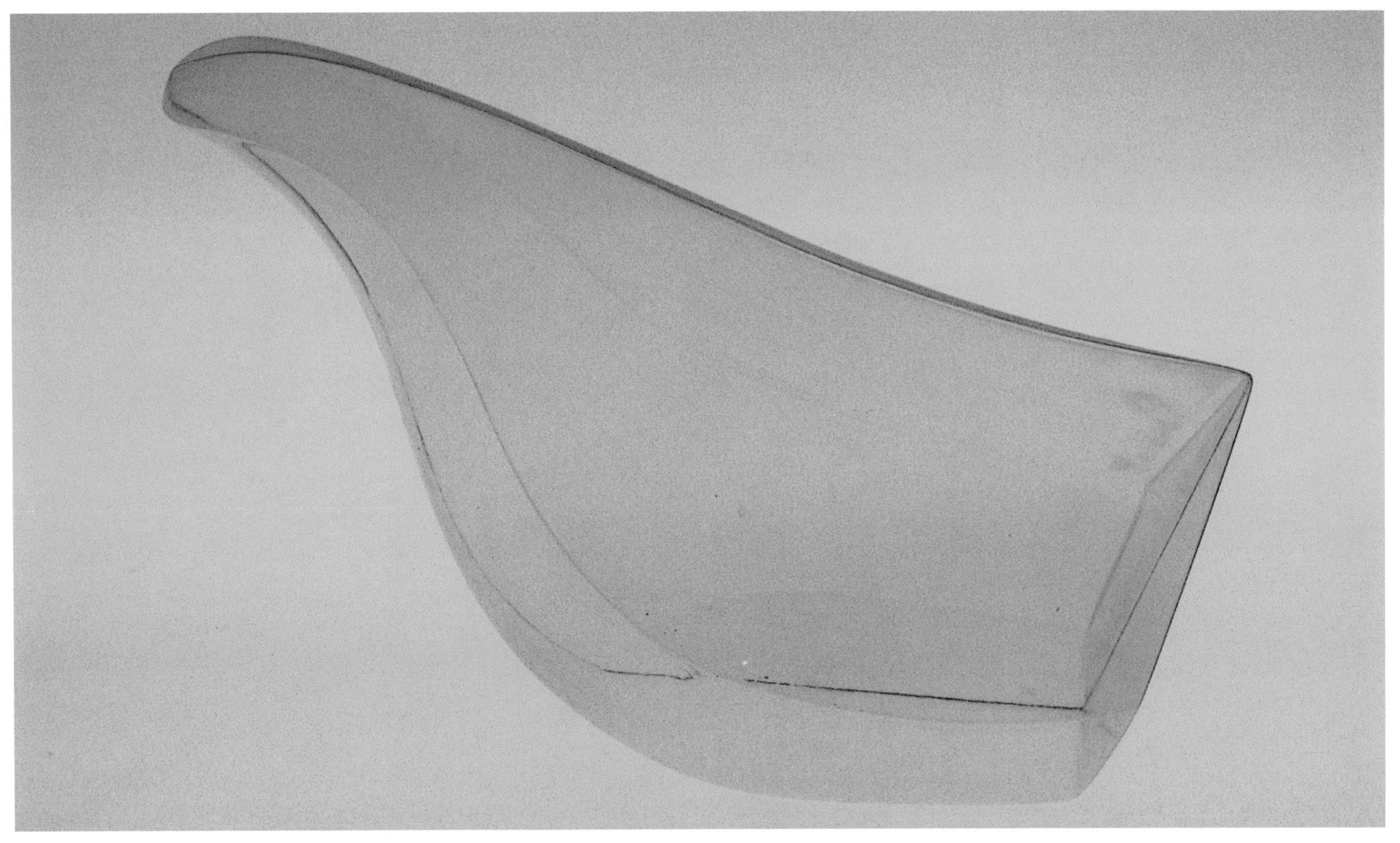

AQUA
DOVE

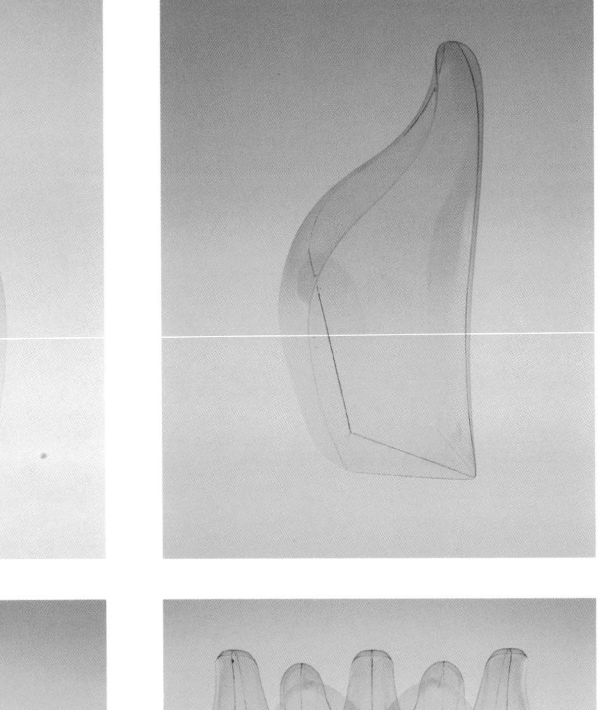

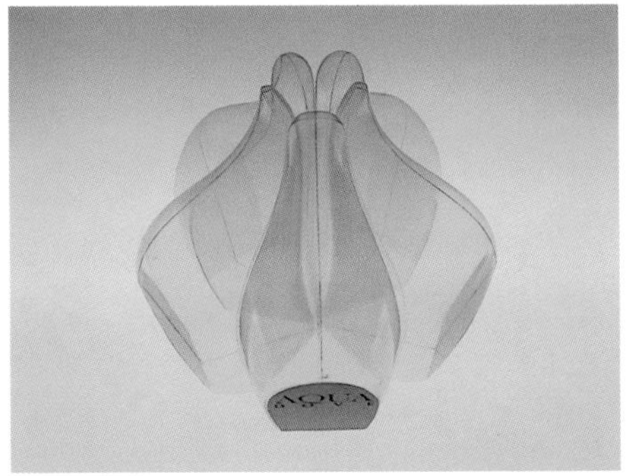

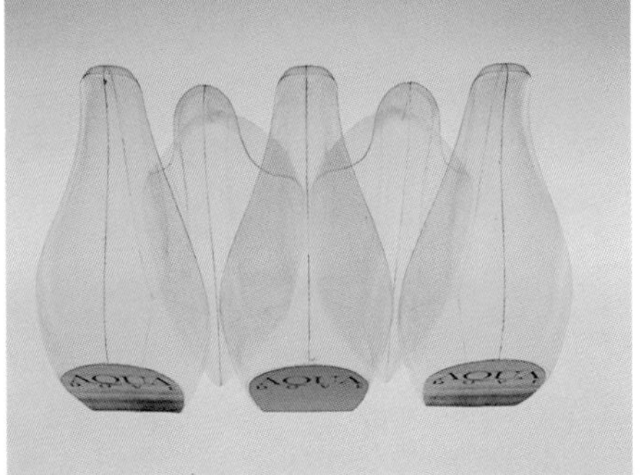

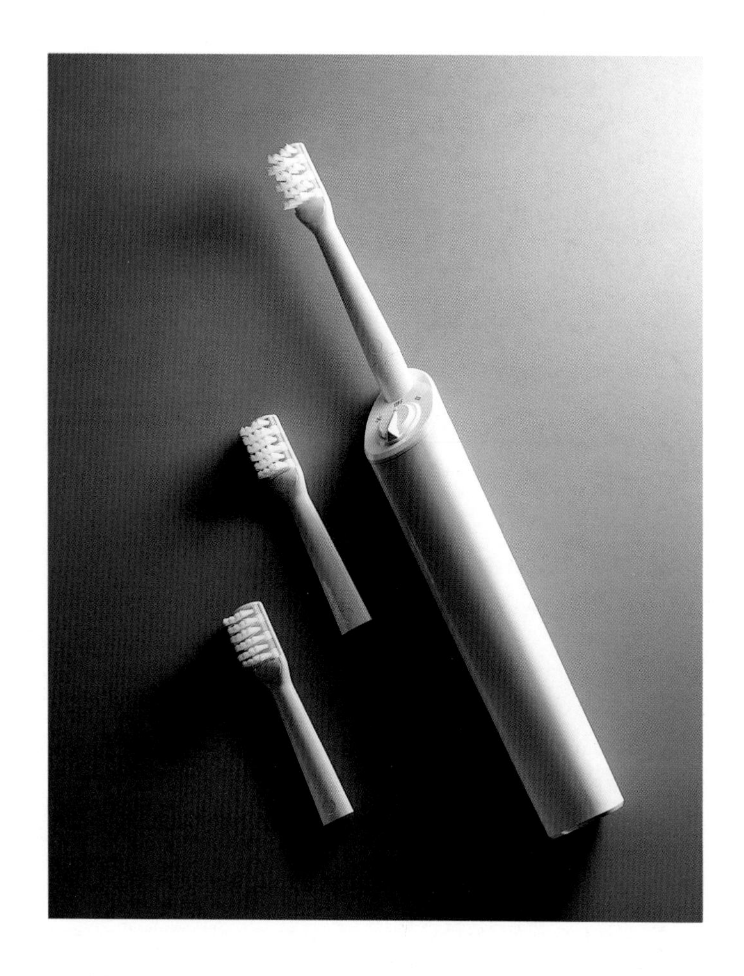

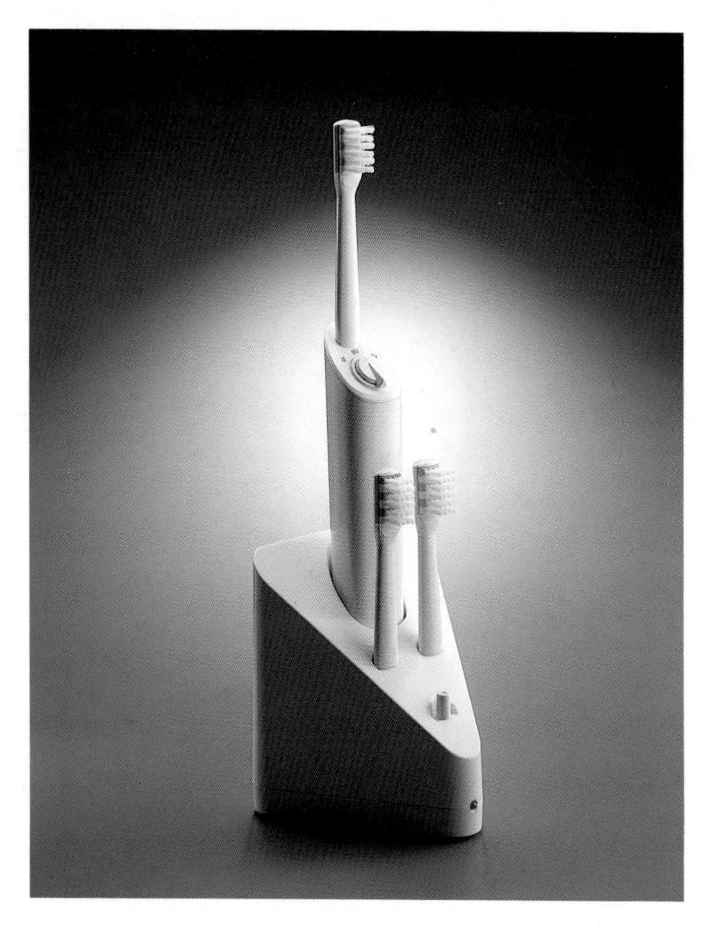

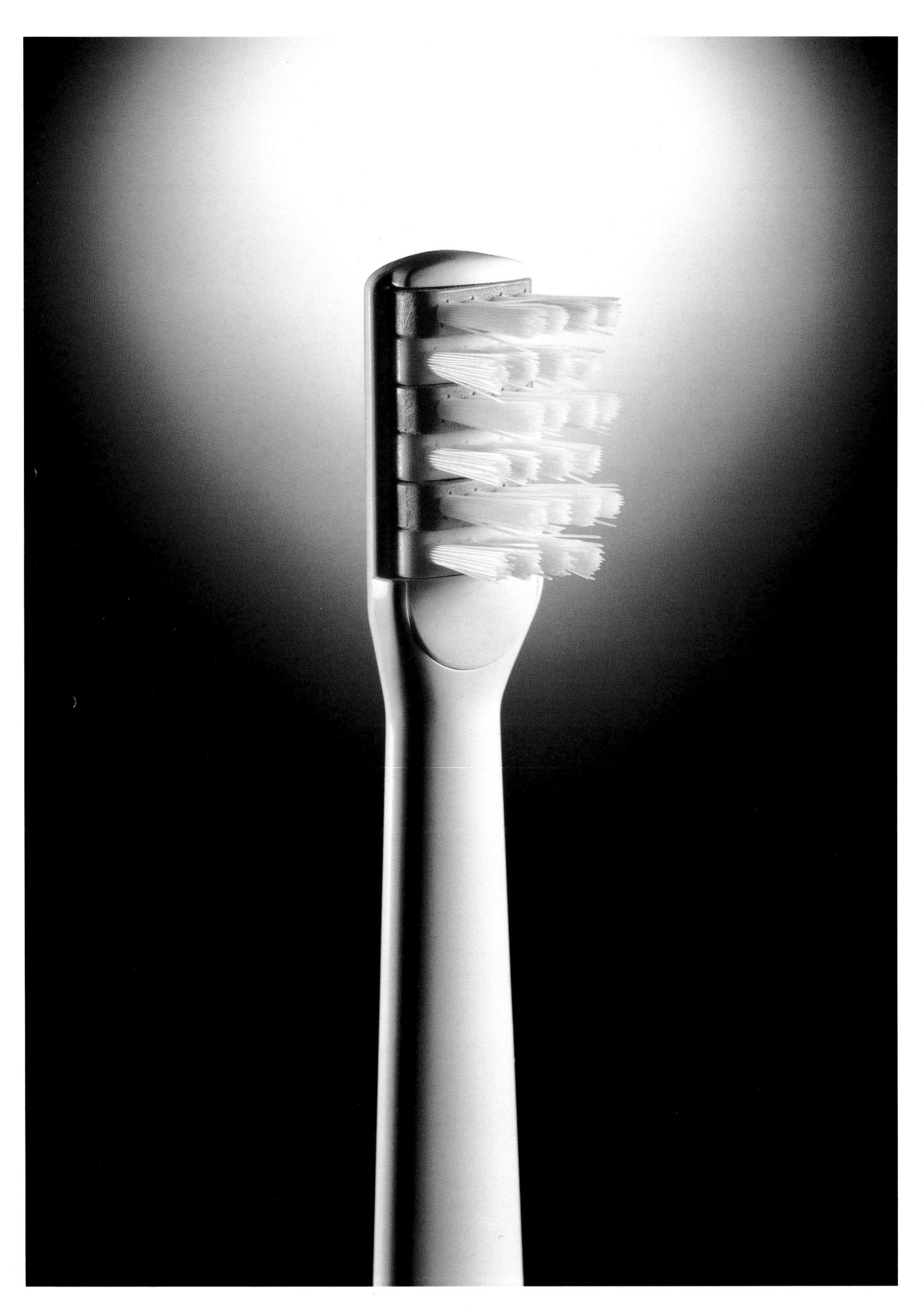

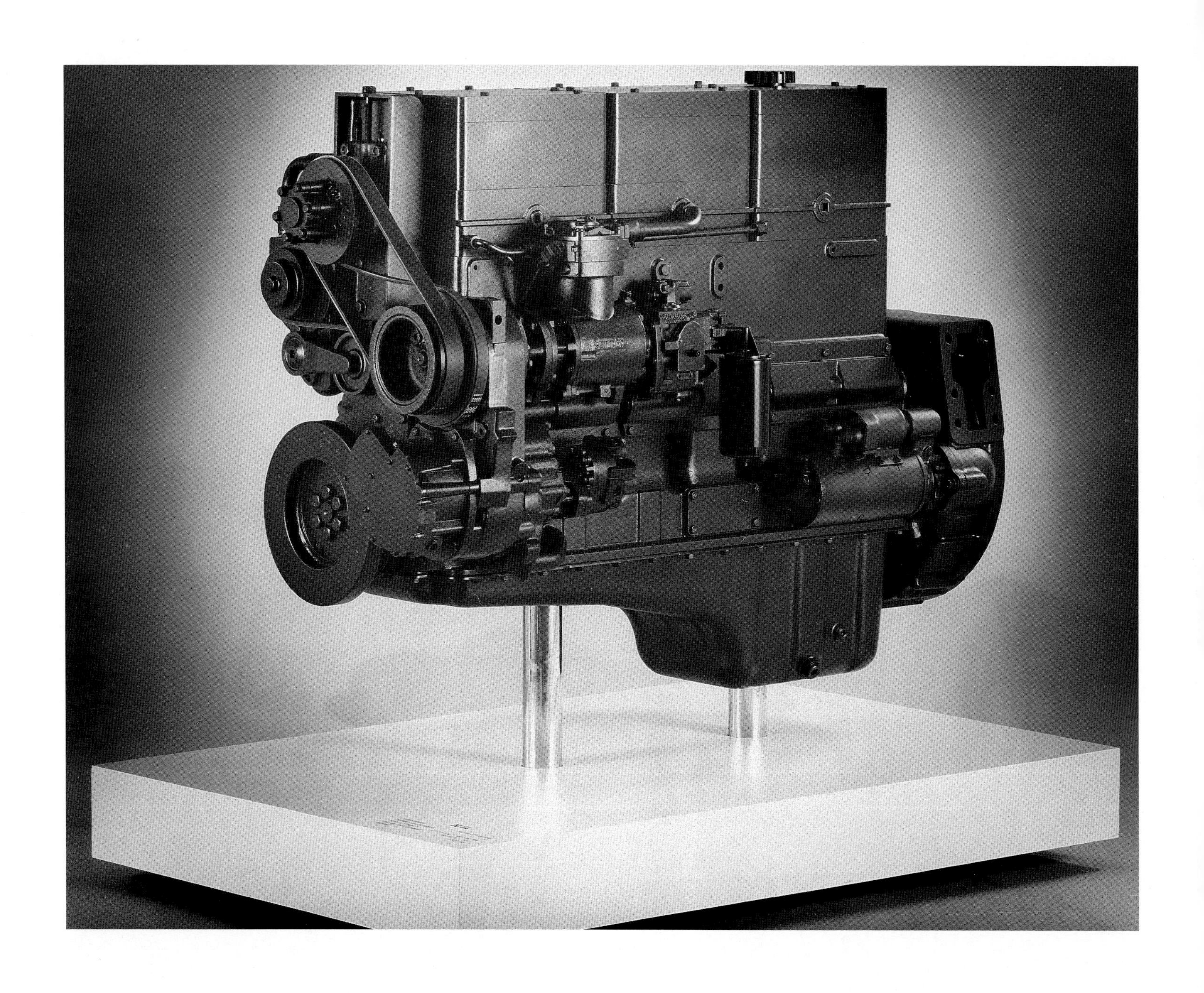

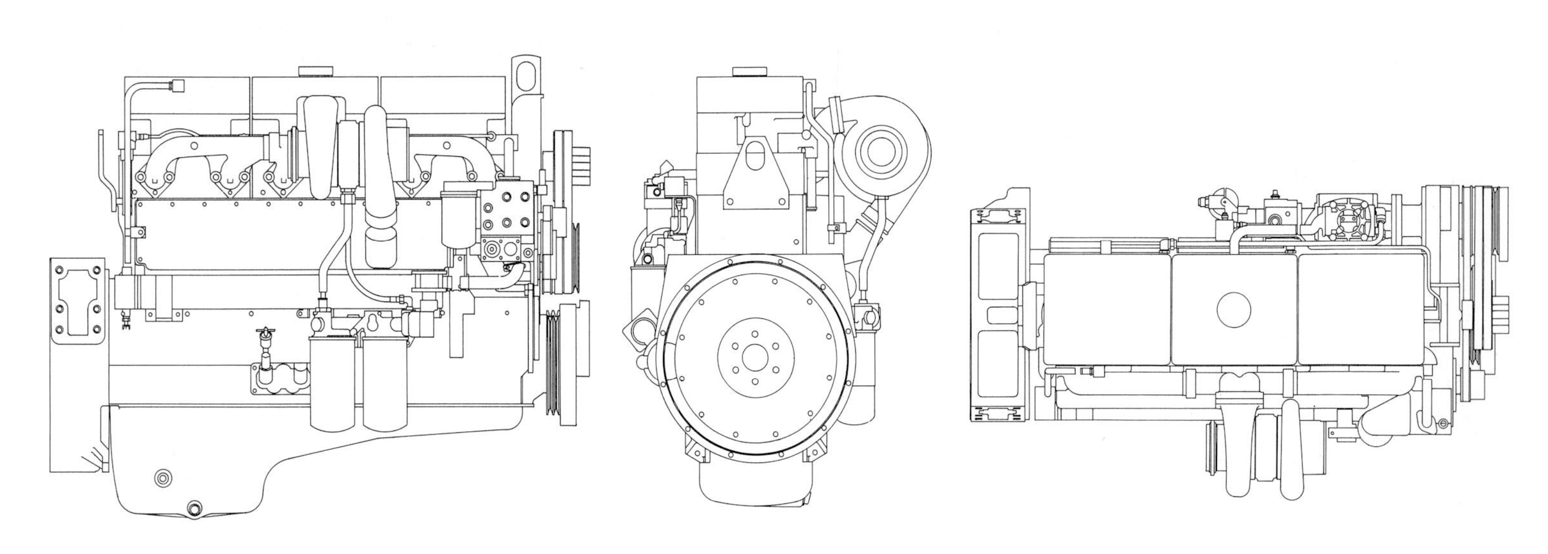

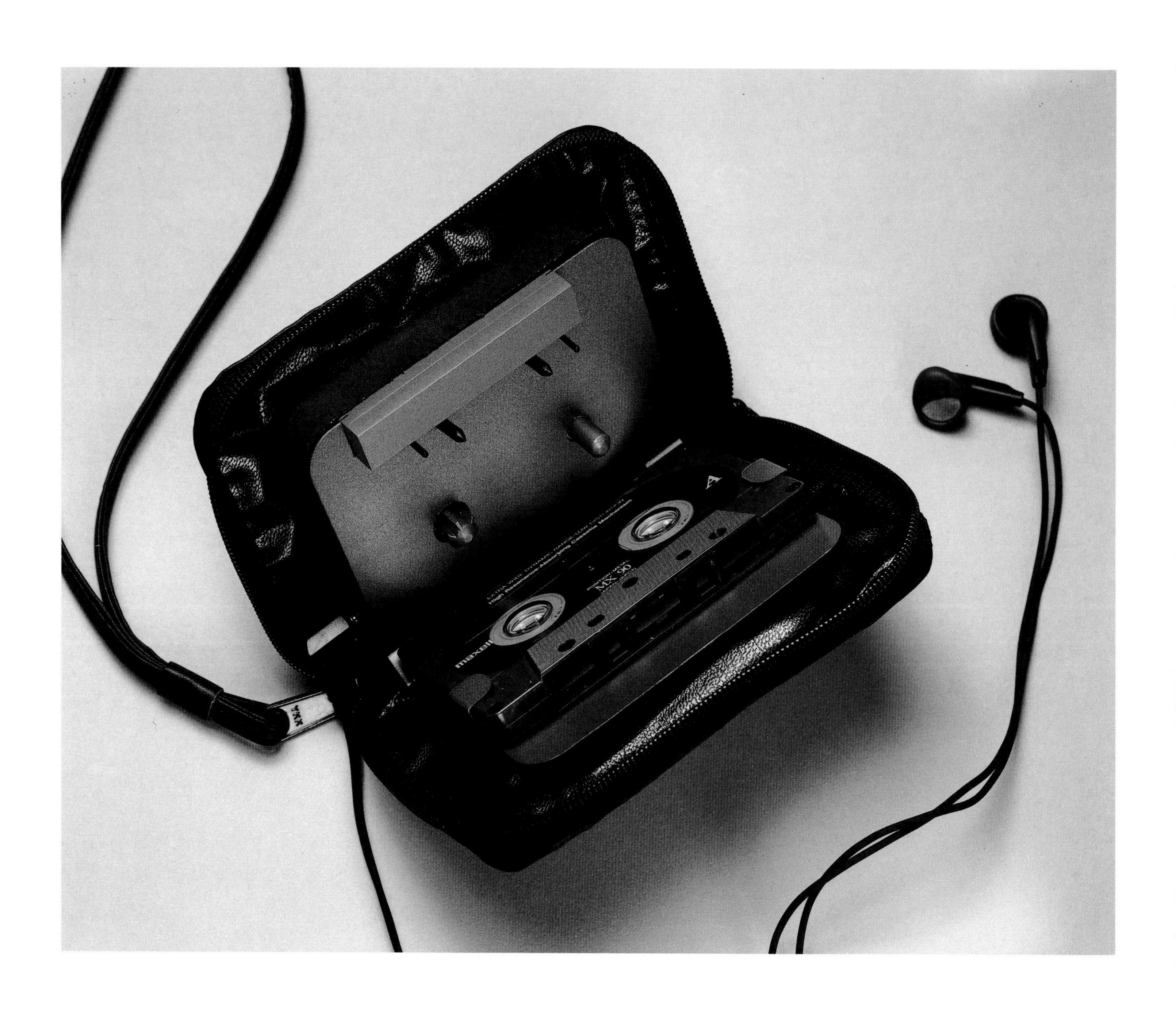

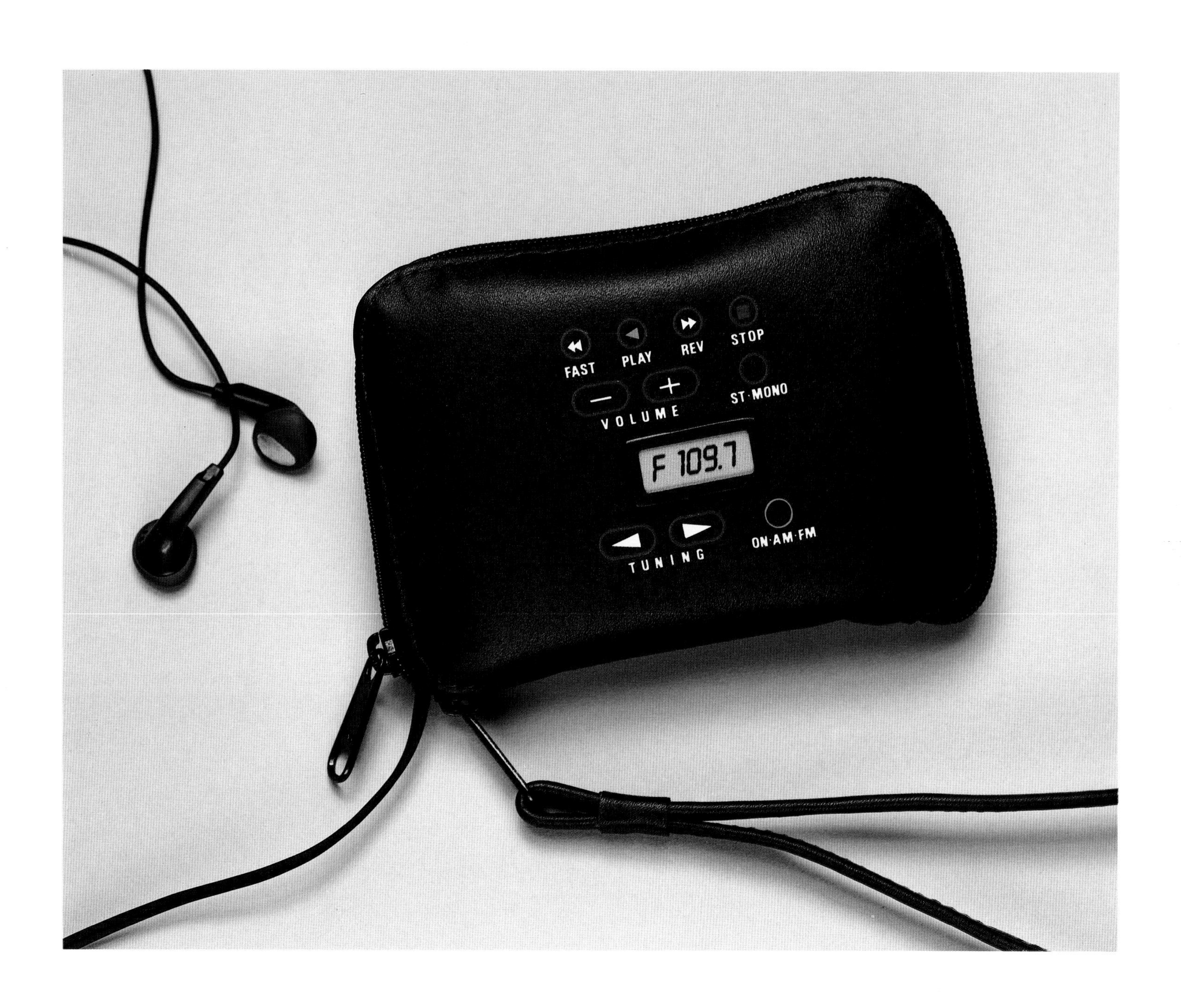
FAST
PLAY
REV
STOP
ST·MONO
VOLUME
F 109.7
ON·AM·FM
TUNING

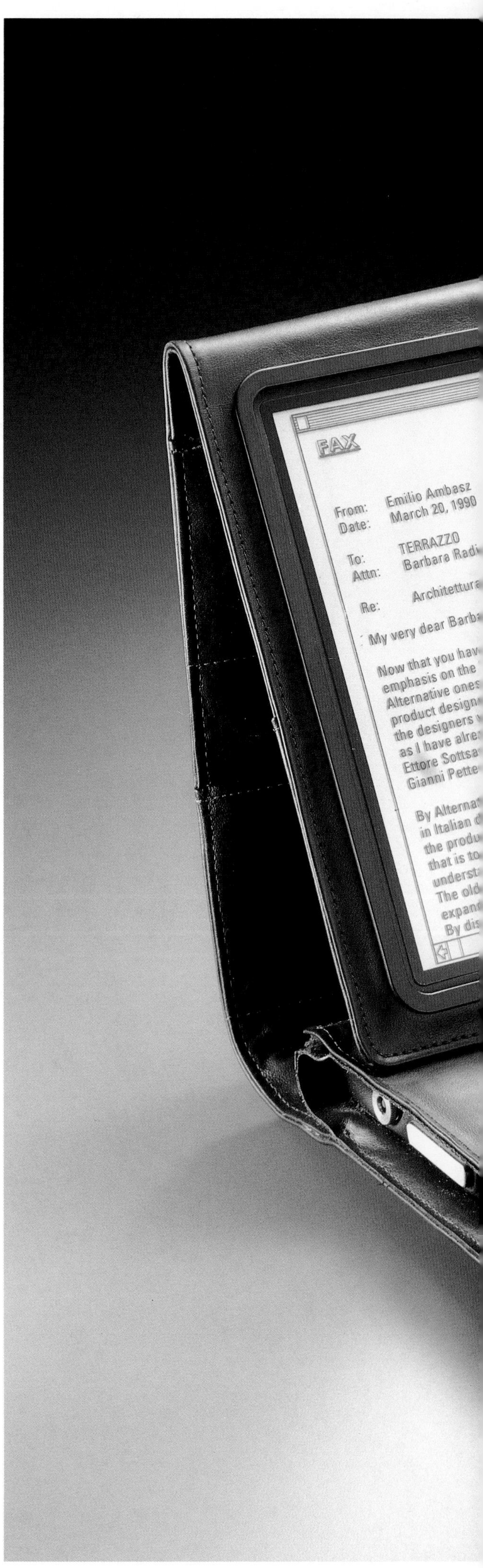
FAX
From: Emilio Ambasz
Date: March 20, 1990
To: TERRAZZO
Attn:
Re:

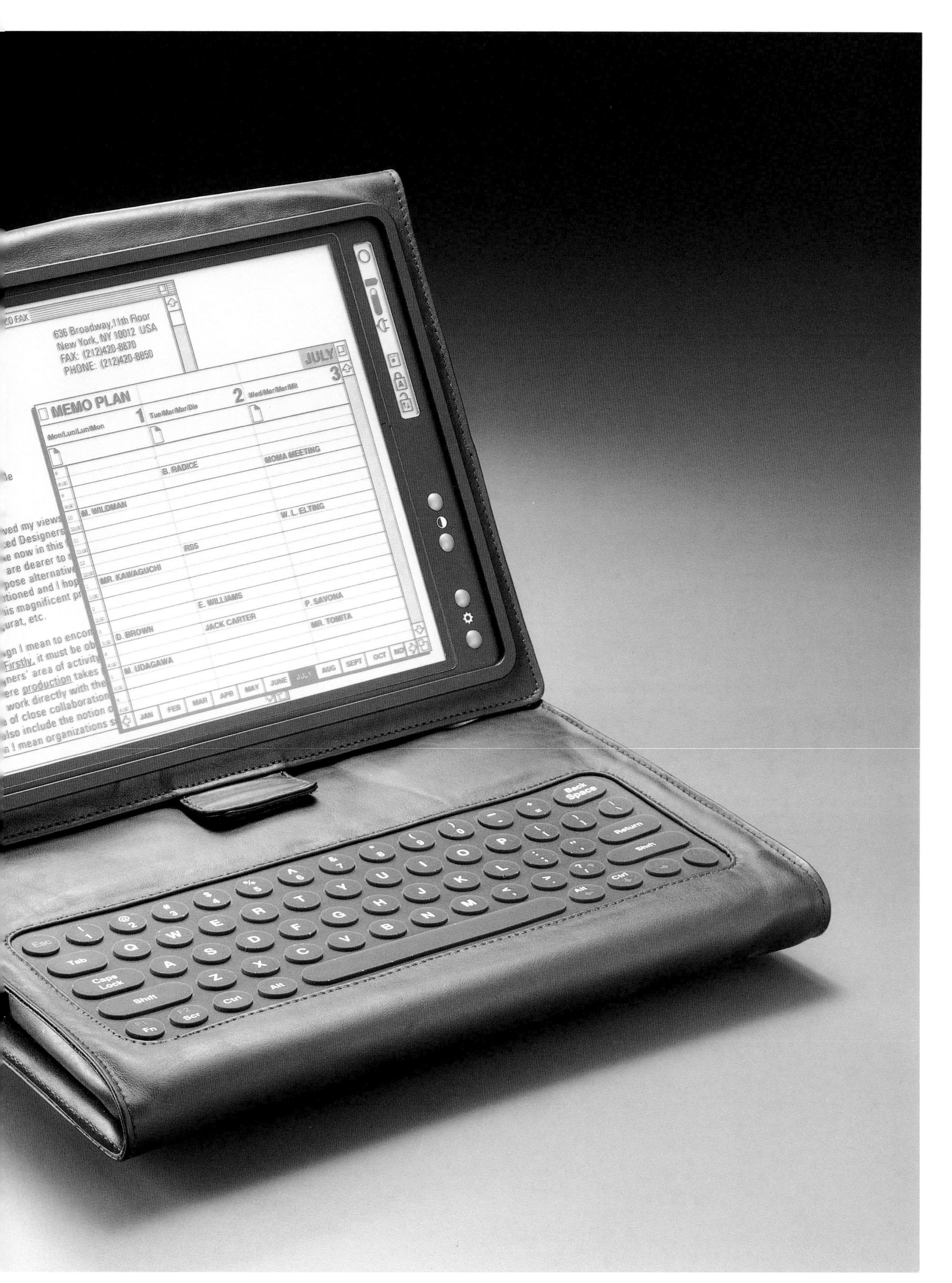
636 Broadway,11th Floor
New York, NY 10012 USA
FAX: (212)420-8670
PHONE: (212)420-8850
JULY
MEMO PLAN
B. RADICE
MOMA MEETING
M. WILDMAN
W. L. ELTING
RSS
MR. KAWAGUCHI
E. WILLIAMS
P. SAVONA
JACK CARTER
MR. TOMITA
D. BROWN
M. UDAGAWA
JAN FEB MAR APR MAY JUNE JULY AUG SEPT OCT

Handkerchief TV

VOL
MODE
–
+

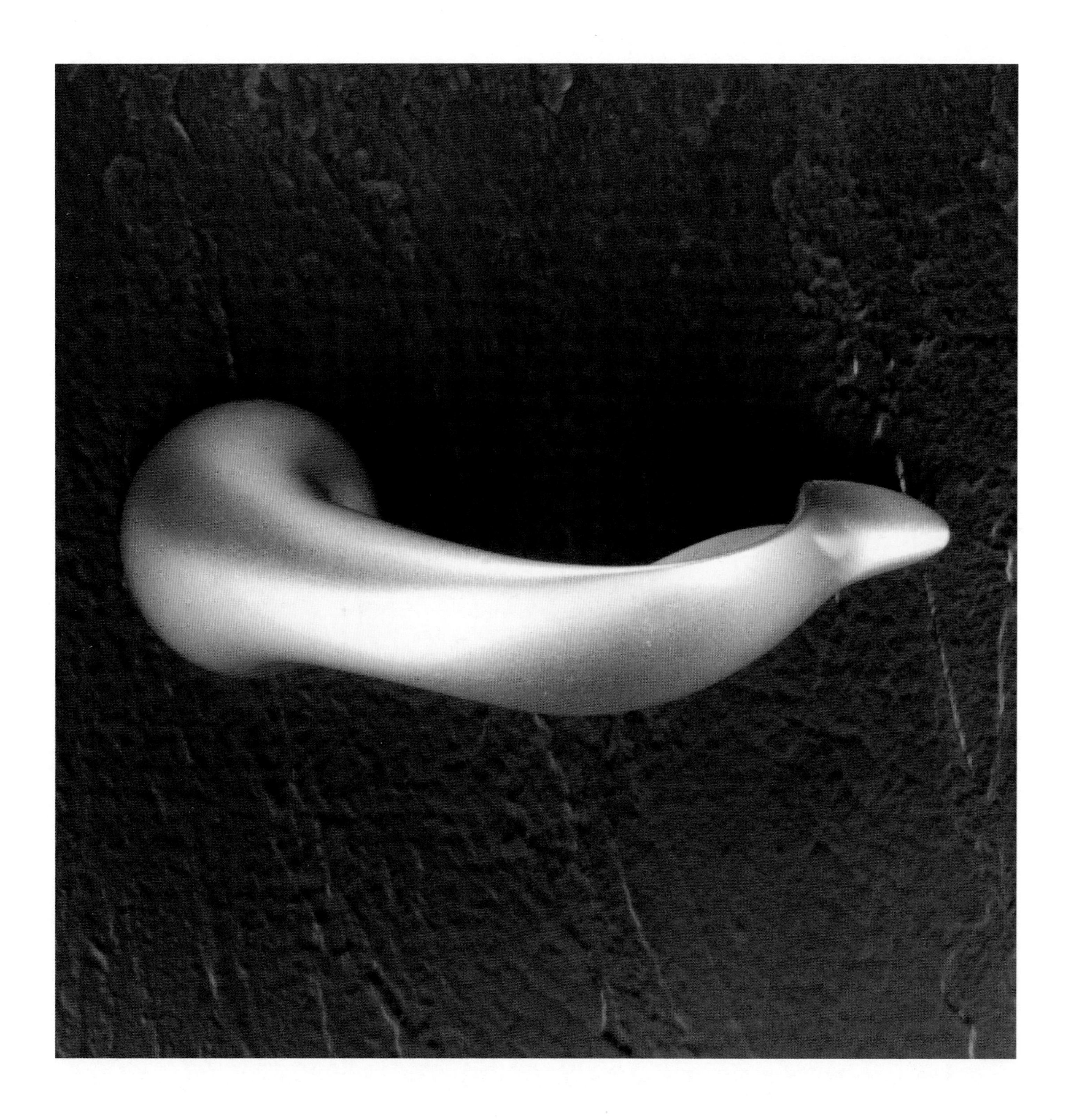

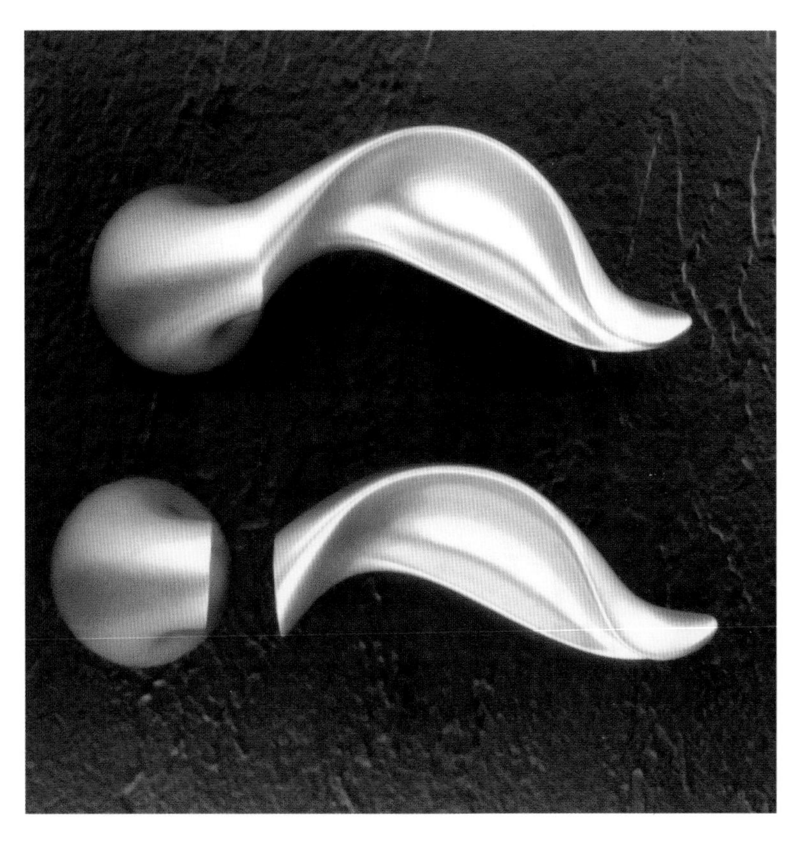

Magic Wand Automatic Roller Pen

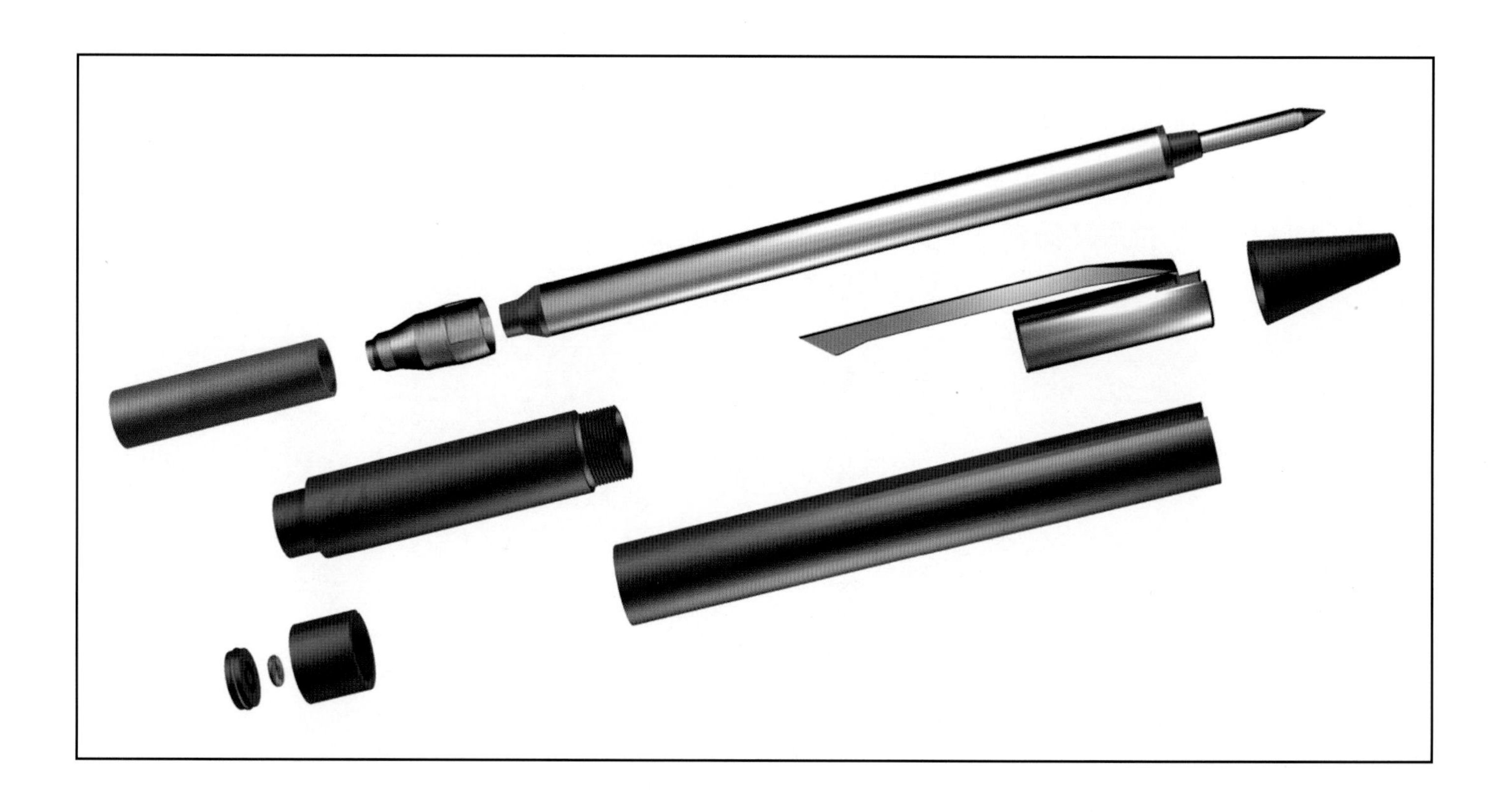

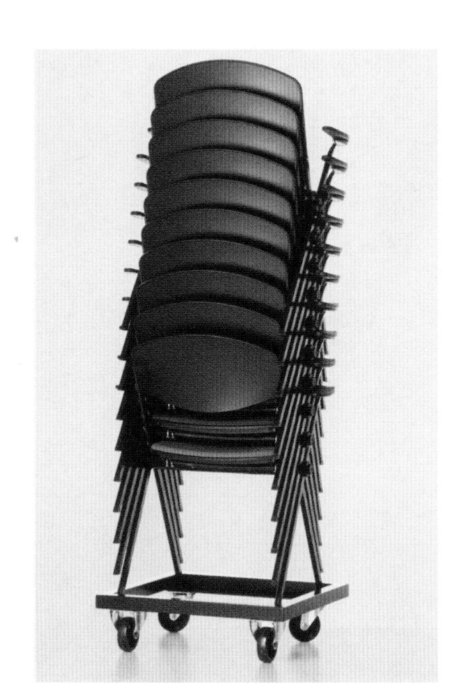

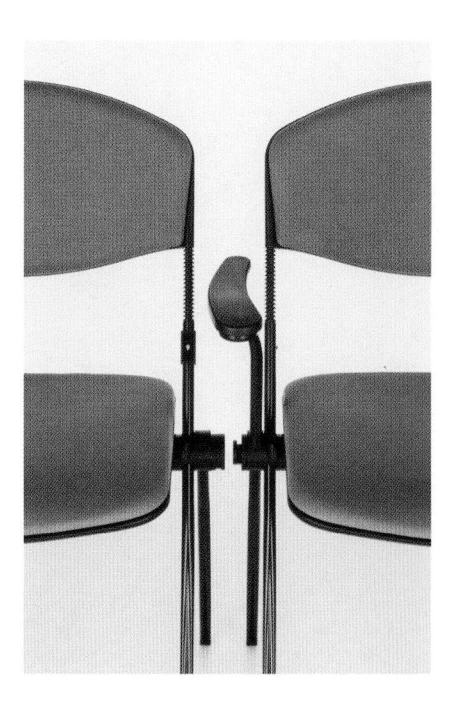

IBM

IBM

IBM
Progetto
IBM
comput
er
system,
realizzat
o in
Milano -
Italia -
2000 -
Portatile
compon
ibile.
Abbiam
o

IBM

IBM

Dior
j'adore
When you integrate hardware
with software, anything's possible
Microso

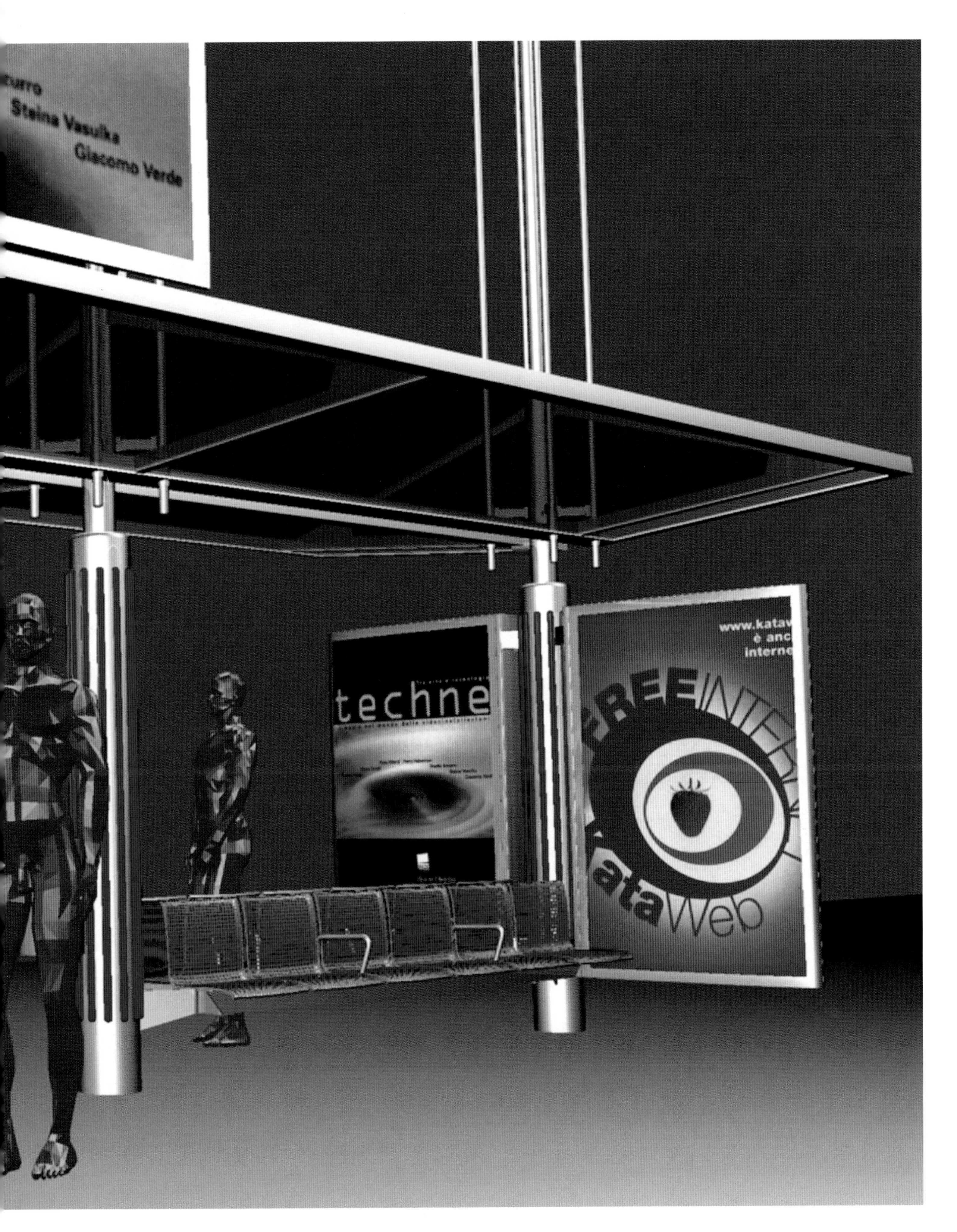
Steina Vasulka
Giacomo Verde
techne
FREE

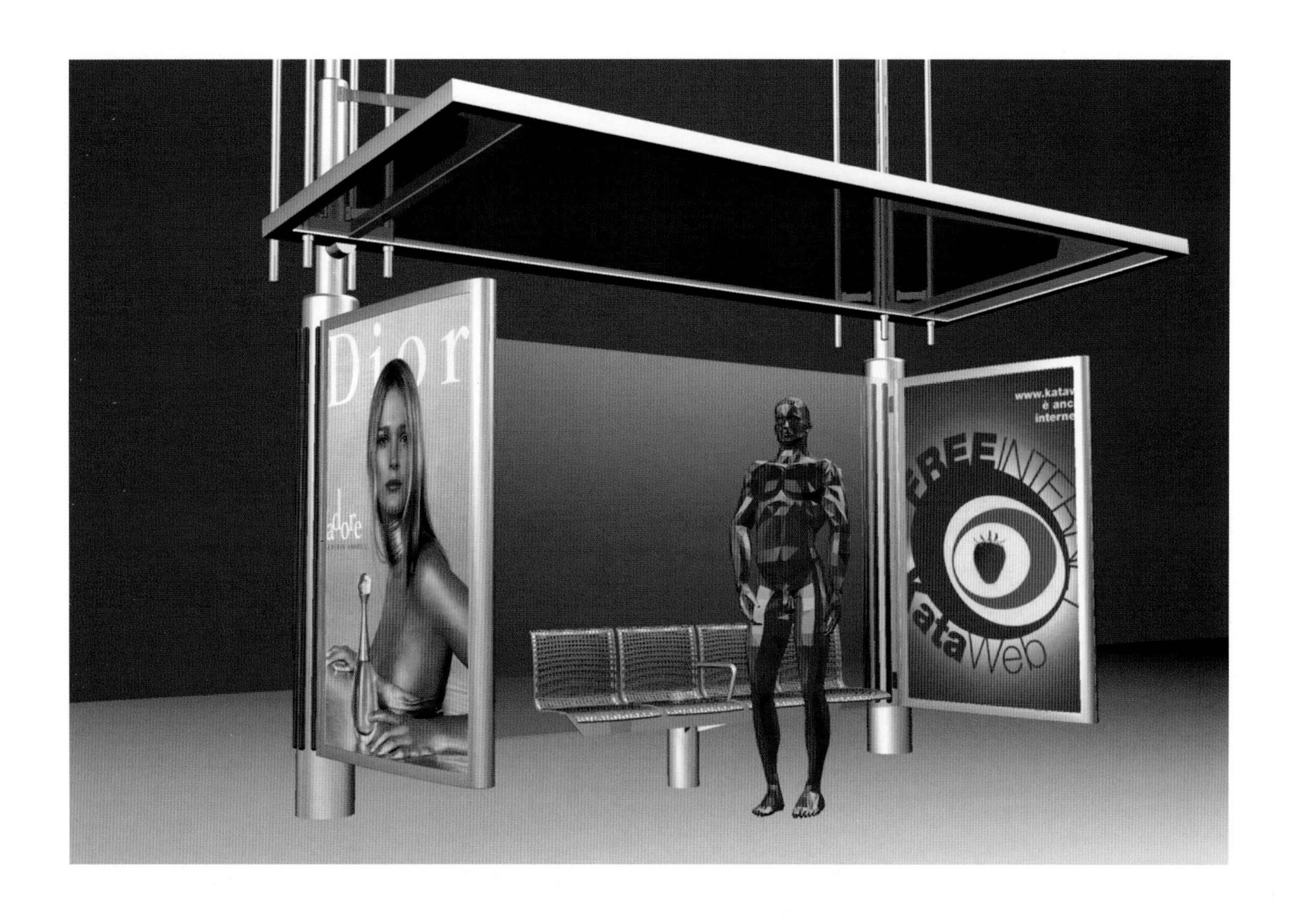
Dior
adore
FREE

Dior
adore
FREE

Dior

Dior
adore
When you integrate hardware
with software, anything's possible
Hardware Design Group

pure

Tra arte e tecnologia
techne

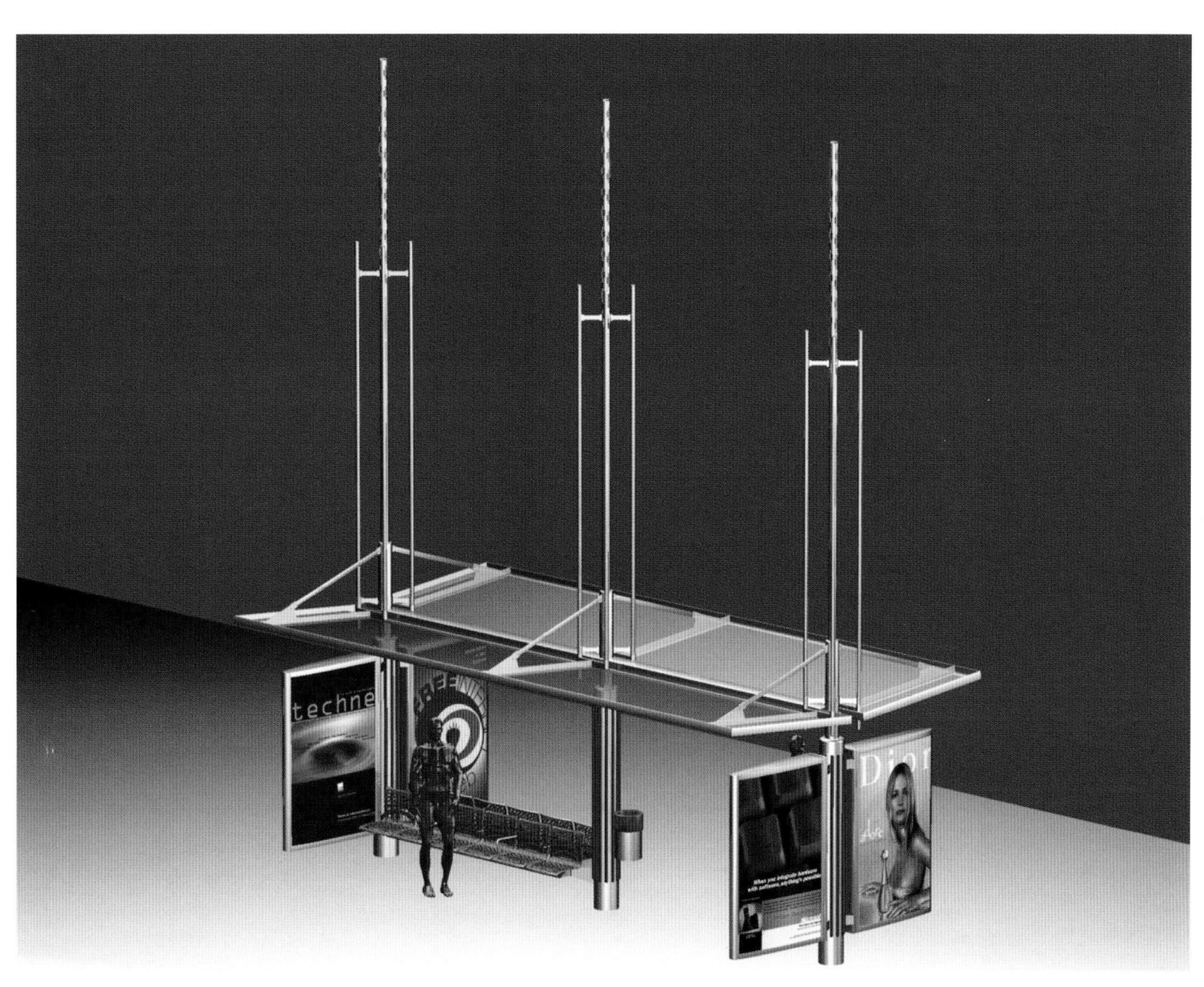
techne
Dior

Dior
pure

Tra arte e tecnologia
techne
Viaggio nel mondo delle videoinstallazioni
Piero Gilardi
Perry Hoberman
Studio Azzurro
Dior
techne

Stacker Contract Chair

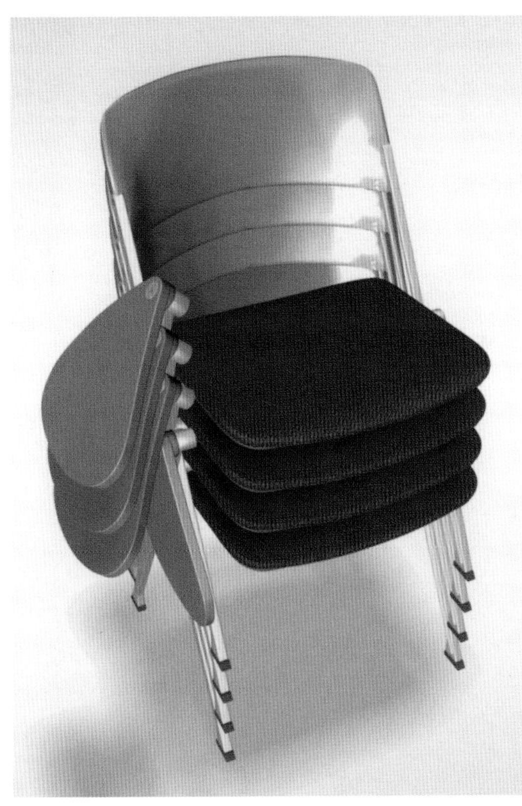

Cummins Diesel Engine

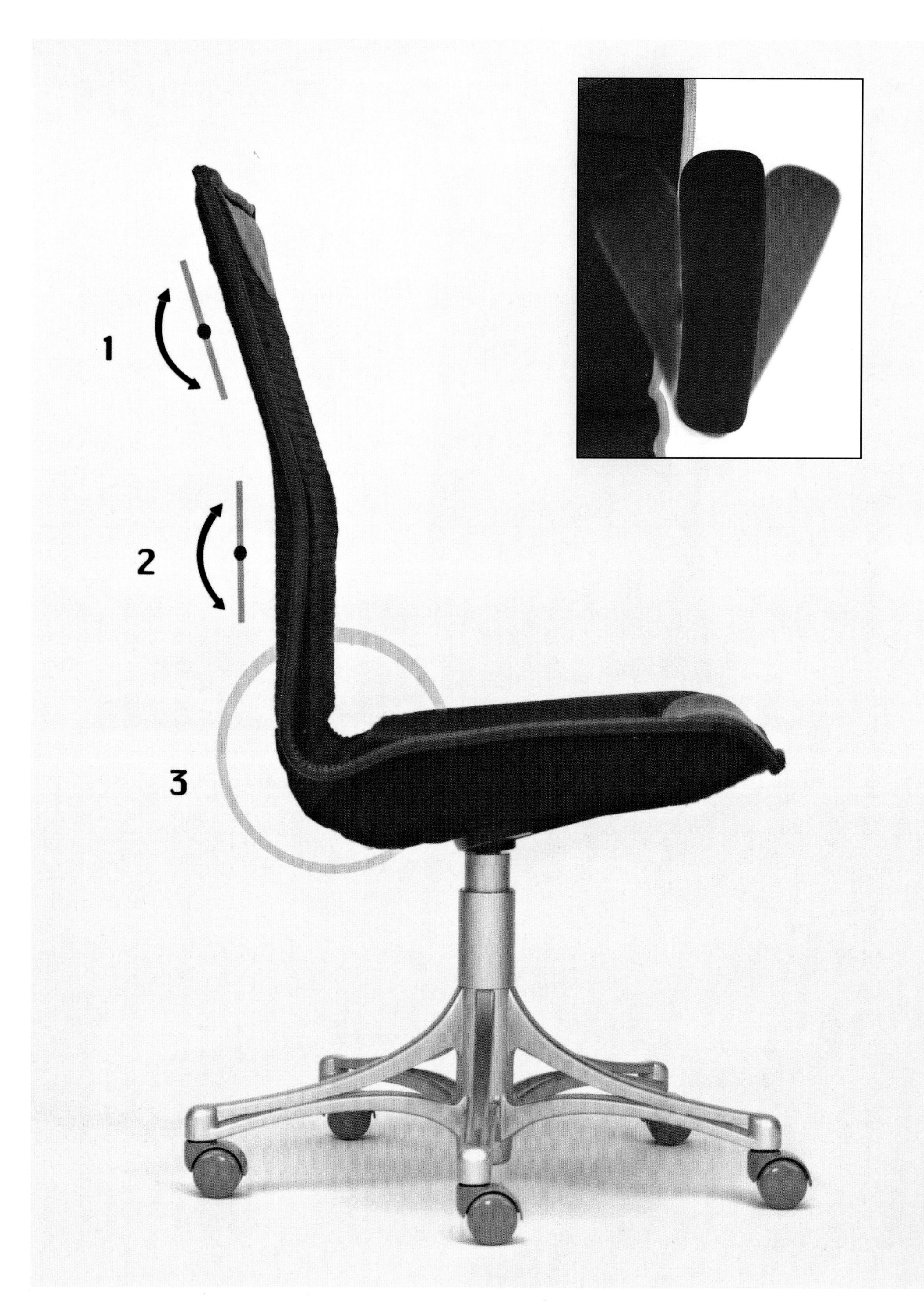
1
2
3

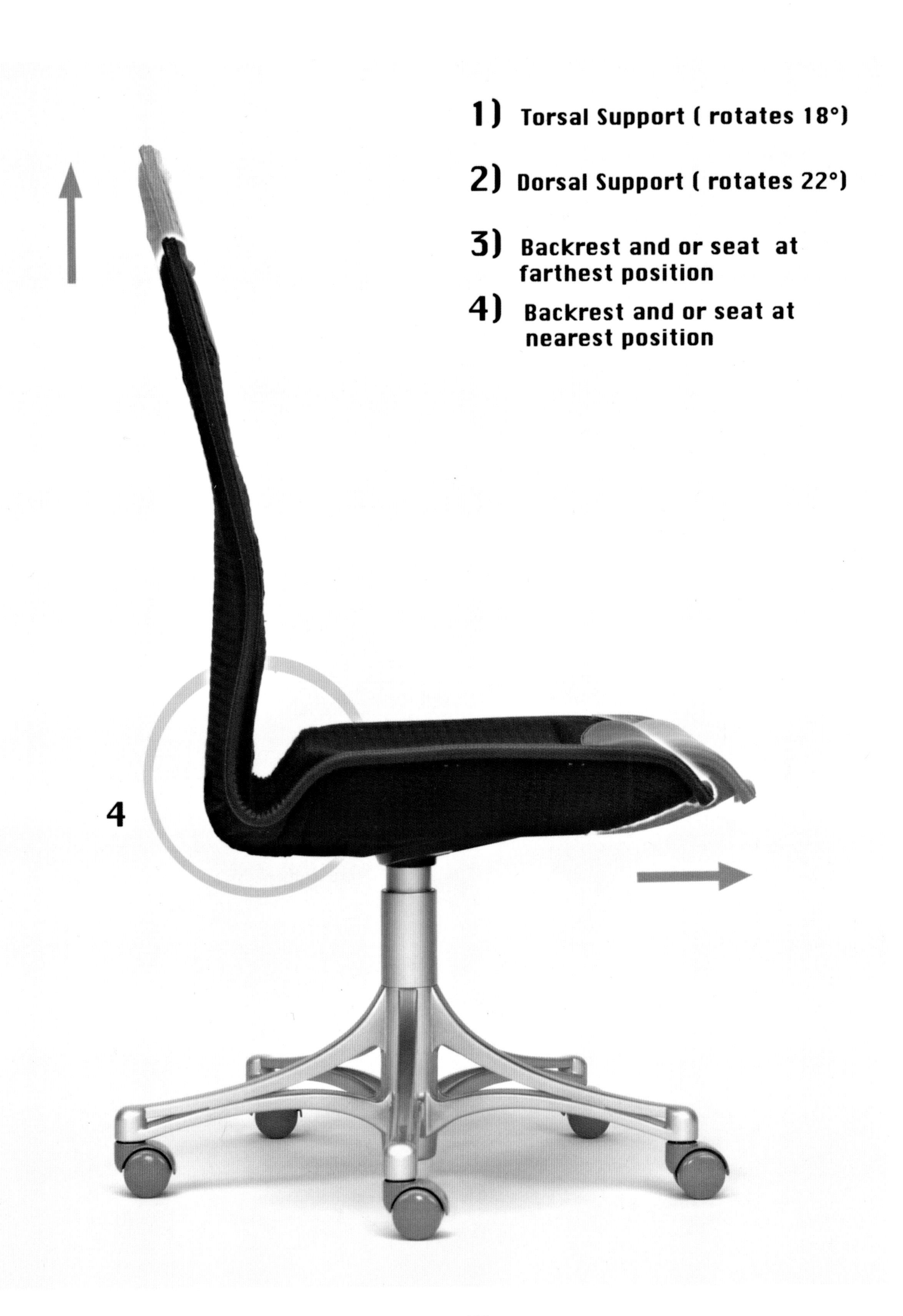
1) Torsal Support (rotates 18°)
2) Dorsal Support (rotates 22°)
3) Backrest and or seat at farthest position
4) Backrest and or seat at nearest position
4

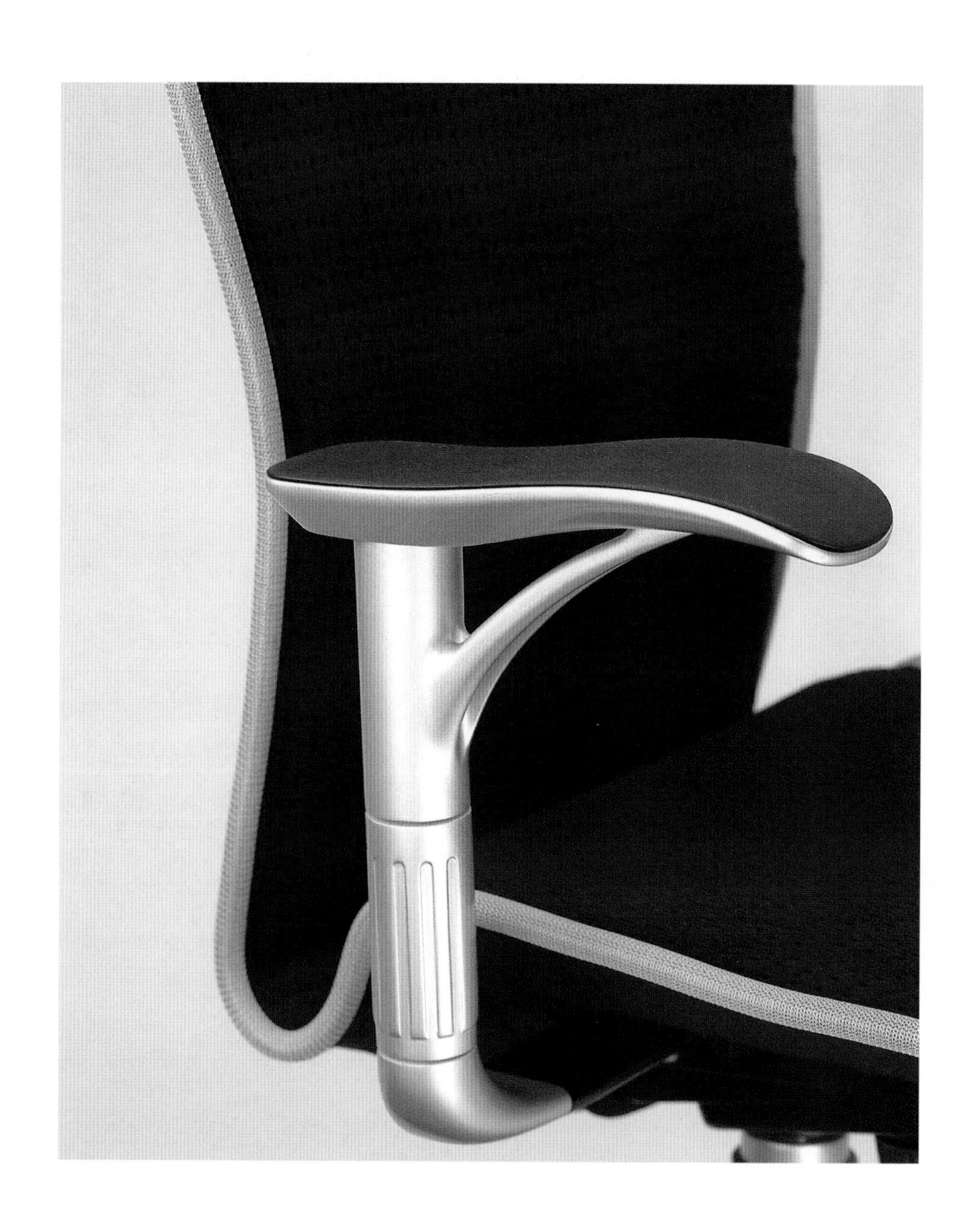

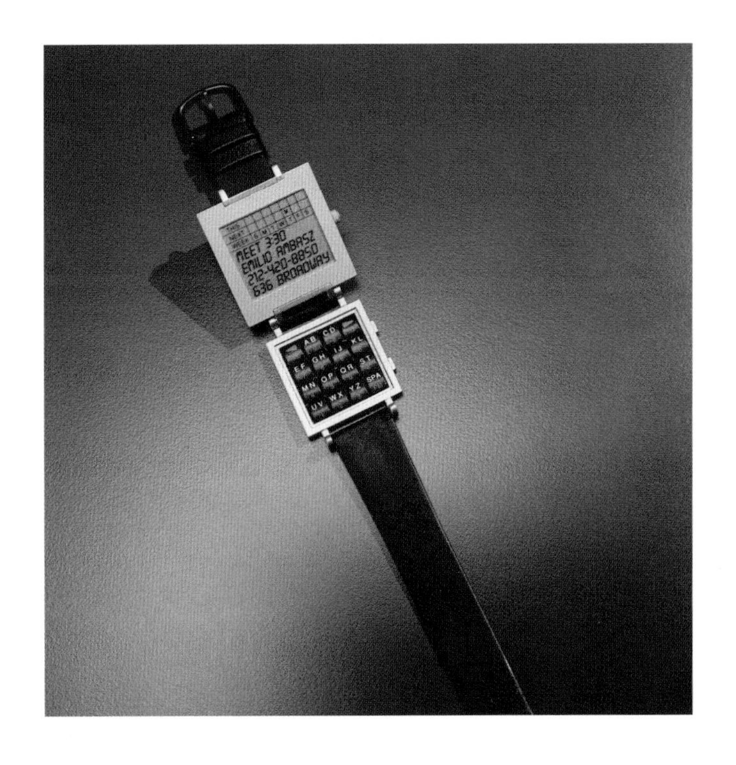
MEET 3:30
EMILIO AMBASZ
212-420-8850
636 BROADWAY
AB
CD
KL
IJ
GH
EF
ST
QR
OP
MN
SPA
YZ
WX
UV

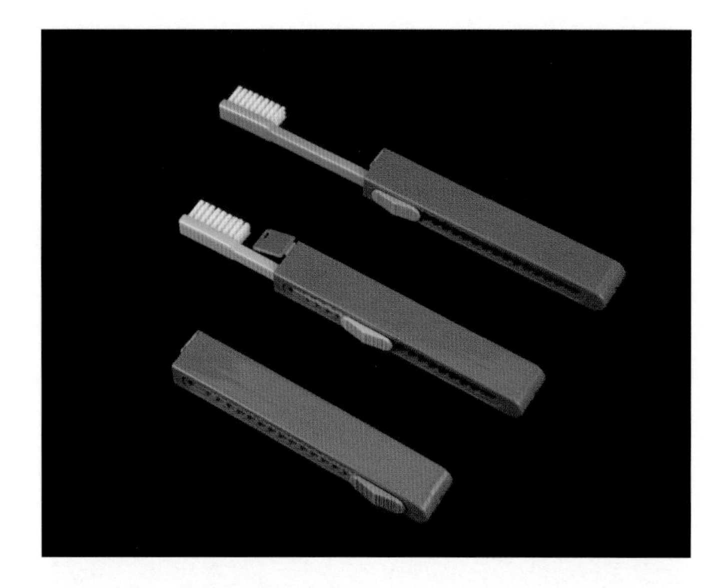

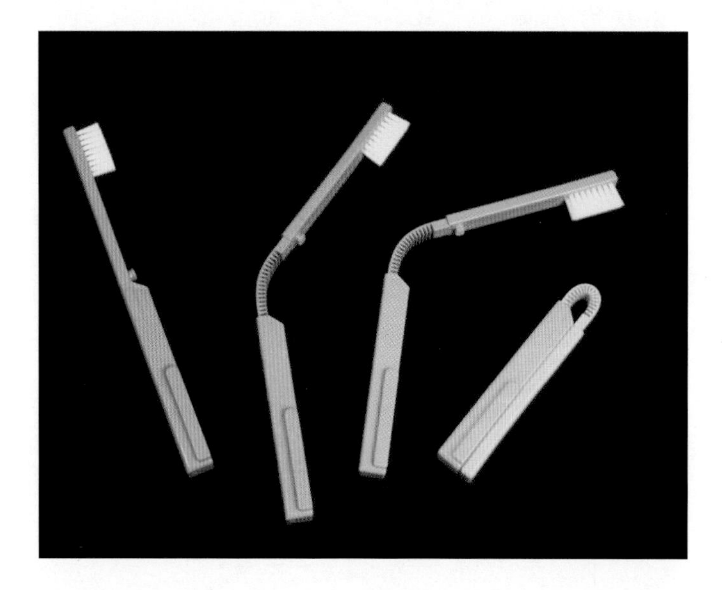

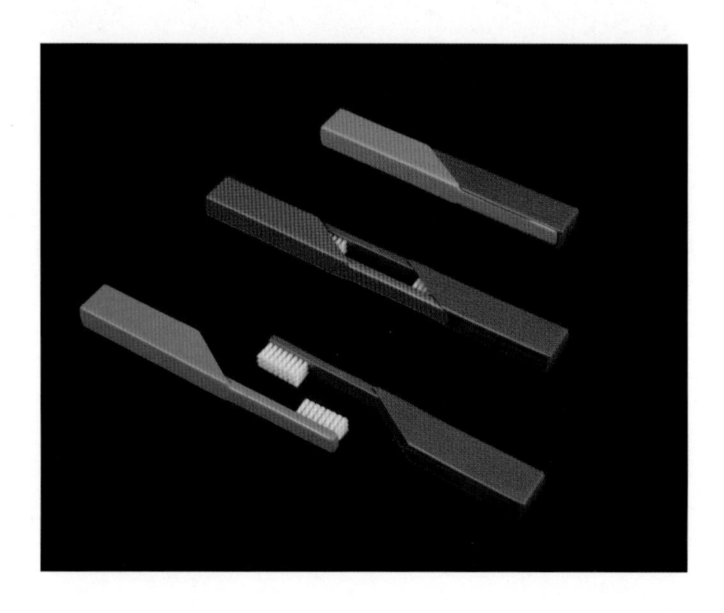

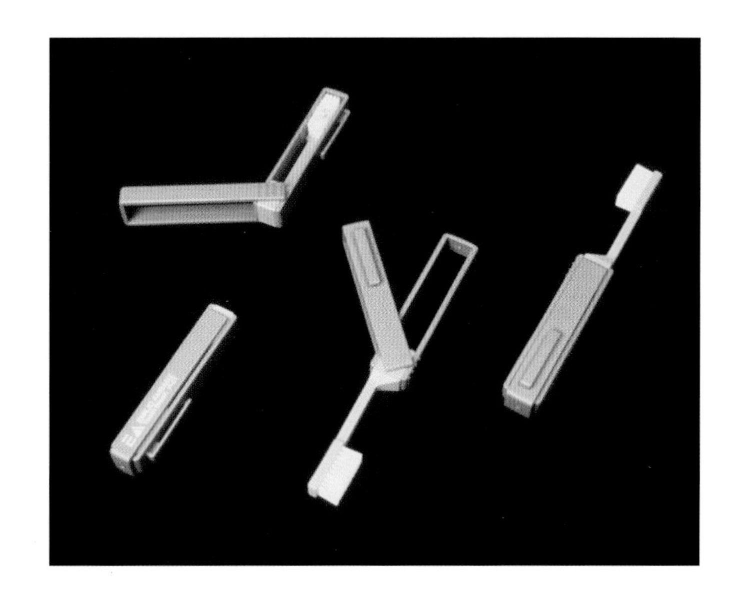

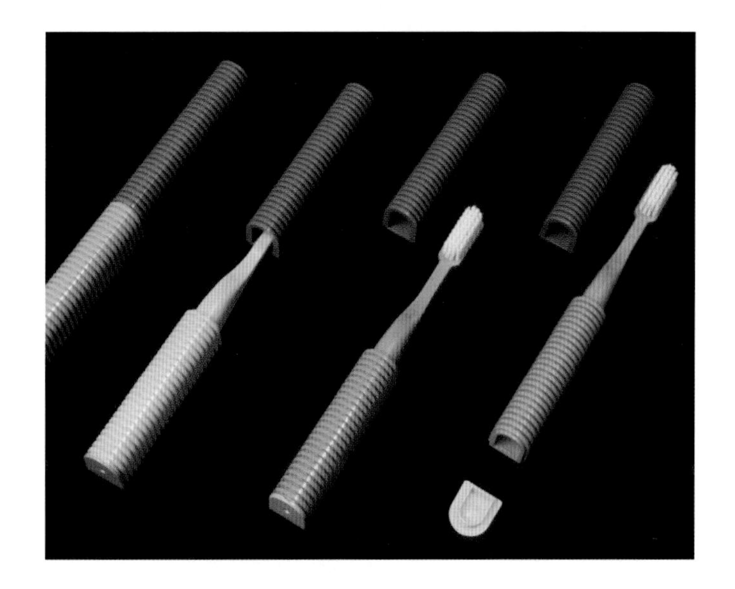

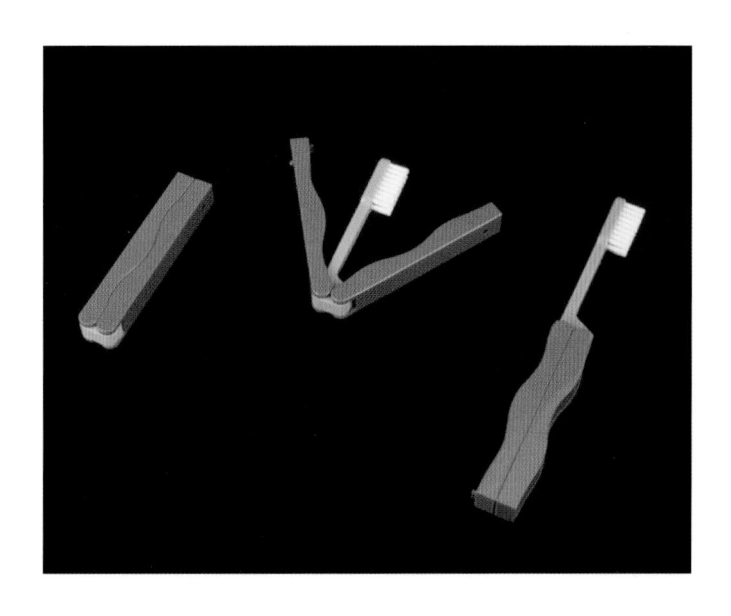

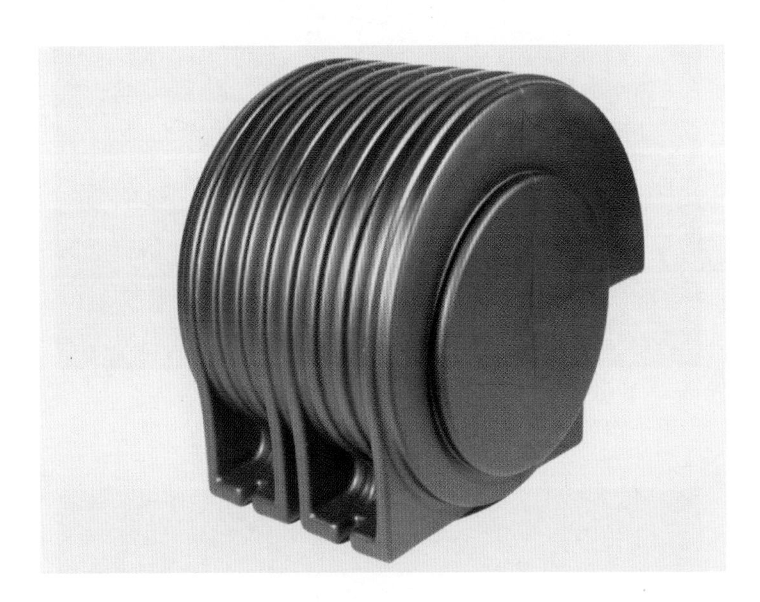

EMILIO AMBASZ DESIGNS™

Geigy Graphics on exhibition april 1967
princeton university school of architecture

Geigy Graphics on exhibition april 1967
princeton university school of architecture

Geigy Graphics on exhibition april 1967
princeton university school of architecture

Geigy Graphics on exhibition april 1967
princeton university school of architecture

Surface & Ornament

Formica Corporation invites you to submit entries in its two COLORCORE™ "Surface & Ornament" design competitions in 1983-84.

COLORCORE laminate is a new surfacing material from Formica Corporation. It is the first laminate with integral solid color. This feature eliminates the dark line associated with laminate applications. Additionally, dimensional and graphic effects are possible with routed channels which remain the same color as the surface.

"Surface & Ornament" is composed of two independent competitions which invite the design community to explore the potential of COLORCORE. Over $80,000 in prizes will be awarded. Judging will be based on overall excellence, technique and inventiveness in demonstrating the unique characteristics of COLORCORE. Colors are limited to original 12 COLORCORE colorways in Competition I.

COMPETITION I (CONCEPTUAL): Open to all architects, designers, fabricators and students, with students having their own category. The purpose is to design an object no larger than 64 cubic feet surfaced with COLORCORE. Designs can be either residential or commercial including such items as TV cabinets, office work stations, dining tables or any decorative or useful object. Prizes are as follows: Professionals—1st Prize $10,000; 2nd Prize $5,000; 3rd Prize $2,000; 4th Prize $1,000. Students—1st Prize $5,000 plus a $5,000 contribution to the student's school. Citations will also be awarded.

Scale models of winning entries will be built and exhibited during NEOCON XV, along with invited designs by the following prominent designers and architects: Emilio Ambasz, Ward Bennett, Frank O. Gehry, Milton Glaser, Helmut Jahn, Charles W. Moore, Stanley Tigerman, Venturi, Rauch and Scott Brown, Massimo and Lella Vignelli and SITE Inc.

Publication of the designs and a traveling exhibit of winning projects are also planned.

Judges are: Charles Boxenbaum, Joe D'Urso, Paul Segal, William Turnbull from Formica Corporation's Design Advisory Board. Other judges are: Niels Diffrient, industrial designer; David Gebhard, University of California, Santa Barbara; and Robert Maxwell, Dean, School of Architecture, Princeton University.

Entries must be postmarked by February 15, 1983. Judging will take place March 15, 1983. Winners will be notified by April 1 and publicly announced at NEOCON XV. For deadline and full details, please write Formica Corporation immediately.

COMPETITION II (BUILT): Open to all professional architects and designers for completed room installations, and in-production product designs utilizing COLORCORE. Prizes are as follows: In each of three categories—Product Design, Residential and Contract installations: 1st Prizes of $15,000 and 2nd Prizes of $5,000. Citations will also be awarded.

Judges are: Alan Buchsbaum and John Saladino from Formica Corporation's Design Advisory Board. Other judges are: Jack Lenor Larsen, James Stewart Polshek, Dean, School of Architecture, Columbia University; Andrée Putman, interior designer; Laurinda Spear, Arquitectonica; Robert A.M. Stern.

Documentation of completed work must be submitted in a series of 35mm slides with complete description of project and why COLORCORE was used. Current projects are eligible. Judging will take place March 15, 1984. For deadline and full details, please write Formica Corporation.

DEADLINES TIMETABLE	COMPETITION I	COMPETITION II
Earliest date for mailing entries	10-15-82	3-1-83
Last postmark date for mailing of inquiries	12-31-82	12-1-83
LAST POSTMARK DATE FOR MAILING ENTRIES	2-15-83	2-15-84
Jury Convenes	3-15-83	3-15-84
Announcements To Winners (confidentially)	4-1-83	4-2-84
To Public	NEOCON XV	NEOCON XVI
Exhibitions	NEOCON XV	NEOCON XVI

For free samples, call toll-free number 1-800-543-3000. Ask for Operator #375. In Ohio call: 1-800-582-1396.

Entrants are strongly urged to call for "Rules and Regulations" manual and samples.

Address entries or requests for information to: COLORCORE "Surface and Ornament" Competition, Formica Corporation, One Cyanamid Plaza, Wayne, N.J. 07470.

Design by Emilio Ambasz for Formica Corporation.

Axis Poster

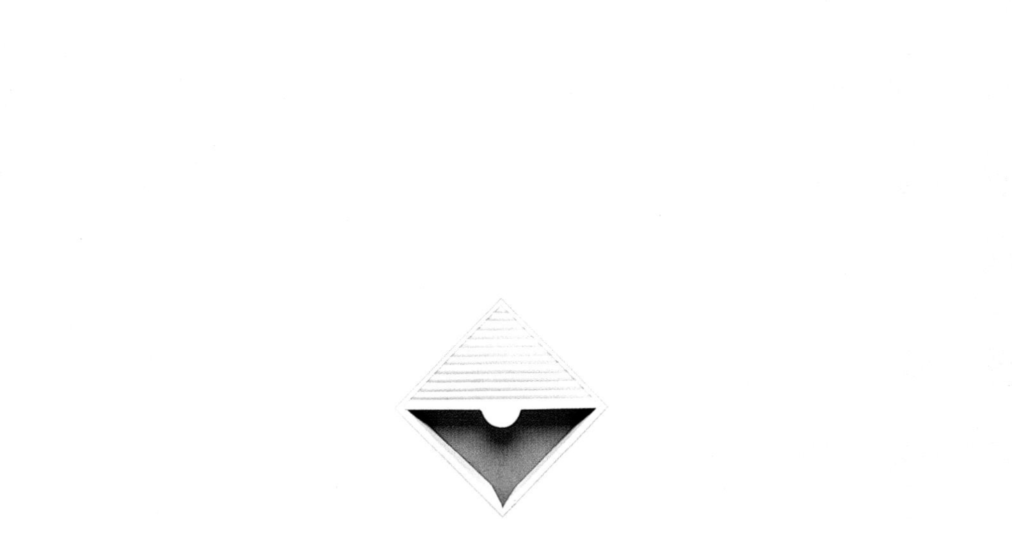

EMILIO AMBASZ

ARCHITECTURAL • INDUSTRIAL • GRAPHIC • EXHIBIT DESIGN

April 16, 1985 to May 6, 1985

AXIS
4F AXIS Building
5-17-1, Roppongi
Minato-ku, Tokyo

1985年4月16日(火)より5月6日(月)まで
11:00AM〜6:00PM(最終日3:00PM)
入場無料

アクシスギャラリー(アクシスビル4F)
東京都港区六本木5-17-1
TEL.03(587)2781

ORGANIZER
AXIS

PATRONS
Embassy of the United States of America
Argentine Embassy
Italian embassy
The International Council of
The Museum of Modern Art

COLLABORATOR
Bridgestone

COOPERATORS
Itoki
Shokobijutsu
Toray Industries
Nuno
TL Yamagiwa Laboratory

Argentine born Emilio Ambasz is very active in various fields as architect, industrial designer, graphic designer, and curator. His New York office engages mainly in architectural projects whose innovative technology and lyrical qualities have won him many awards and worldwide recognition. In addition to acting as Chief Design Consultant for Commins Engine Company, a company reknown for its enlightened support of architecture and design, Mr. Ambasz maintains an office in Bologna, Italy's combination of traditional craftsmanship and sophisticated engineering has enabled him to design as well as develop a number of influential products, ranging from furniture to lighting and electronic equipment. A writer of articles and books, Mr. Ambasz is also active in cultural endeavors related to architecture and design both as former Curator of Design at the Museum of Modern Art (NYC) and as current President of The Architectural League.

エミリオ・アンバースはアルゼンチンに生まれ、建築家、工業デザイナー、グラフィックデザイナー、キューレーターとして多くの領域で精力的な活動をしている。彼はニューヨークの事務所では、主に建築のプロジェクトに従事し、革新的な技術と叙情的な質により、多くの賞と世界的な評価を受けている。そして、建築やデザインを積極的に奨励する企業として知られるカミンズエンジン社の主席デザインコンサルタントとしても活動している。また彼はボローニアにも事務所を持ち、イタリアの伝統的な職人芸と洗練された工業技術により、家具から照明器具や電気製品までの、数多くの影響力ある製品の開発、デザインをおこなっている。現在、アンバース氏はアーキテクチュラル・リーグの代表をしているのみならず、ニューヨーク近代美術館のデザイン部門の元キューレーターとして、また論文や本の著述を通して、建築やデザイン部門での文化的な貢献をしている。

主催
アクシス

後援
アメリカ大使館
アルゼンチン大使館
イタリア大使館
ニューヨーク近代美術館国際協議会

協賛
ブリヂストン

協力
イトーキ
商工美術
東レ
布
TLヤマギワ研究所

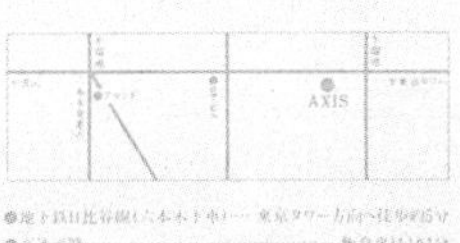

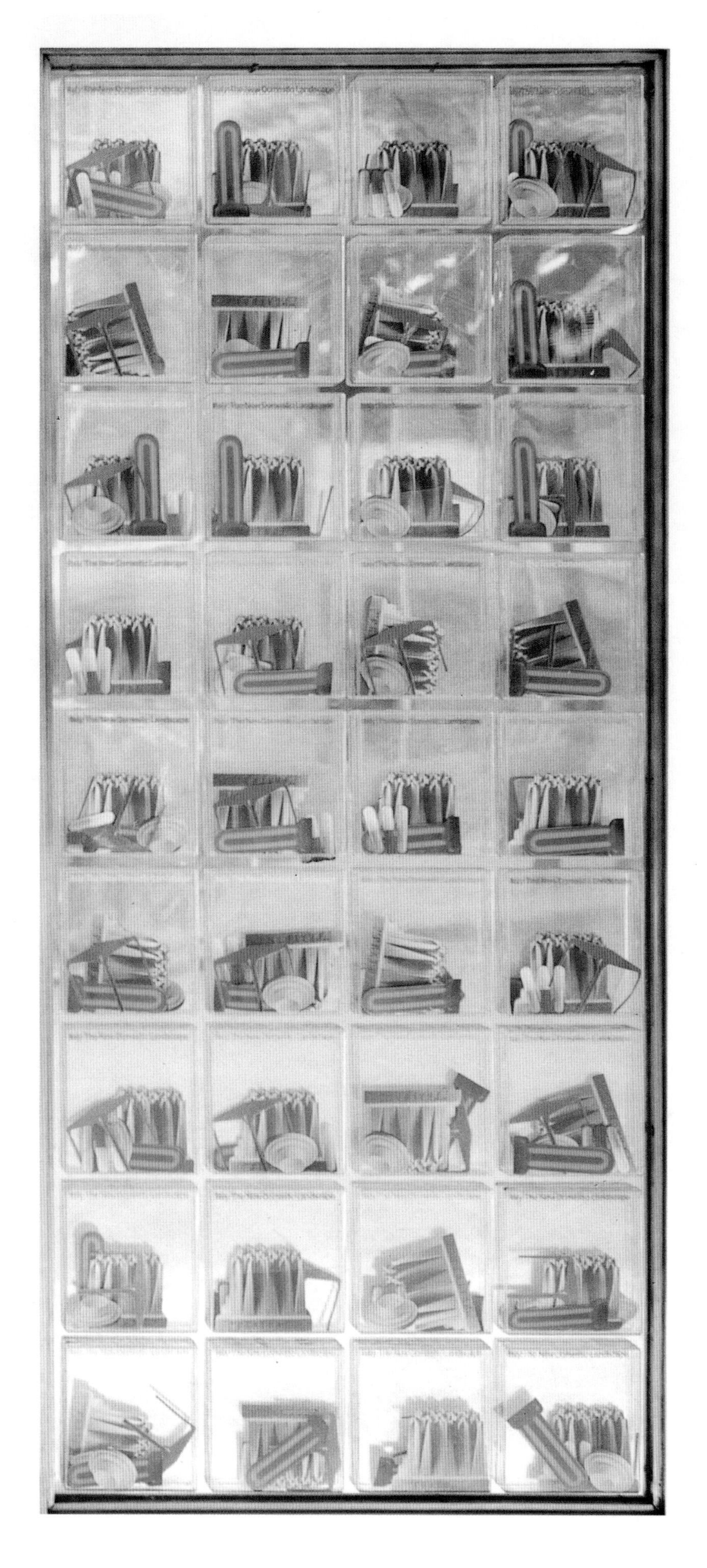

Italy: The New Domestic Landscape
Achievements and Problems of Italian Design
The Museum of Modern Art, New York

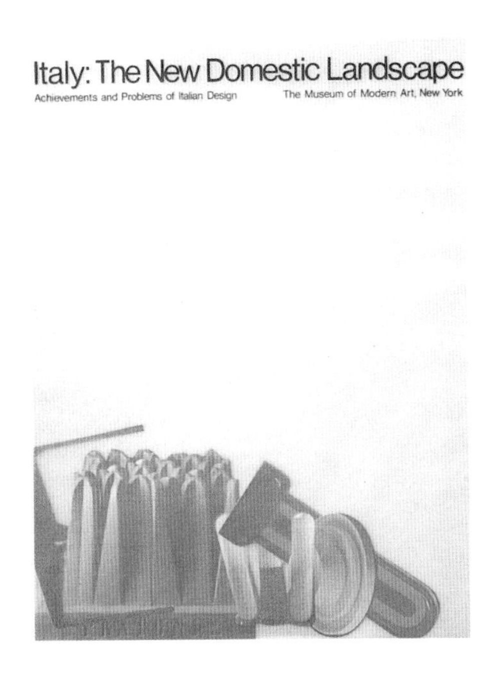
Italy: The New Domestic Landscape
Achievements and Problems of Italian Design
The Museum of Modern Art, New York

Italy: The New Domestic Landscape
Achievements and Problems of Italian Design
The Museum of Modern Art, New York

Italy: The New Domestic Landscape
Achievements and Problems of Italian Design
The Museum of Modern Art, New York

La Jolla Exhibition Poster

EMILIO AMBASZ

Architecture, Exhibition, Industrial And Graphic Design

La Jolla Museum of Contemporary Art	June 11–August 6, 1989
Musée des Arts Décoratifs de Montréal	October 1, 1989 – January 6, 1990
Akron Art Museum	January 26 – March 25, 1990
The Art Institute of Chicago	May 1 – July 2, 1990
Laumeier Sculpture Park	September 9 – November 11, 1990
Des Moines Art Center	April 27 – June 23, 1991
The Queens Museum	October 5 – December 1, 1991

This exhibition has been organized by the La Jolla Museum of Contemporary Art and made possible by a grant from the Graham Foundation for Advanced Studies in the Fine Arts; generous contributions from Mr. and Mrs. Rea A. Axline and Dr. and Mrs. Jack M. Farris; and by grants from the National Endowment for the Arts, a federal agency; and the California Challenge Program of the California Arts Council. Additional project support was received from the San Diego Design Center.

MYCAL GROUP

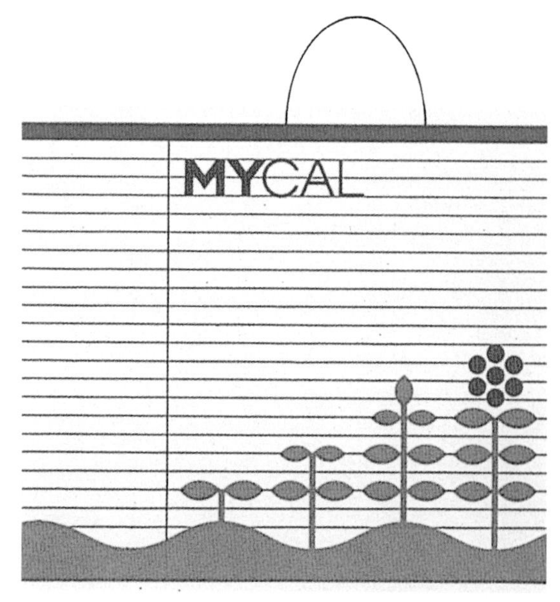

EMILIO AMBAS
AXIS

Emilio Ambasz

Biography

Emilio Ambasz, born in 1943 in Argentina, studied at Princeton University. He completed the undergraduate program in one year and earned, the next year, a Master's Degree in Architecture from the same institution. He served as Curator of Design at the Museum of Modern Art, in New York (1970-76), where he directed and installed numerous influential exhibits on architecture and industrial design, among them "Italy: The New Domestic Landscape," in 1972; "The Architecture of Luis Barragan," in 1974; and "The Taxi Project," in 1976.

Mr. Ambasz was a two-term President of the Architectural League (1981-85). He taught at Princeton University's School of Architecture, was visiting professor at the Hochschule für Gestaltung in Ulm, Germany, and has lectured at many important American universities. Mr. Ambasz' large number of prestigious projects include the Mycal Sanda Cultural Center in Japan, the Museum of American Folk Art in New York City; and an innovative design of the Conservatory at the San Antonio Botanical Center, Texas, which was inaugurated in 1988. Among his award winning projects are the Grand Rapids Art Museum in Michigan, winner of the 1976 Progressive Architecture Award; a house for a couple in Cordoba, Spain, winner of the 1980 Progressive Architecture Award; and for the Conservatory at the San Antonio Botanical Center in Texas he has awarded the 1985 Progressive Architecture Award, the 1988 National Glass Association Award for Excellence in Commercial Design, and the highly esteemed 1990 Quaternario Award for high technological achievement.

His Banque Bruxelles Lambert in Lausanne, Switzerland, a bank interior, received the 1983 Annual Interiors Award, as well as a Special Commendation from the jury. His design for the Banque Bruxelles Lambert branch in Milan, Italy, and his design for their New York City branch at Rockefeller Center have been completed. He won the First Prize and Gold Medal in the closed competition to design the Master Plan for the Universal Exhibition of 1992, which took place in Seville, Spain, to celebrate the 500th anniversary of America's discovery. This project was also granted the 1986 Architectural Projects Award; from the American Institute of Architects/New York.

The headquarters he designed for the Financial Guaranty Insurance Company of New York won the Grand Prize of the 1987 International Interior Design Award of the United Kingdom, as well as the 1986 IDEA Award from the Industrial Designers Society of America (IDSA). He won First Prize in the 1986 closed competition for the Urban Plan for the Eschenheimer Tower in Frankfurt, West Germany. The magazine Progressive Architecture in 1987 and the American Institute of Architects, New York, in its 1986 Architectural Projects Award, cited for awards the Mercedes Benz Showroom design. Mr. Ambasz represented America at the 1976 Venice Biennale. He has been the subject of numerous international publications as well as museum and art gallery exhibitions, principal among them the Leo Castelli Gallery, the Corcoran Gallery, the Museum of Modern Art, New York and the Philadelphia and Chicago Art Institutes. An exhibition entitled "Emilio Ambasz: 10 Years of Architecture, Graphic and Industrial Design," was held in Milan in the fall of 1983, traveling to Madrid in May '84, and Zurich in the fall of '84.

The Axis Design and Architecture Gallery of Tokyo dedicated a special exhibition of his work in April 1985. In 1986, the Institute of Contemporary Art of Geneva, Switzerland at "Halle Sud" and, in 1987, the "Arc-en-Ciel" Gallery of the Centre of Contemporary Art in Bordeaux, France also presented one-man shows of his work.

In 1989, a retrospective of Mr. Ambasz' architectural designs, "Emilio Ambasz: Architecture," was held at The Museum of Modern Art, New York; and a second traveling exhibition, "Emilio Ambasz: Architecture, Exhibition, Industrial and Graphic Design," was held in June of 1989 at the San Diego Museum of Contemporary Art, traveling to, among others, the Musee des Arts Decoratifs de Montreal, the Akron Art Museum in Ohio, the Art Institute of Chicago in Illinois, and the Laumeier Sculpture Park in St. Louis. Another retrospective of Mr. Ambasz' complete works was held in 1993 at the Tokyo Station Gallery in Japan, and in 1994 at the Centro Cultural Arte Contemporaneo in Mexico City, as well as in other locations in South America and Europe. Major international publications, Domus, ON, Space and Design, Architectural Record and Architecture plus Urbanism, among others, have dedicated special issues to his architectural work. In 1989, Rizzoli International Publications printed a monograph of Mr. Ambasz' work to coincide with his exhibition at The Museum of Modern Art. In 1993, Rizzoli International Publications printed a second monograph chronicling Mr. Ambasz' complete works.

Mr. Ambasz also holds a number of industrial and mechanical design patents. Since 1980 he has been the Chief Design Consultant for the Cummins Engine Co., a company internationally celebrated for its enlightened support of architecture and design. Mr. Ambasz has received numerous industrial design awards. Included among them are the Gold Prize awarded for his co-design of the Vertebra Seating System by the IBD (USA) in 1977, the SMAU Prize (Italy) in 1979, and the coveted Compasso d'Oro (Italy) in 1981. In 1991 Mr. Ambasz was again awarded the Compasso d'Oro (Italy) prize for his new seating design, Qualis. The Vertebra chair is included in the Design Collections of the Museum of Modern Art, New York, and the Metropolitan Museum of Art, New York. The Museum of Modern Art, New York, has also included in its Design Collection his 1967 3-D Poster Geigy Graphics and his Flashlight, a design also cited for awards by the Compasso d'Oro (Italy) in 1987 and the IDSA in 1987. His design for Cummins' N14 Diesel Engine won the 1985 Annual Design Review from Industrial Design magazine. This publication also awarded him similar prizes in 1980 for his Logotec spotlight range (which also received the 1980 IDSA Design Excellence Award), in 1983 for the Oseris spotlight range, as well as in 1986 for his design of Escargot, an air filter designed for Fleetguard Incorporated.

In 1987, the Industrial Designer's Society of America granted its Industrial Design Excellence top award for his Soffio, a modular lighting system. In 1988, the IDSA awarded him the same top honor and in 1989, the ID Designer's Choice Award for design excellence and innovation was given for AquaColor, a water color set. In 1992, IDSA again awarded Mr. Ambasz top honors for his innovative design of the Handkerchief Television, as well as awards for his Sunstar Toothbrushes and the Soft Notebook, and in 2000, the IDSA/Business Week GOLD AWARD for Design Excellence went to Mr. Ambasz for Saturno, an innovative street lighting system. The Tenth Biennial of Industrial Design (BIO 10, Ljublianna, 1984) granted Mr. Ambasz their Jury Special Award "for his many contributions to the design field."

In 1997, Ambasz received the Vitruvius Award from the Museo Nacional de Bellas Artes for the innovative quality in his work. Recently, the Mycal Cultural Center in Shin-Sanda, constructed within close proximity to the epicenter of the devastating Kobe earthquake, received a special award from the Japanese Department of Public Works. Its high quality of construction and structural integrity allowed it to withstand this overwhelming natural disaster. This building, the Mycal Cultural Center at Shin Sanda, Japan, received the 2000 Saflex Design Award.

In 1999, his design for Cummins Signature 600 Engine received a Industrial Designers Society of America/Business Week Award for Design Excellence and in 2000 he received from the same group the Gold Award for his design of Saturno, an urban lighting pole system, which also received the highly regarded Compasso d'Oro Award in 2001. The very prestigious American Institute of Architects/Business Week Architectural Award was granted in 2000 to Mr. Ambasz for his design of the Fukuoka Prefectural and International Hall in Fukuoka, Japan. The same building received in 2001 the First Prize for environmental architecture from the Japanese Institute of Architects.

This volume was printed for Elemond S.p.a.
by Martellago Mondadori Printing S.p.a.
via Castellana 98 Martellago (VE) in the year 2001